PALGRAVE STUDIES IN THEATRE AND PERFORMANCE HISTORY is a series devoted to the best of theatre/performance scholarship currently available, accessible, and free of jargon. It strives to include a wide range of topics, from the more traditional to those performance forms that in recent years have helped broaden the understanding of what theatre as a category might include (from variety forms as diverse as the circus and burlesque to street buskers, stage magic, and musical theatre, among many others). Although historical, critical, or analytical studies are of special interest, more theoretical projects, if not the dominant thrust of a study, but utilized as important underpinning or as a historiographical or analytical method of exploration, are also of interest. Textual studies of drama or other types of less traditional performance texts are also germane to the series if placed in their cultural, historical, social, or political and economic context. There is no geographical focus for this series and works of excellence of a diverse and international nature, including comparative studies, are sought.

The editor of the series is Don B. Wilmeth (EMERITUS, Brown University), Ph.D., University of Illinois, who brings to the series over a dozen years as editor of a book series on American theatre and drama, in addition to his own extensive experience as an editor of books and journals. He is the author of several award-winning books and has received numerous career achievement awards, including one for sustained excellence in editing from the Association for Theatre in Higher Education.

Also in the series:

Undressed for Success by Brenda Foley
Theatre, Performance, and the Historical Avant-garde by Günter Berghaus
Theatre, Politics, and Markets in Fin-de-Siècle Paris by Sally Charnow
Ghosts of Theatre and Cinema in the Brain by Mark Pizzato
Moscow Theatres for Young People by Manon van de Water
Absence and Memory in Colonial American Theatre by Odai Johnson
Vaudeville Wars: How the Keith-Albee and Orpheum Circuits Controlled the Big-Time and Its Performers by Arthur Frank Wertheim
Performance and Femininity in Eighteenth-Century German Women's Writing by Wendy Arons
Operatic China: Staging Chinese Identity across the Pacific by Daphne P. Lei
Transatlantic Stage Stars in Vaudeville and Variety: Celebrity Turns by Leigh Woods
Interrogating America through Theatre and Performance edited by William W. Demastes and Iris Smith Fischer
Plays in American Periodicals, 1890–1918 by Susan Harris Smith
Representation and Identity from Versailles to the Present: The Performing Subject by Alan Sikes
Directors and the New Musical Drama: British and American Musical Theatre in the 1980s and 90s by Miranda Lundskaer-Nielsen
Beyond the Golden Door: Jewish-American Drama and Jewish-American Experience by Julius Novick
American Puppet Modernism: Essays on the Material World in Performance by John Bell
On the Uses of the Fantastic in Modern Theatre: Cocteau, Oedipus, and the Monster by Irene Eynat-Confino
Staging Stigma: A Critical Examination of the American Freak Show by Michael M. Chemers, foreword by Jim Ferris
Performing Magic on the Western Stage: From the Eighteenth-Century to the Present edited by Francesca Coppa, Larry Hass, and James Peck, foreword by Eugene Burger
Memory in Play: From Aeschylus to Sam Shepard by Attilio Favorini

Danjūrō's Girls: Women on the Kabuki Stage by Loren Edelson
Mendel's Theatre: Heredity, Eugenics, and Early Twentieth-Century American Drama by Tamsen Wolff
Theatre and Religion on Krishna's Stage: Performing in Vrindavan by David V. Mason
Rogue Performances: Staging the Underclasses in Early American Theatre Culture by Peter P. Reed

Rogue Performances

Staging the Underclasses in Early American Theatre Culture

Peter P. Reed

ROGUE PERFORMANCES
Copyright © Peter P. Reed, 2009.

All rights reserved.

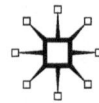

First published in 2009 by
PALGRAVE MACMILLAN®
in the United States—a division of St. Martin's Press LLC,
175 Fifth Avenue, New York, NY 10010.

Where this book is distributed in the UK, Europe and the rest of the world, this is by Palgrave Macmillan, a division of Macmillan Publishers Limited, registered in England, company number 785998, of Houndmills, Basingstoke, Hampshire RG21 6XS.

Palgrave Macmillan is the global academic imprint of the above companies and has companies and representatives throughout the world.

Palgrave® and Macmillan® are registered trademarks in the United States, the United Kingdom, Europe and other countries.

ISBN: 978-0-230-60792-7

Library of Congress Cataloging-in-Publication Data

Reed, Peter P.
    Rogue performances : staging the underclasses in early American theatre culture / Peter P. Reed.
        p. cm.—(Palgrave studies in theatre and performance history)
    Includes bibliographical references.
    ISBN 978-0-230-60792-7 (alk. paper)
       1. American drama—18th century—History and criticism.
   2. American drama—19th century—History and criticism. 3. Theater and society—United States—History—18th century. 4. Theater and society—United States—History—19th century. 5. Poor in literature.
   6. Working class in literature. 7. Rogues and vagabonds in literature.
   I. Title.

PS341.R44 2009
812'.209352624—dc22   2008042333

A catalogue record of the book is available from the British Library.

Design by Newgen Imaging Systems (P) Ltd., Chennai, India.

First edition: July 2009

10 9 8 7 6 5 4 3 2 1

For my friends and family.

Contents

List of Illustrations		ix
Acknowledgments		xi
1	Atlantic Underclasses and Early American Theatre Culture	1
2	Gallows Performance, Excarceration, and *The Beggar's Opera*	27
3	Algerians, Renegades, and Transnational Rogues in *Slaves in Algiers*	53
4	Treason and Popular Patriotism in *The Glory of Columbia*	79
5	Pantomime and Blackface Banditry in *Three-Finger'd Jack*	101
6	Class, Patronage, and Urban Scenes in *Tom and Jerry*	127
7	Slave Revolt and Classical Blackness in *The Gladiator*	151
Epilogue: Escape Artists and Spectatorial Mobs		175
Notes		189
Works Cited		225
Index		243

Illustrations

2.1 Exhortation to Young and Old [...] Occasioned by the Unhappy Case of Levi Ames, Executed on Boston-Neck, October 21st, 1773, for the Crime of Burglary (1773), broadside; Historical Society of Pennsylvania. 47

3.1 "Dramatis Personae," Susanna Haswell Rowson, *Slaves in Algiers; or, a Struggle for Freedom* (1794); Historical Society of Pennsylvania. 59

4.1 Philadelphia Effigy Parade of Benedict Arnold, September 30, 1780; *Americanischer Haus- und Wirthschafts-Calender auf das 1781ste Jahr Christi* (Philadelphia, 1780); Historical Society of Pennsylvania. 85

5.1 Hand-colored etching entitled "Mr. Smith as Obi, in Three-Fingered Jack, drawn and etched by J. Findlay," n.d. Billy Rose Theatre Division, The New York Public Library for the Performing Arts, Astor, Lennox, and Tilden Foundations. 108

5.2 Isaac Mendes Belisario, "Jaw-bone, or House John Canoe" (1837); Lithograph in Belisario, Sketches of Character. Yale Center for British Art, Paul Mellon Collection. 119

7.1 James and Eliphalet Brown, "Dancing for Eels; A scene from the new play of *New-York As It Is*, as played at the Chatham Theatre, N.Y." (1848); Library of Congress Prints and Photographs Division [LC-USZC4-632]. 170

Acknowledgments

It is a pleasure to thank the many people who have offered material support, intellectual feedback, encouragement, and companionship. This project has received tremendous institutional support from beginning to end. Florida State University granted material support with University Fellowships and Dissertation research grants. My friends at Florida State offered years of encouragement and intellectual support. I remain grateful to Karen Bearor, Ralph Berry, Helen Burke, Leigh Edwards, Andrew Epstein, Barry Faulk, Ray Fleming, Leon Golden, Hunt Hawkins, Joseph McElrath, Dennis Moore, Jim O'Rourke, Dan Vitkus, Eric Walker, Candace Ward, Nancy Warren, and many others for their collegial support and early encouragement. Special thanks to my hallmates off the skybox. Deborah Brock, Stephanie Cameron, Carolyn Hall, and Scott Kopel are enablers of the best kind; they have each made my career possible in the most crucial and material of ways. The support continues at the University of Mississippi, and I am grateful to the Office of Research and Sponsored Projects for a Faculty Research Fellowship. In the Department of English, Kara Hobson and Jamie Dakin deserve awards for their resourceful help. My new colleagues have given me a kind welcome to Oxford and offered stimulating intellectual community. Thanks especially to Ben Fisher, Adam Gussow, Jaime Harker, Katie McKee, Patrick Quinn, Karen Raber, Annette Trefzer, and Jay Watson for support and feedback. Special gratitude to Kacy Tillman for selfless last-minute research assistance and to Sarah Wurgler Walden and Dan Walden for their help in the final stages of book completion.

This project has benefited immensely from generous access to archival resources. The Harvard Theatre Collection provided unparalleled and still unplumbed depths of archival resources; a generous Stanley J. Kahrl Visiting Fellowship in Theatre History made my research there possible. Betty Falsey, Rachel Howarth, and Fredric Woodbridge Wilson assisted in numerous ways, initiating me into the secrets of the old card catalogues. Annette Fern was an always cheerful and helpful presence. The Houghton Library also provided invaluable access to printed materials. A visiting fellowship at the American Antiquarian Society sponsored by the Northeast Modern Language Association gave me access to an extravagantly broad and rich array of materials. Joanne Chaison, Ellen Dunlap, Paul Erickson, Elizabeth Pope, and Caroline Sloat helped make my time in Worcester enjoyable and productive. The Library Company of Philadelphia's NEH Post-Doctoral Fellowship has provided invaluable resources and a terrific intellectual community. Linda August, Jim Green, Connie King, Phil Lapsansky, Linda Wisniewski, John C. Van Horne, Wendy Woloson, and many others offered creative and incisive feedback and constant assistance. The Cassatt House snuggery offered an ideal place to recover from archival binges.

Also in Philadelphia, R. A. Friedman at the Historical Society of Pennsylvania and John Pollack at the University of Pennsylvania Rare Book and Manuscript Library gave generously of their resources and assistance. The resources and community of the McNeil Center for Early American Studies have proven priceless, and Dan Richter has earned my admirations for his generous ways and academic dedication. I owe thanks as well to the Yale Center for British Art, the Huntington Library, the Billy Rose Theatre Collection of the New York Public Library, the Library of Congress, the Victoria and Albert Museum, and the British Library for valuable archival access and reproductions. I am grateful to the Johns Hopkins University Press and the University of Alabama Press for permission to reprint portions of my articles previously published in *Theatre Journal* and *Theatre History Studies*.

In my vagrant wanderings, I have found extraordinary communities of scholars and friends. My deepest gratitude goes to Heather Nathans, who gives selflessly, cheerfully, and constantly. Elizabeth Maddock Dillon has offered good thinking and warm hospitality at just the right moments. I am also grateful for the insightful and critical comments of Jeffrey Richards and Odai Johnson at key junctures. Jean Graham Jones, Rhona Justice Malloy, David Saltz, and numerous insightful readers have helped shape my work into published form. Don Wilmeth has also been particularly supportive and helpful of this project at every stage. I remain grateful for the friendship and feedback of my fellow travelers Tiffany Brooks, Vincent Brown, Andrew Burstein, Jo Cohen, Ken Cohen, Jonathan Chu, Maura D'Amore, Jeannine DeLombard, Matt Garrett, Jenna Gibbs, Hunt Howell, Jennifer Hughes, Nancy Isenberg, Tatia Jacobson Jordan, Jeffrey Kaja, Dan Kilbride, Adrienne Macki, Will Mackintosh, Michelle Craig McDonald, Roderick McDonald, Lucia McMahon, Peter Messer, Katie Paugh, Matthew Pethers, Chris Phillips, Stacey Robertson, and Jason Shaffer. My apologies to any whom I inadvertently neglect.

From start to finish, Rip Lhamon has given robust support, unwavering belief, and kind hospitality. The invitations to Vermont always seem to come at just the right time. Special thanks to Mom and Dad, Aunt Patty, Carol, Andy, and Lu for holiday breaks and moral support. Finally, thanks to TRG, Go Vertical Philly, Jules, and my climbing friends, who remind me of the value of a good belay partner.

1. Atlantic Underclasses and Early American Theatre Culture ∾

In December 1775, outside London's Covent Garden theatre, rogue characters staged a scene of underclass defiance. Although taking place in London, the event reveals the contours of circumatlantic theatre culture. It started with a young girl's struggle with parental authority: a teenage actress named Ann Brown eloped and fled her apparently controlling father. Mr. Brown, opposed to the fast-living theatre world, forcibly carried her away from London in a coach. Ann put her histrionic skills to good use as they rolled through an outlying village, alerting onlookers that her father "was carrying her away by force, in order to ship her for America."[1] At that moment, her struggle became a public performance. The young actress's improvisation likens her father's enforcement of parental authority to criminal transportation or the press gangs that unscrupulously sent Britain's less fortunate along circumatlantic routes of forced labor and inter-imperial conflict. Responding to her pleas, onlookers freed the actress and returned her to the care of an aunt more sympathetic to the stage. Ann Brown returned to act at Covent Garden during the 1776–77 season; this scene, however, is only the opening act in a more spectacular drama.

Brown had built a reputation in the late 1770s, roguishly crossing gender lines on stage; she played the two lead roles in John Gay's *The Beggar's Opera*, performing as Polly at Covent Garden and as her highwayman lover Macheath in cross-dressed productions at the Haymarket Theatre. The parts and her theatrical versatility had made Brown an "amazing favorite" with her public, according to biographers.[2] Those rogue roles also inflected offstage life, and Brown's teenage rebellion replays and improvises on her most popular onstage parts. Like Polly Peachum in *The Beggar's Opera*, Brown defied her father to pursue an undesirable match. Her act also echoes Gay's lesser-known sequel *Polly*, which sends its female lead on a cross-dressed Caribbean search for her husband Macheath. As Polly, Ann replays Gay's fugitive dramas of captivity and Atlantic mobility that had regained popularity in the 1770s (*Polly*, although published in 1729, saw its first performance in 1777). As in those plays, Ann's rebellion centers on the problems and possibilities of female agency in the Atlantic world, contesting restraints and asserting her youthful independence. Life imitates art as character types and plot lines spill over the confines of the London stage. With her theatrical training, Brown may even have consciously adapted Gay's characters as models for her rebelliously theatrical escape.

Brown's story reached its sensational climax later that same year when her dilemma called up the very outlaw gang that had defined her stage career. In the autumn of 1776, the conflict made the newspapers again:

> Friday night Mr. Brown, the coal-merchant, whose daughter eloped from him some months since, attempted to seize her, as she stopped in a coach at the end of the play-house passage in Bow-street. The little Syren was accompanied by her aunt, who made a great outcry, and told the populace Mr. Brown was mad; the alarm presently reached the play-house, and the theatrical garrison sallied out in great numbers, headed by Messrs. M—— ——, B—— ——, W—— ——, and S—— ——, to relieve the distressed damsel. The thieves in the Beggar's Opera, armed with pistols, &c. made a most formidable appearance, and the crowd was so numerous, that for a considerable time the street was impassable. At length, however, the lady was handed into the play-house in triumph, and, notwithstanding her great agitation of spirits, performed the part of Polly greatly to the satisfaction of a very numerous and brilliant audience, who received her with repeated shouts of applause.[3]

The scene is complexly theatrical. The girl's aunt preempts the typical misogynist charge of hysteria, histrionically accusing Mr. Brown of madness. A "theatrical garrison" costumed as a gang of thieves reclaims one of their own from the grips of parental authority, assisted by a street-clogging audience-turned-mob. As the newspaper report suggests, it must have seemed as if the imaginative world of *The Beggar's Opera* had spilled over the boundaries of the stage. The London streets transform, however briefly, into the domain of a gang of charismatic outlaws who chronically flaunted their escapes from the law. The scene embodies an inversive, carnivalesque playworld in which the low mimicked and mocked the high to popular acclaim. Just down the street from the Bow Street offices of Henry and John Fielding's thief-catching operations, the stage materializes a world in which rogues and their gangs, assisted by the riot-ready eighteenth-century mob, successfully make their stand.

Brown's fear of transportation to America situates this story in the circumatlantic routes that shaped English and early American theatre. Any view of Philadelphia, New York City, or Boston theatricals inevitably finds itself tracking London and provincial English theatre, and this is no exception. Atlantic theatre follows the same routes that threaten to carry a young actress away from her admiring audiences. Scripts, actors, and theatre's material resources move along circumatlantic trajectories; they pass among theatrical sites, creating common cultures of Atlantic theatricality. As a result, widely shared modes of performance permeate late-eighteenth-century Anglo-Atlantic culture, articulating public senses of subjectivity and agency. Acting in this thickly textured culture of conventional social performance and formal theatre, Ann Brown's rescuers do not bother with fine distinctions between the "imaginary" world of the stage and the "reality" of offstage actions. Actors exit the playhouse and enter street scenes. Street performances also feed into stage acts; after her rescue, Brown quickly transitioned into her onstage performance of Polly, "notwithstanding her great agitation of spirits."[4] Macheath and his gang freely traverse the "fourth wall" (and the playhouse's other walls as well) that later conventions imagine dividing the play from its contexts. The "shimmering liquid play on the themes of self-expression and self-concealment" that Terry Castle finds in the masquerade are, indeed, "exemplary phenomena of the period."[5]

Both everyday and onstage varieties of performance articulate, revise, and reinforce one another in a theatricalized Atlantic culture. The scene speaks of theatre's position in society, but also of the roles assigned to (or assumed by) the low. The magical appearance of the outlaw gang suggests that they perform a particularly noteworthy brand of theatricality. Their mobile roguishness seeps into the fabric of Anglo-Atlantic culture, providing models for real behavior and theatrical templates for other stage acts. Macheath and his gang perform a show of force, enacting the potential of low culture to resist authority through vernacular forms of community policing and collective expression. In that respect, the gang seems an imagined and nostalgic cohort, carrying out the primitive rebellion of Eric Hobsbawm's "social banditry" operating through grassroots community support.[6] Indeed, the London troupe wrests control of the script from managers and playhouse conventions, producing an impromptu paratheatrical act.

The players, of course, are not really Macheath's gang, nor do they precisely perform underclass roles. Instead, they play at performance itself, enacting those roles twice removed. The acting-troupe-turned-theatrical-gang recruits stage forms, rather than outlawry proper, to its antiauthoritarian ends. They display "formidable appearance" rather than actual criminal activity. Their rogue rescue of Ann Brown represents above all a reenactment of cultural memory, a self-conscious "surrogation," to use Joseph Roach's term, of earlier stage performances.[7] Macheath and his gang, busted out of the playhouse, signal the persistence of an informal Atlantic cultural repertoire of theatrical criminality. Aware of their own stage origins, they transfer theatre to the street in an act that Richard Schechner might call "restored behavior."[8] Theatricality infiltrates and shapes the "presentation of self in everyday life" and performatively produced acting bodies.[9]

The deliberate theatricality of the rescue also suggests that (however spontaneous and insurgent it may seem) the gang hardly stages an artless defense of low prerogatives. If the gang shows force for customary rights, they claim the customs of theatre and its audiences. They act, along with their mob, in urban spaces shaped and marked by the traffics of celebrity and publicity. Those London onlookers in 1776 watched a highly mediated scene shaped by the means of communication and print representation. Customary forms appear in a thoroughly commercialized world with its own shared and transitive forms of community and communication. The gang rescue happens live on the street, but it also reaches a far wider public through newspaper accounts. Those forms of print culture disseminate performance vigorously in the eighteenth century. The column in the *Morning Chronicle* that recounts the street scene also briefly reviews Brown's ensuing performance in *The Beggar's Opera*, further reinforcing the scene's staginess. A pervasive theatricality holds acting bodies and reenacting texts in wavering tension; print forms publicly reproduce and transmit the theatrical low. Biographies, broadsides, and newspaper accounts magically and perversely conflate actor and character, performatively inscribing roles onto histrionic bodies. The manipulation of text and performance goes both ways, however, and the outlaw gang improvises popular versions of what might otherwise seem stable scripts. The interplay of text and act shows performance's mutually constitutive symbiosis with other forms of representation. Print culture, the formal stage, and commercial celebrity colluded with informal and customary practices to render the street performance of *The Beggar's Opera*.

Behind it all, Ann Brown's story hints at the presence of the less powerful of the eighteenth-century Atlantic world. Although often invisible and disdained, the low also emerge in spectacular scenes. They become a major theme, even an obsession, of eighteenth- and nineteenth-century Atlantic theatre culture. The stage traffics constantly in rogue acts. Brown's rescue signals the paradoxical power those low characters confer on performance and reenactment. They also point us toward the onstage scenes that reenact and claim such power. Such acts negotiate the status of high culture and low, of live bodies and mediated forms, of modern institutions and preindustrial communities. Most of all, they reveal the stage underclasses emerging, highly scripted but always ready to improvise, as a compelling force in Atlantic and early American theatre culture.

THEATRICAL UNDERCLASSES ON DISPLAY

Rogue Performances examines the stage lives of the theatrical low. This study focuses on plays produced in American theatres from the 1770s to the 1830s and takes as its most immediate contexts American theatre and paratheatricals. Although this study examines performances that occur at American sites, it surveys a broader Anglo-Atlantic landscape. The constant trafficking of Atlantic culture industries means that English contexts, from the rise of London's illegitimate theatre to the discourses of empire, constantly inflect this study. Of the eight plays discussed in detail here, four have "American" authorship. One of those (Susanna Rowson's *Slaves in Algiers*) is by a transatlantic immigrant, and another (Jonas B. Phillips's *Jack Sheppard*) closely follows the script of an English play. Of the other two, Robert Montgomery Bird's *The Gladiator* is the only play in this study that has warranted inclusion in the traditional anthology of "American drama."[10] The last, William Dunlap's *The Glory of Columbia*, seems too derivative or popular for anthologizing, although the tragedy on which Dunlap based it, *André*, rates inclusion. The rest of my chapters examine performances based on English scripts. This is not an accident, since early American theatre (with its British-born and trained actors) trafficked heavily in English conventions. Nor is it a drawback, since this study attends to the ways that American theatre reinvents found materials to produce its own dramatic scenes. Even the less common American-authored play, though certainly marking significant movements toward emergent cultural nationalism, reveals the constant recycling of Atlantic forms. Indeed, such performances confirm the now-conventional notion that American identity is itself contested and under construction in the early national period.[11] Moreover, theatre is live, site-specific, and embodied; material contexts and offstage acts inevitably shape the transcribed desires of scripts. This study reads texts for evidence of early American underclass imagination, but it does not imagine play scripts as singular. Neither does it construe performance as ending at the stage's edge. Formal theatre makes meaning alongside the manifold and diverse worlds of vernacular performance. Theatre happens both inside and outside the playhouse, and the two sites produce counterpoised scenes of "theatrically real" characters that seem solid and embodied, but also masked and scripted, each in their own ways. Performance transfers and transmutes from street to stage and back.

The material, economic, and imaginary forms of class are as Atlantic as other forms of culture. Underclasses broadly defined become and influence theatre, moving in vexed, paradoxical ways. They occupy vigorously contested sites of cultural valuation. They appear simultaneously admired and disdained, premodern and quintessentially modern, and frequently resistant, unruly, and outcast in the face of outside socializing and disciplining urges. Low types appear often disenfranchised and disrespected, but rarely fully disempowered. The stage transforms the outcasts and conscripts of circumatlantic Atlantic modernity into entertainment. Words, songs, dances, gestures, and stances attributed to the low become a sort of collective property, pushed and pulled through an increasingly commercialized entertainment industry. Theatre embodies the low not just as abject, but also as roguish—bad, powerful, and attractive. These transgressive but appealing characters stage magical and spectacular solutions to the dilemmas of class and culture in the Atlantic world. In part, this is a function of theatre's commercialization; the stage reproduces the transgressive and often inversive spaces that Mikhail Bakhtin's work has recognized in the early modern market.[12] Such sites host portable conundrums of class performance. This study centers on a central puzzle of low performance. Even as the stage disciplines the low, it repeats and celebrates them incessantly. Bourgeois culture in many ways distances itself from the "grotesque" (and unruly, violent, even violated) bodies of those perceived as low, as Peter Stallybrass and Allon White have argued.[13] At the same time, it also incessantly restages those bodies. The relentless, even obsessive, processes of becoming and watching such types utterly confound bourgeois disavowal.

This study proposes three broad arguments. First, it argues that circumatlantic modernity centers crucially on what I will call the "theatrical underclasses." The stagey low emerges from and destabilizes the identity formations, collective affiliations, and disciplinary practices of Atlantic modernity. While the phrase does not mean to erase the important distinctions between stage and street, I use theatrical underclasses to convey a dual sense of onstage role and theatrical offstage body. Recent histories of the lower sorts have fleshed out our understanding of lived experiences in the less privileged parts of society, and they certainly pursued independent lives of their own. Nevertheless, the imagined experiences of cross-class contact shape those bodies in crucial ways. The theatricalized lower classes emerge from social encounters and perceived differences as much as from material conditions and collective consciousness. Spectacular displays, both on the boards and in the street, produce them as visible before various audiences and publics.

Second, *Rogue Performances* argues that low performance shaped the forms and genres of theatre in profound ways. The low become theatre through the repetition and rehearsal of various stage conventions. They travelled as the types of ballad opera, blackface pantomime, patriotic spectacles, urban scenes, and heroic melodrama. Those forms stylize their gestures, offering underclass bodies packaged in theatrical conventions. In return, certain forms seem to become "low," relegated in cultural hierarchies to lowbrow sites and popular audiences. Older Atlantic theatrical forms continue to circulate and intersect, producing new forms such as melodrama and blackface minstrelsy. Such theatre can at times degenerate into mere stereotypes, meaningless acts imposed upon absent and passive subjects. However, they can also show subtle connections and even symbiotic relationships with offstage performances of class, and this study attends to connections between the offstage low and onstage performance.

Third, this account argues for a broadly defined "print-performance culture" that produces, shapes, transmits, and archives the low. Print-performance culture shares circumatlantic routes with the subjects it portrays; it transfers roles and scripts across generic and cultural boundaries. It includes the different printed and verbal forms that produce and record traces of performance, and it provides my primary access to the theatrical underclasses. Surviving textual evidence of performance comes in a mongrel mix of different forms. Scripts, newspapers, and broadsides show such acts entering public spaces; diaries and letters document theatre in forms shaped by their respective conventions and audiences. Literary accounts, including play scripts, reshape performance through their respective aesthetic conventions and production practices. Print-performance culture traffics in a mix of lived experiences, external perceptions, and cultural memories associated with the Atlantic low. Its materials imperfectly document the fugitive acts of players and audiences in a performance culture predicated upon its own erasure.

Marx's 1852 *Eighteenth Brumaire of Louis Bonaparte* paints a revealing picture of the stagey and troubling Atlantic underclasses that stand at the center of this study. The commentary provides a useful nineteenth-century retrospective and evidence of the durability of such notions. Marx's commentary famously defines the unstable, unruly remnants of the revolution as the "lumpenproletariat." The ragged, mobile sub-proletariat poses problems for class analysis. Marx pictures the politically unreliable dregs of society as partly discernible by their chaotic, unpredictable relationships with other classes:

> Alongside decayed *roués* with dubious means of subsistence and of dubious origin, alongside ruined and adventurous offshoots of the bourgeoisie, were vagabonds, discharged soldiers, discharged jailbirds, escaped galley slaves, swindlers, mountebanks, *lazzaroni*, pickpockets, tricksters, gamblers, *maquereaus*, brothel keepers, porters, *literati*, organ grinders, ragpickers, knife grinders, tinkers, beggars—in short, the whole indefinite, disintegrated mass, thrown hither and thither, which the French term *la bohème*.[14]

They display no clear relationship to the means of production and adhere to no coherent class ideology. The lumpenproletariat becomes the leftover, cast-off, unaccounted-for remains of class struggle that the revolution must comprehend and police.[15] The *Eighteenth Brumaire* displays an urge to homogenize and classify, to name and control social types.

Essentially, Marx encounters the old problem of acting and authenticity in a political guise. The analysis envisions the lumpenproletariat as a theatrical class, a cast of masked, costumed, and artificed characters. Marx's lumpen characteristics—"dubious" means of subsistence, downward mobility of "ruined" fortunes, instability, rootlessness, deception, and duplicity—might well describe a cast of theatrical rogues. The *Eighteenth Brumaire*'s labeling and libeling suggests that the lumpenproletariat stands in the troubling theatrical space between representation and self-representation, between agency and passivity. In the end, Marx argues that such types "cannot represent themselves"; they must passively await representation.[16] Theatre's rogue antiheroes and their supporting casts imagine the menaces, motivations, and consequences of underclass agency and self-representation. Both acting and acted for, troubling both authenticity and imitation, their staginess provokes Marx's wariness. Even so, the passage exhibits something of the fascination with which audiences greeted numerous stage rogues and their outcast cohorts.

Quoting Hegel, Marx introduces the lumpenproletariat as a reenactment of history in which the actors and scenes appear "the first time as tragedy, the second time as farce."[17] To Marx, the revolution's troubling theatricality (and the retreat of class conflict from high drama to low comedy) represents a serious failing. *Rogue Performances* understands its characters as motley, mobile, and theatrical, but I hardly share Marx's disappointment at their disorder. I do not aim to rehabilitate the stage low into respectability, and cannot consistently imagine them as revolutionary or even class-conscious. The stagey underclasses (both inside and outside of playhouses) produce varying kinds of frictions in the machinery of capitalism, colonialism, and nationalism. They frequently seem out of place, forcibly displaced, in fugitive flight. They collectively appear, like Marx's lumpenproletariat, as extravagantly mobile, creating and claiming spaces in which to act. They circulate in unruly patterns that defy the ordering practices of power. Although rarely confronting authority in direct combat, their acts at times shift the terrain and achieve occasional and partial—though still significant—victories. If they constitute a many-headed hydra, the beast shows itself rarely, appearing mostly in the theatrical imaginations of the Atlantic world's more privileged classes.

The stage invokes such characters repeatedly in fascination, if only to discipline and defeat them. Although scripting rarely allows rogues and mobs to triumph, or even to substantially alter the social order in which they acted, they remain popular. Coded and stylized distortions of the low enact scenes frequently created or appropriated—but seldom fully controlled—by those in positions of power. Rogue performances appear rebellious, unruly, or absurdly comedic. They also frequently show off a self-conscious theatricality, as if aware of acting their part in Atlantic scenes. They produce complex and frequently masked acts of resistance, at times showing the system as artificial and constructed. Atlantic and early American theatre's anxious compulsion to stage the low rehearses triumphs of authority and order while demonstrating at the charismatic appeal of rogue performances.

* * *

Early Americans encountered the lower sorts in a variety of ways. Valuable recent scholarship has recovered the lived experiences and public visibility of America's less privileged, providing richly detailed contexts for theatrical scenes. Early American culture was quite literally built by and on the bodies of the circumatlantic underclasses. Others "interpreted, regulated, and controlled" those bodies, but they also belonged to people who acted out, who, as Simon P. Newman writes, "refused to surrender all control."[18] Those men and women lived, worked, played, and fought near and among other parts of society. They labored as servants, apprentices, artisans, mechanics, laborers, and slaves. They engaged in illicit and quasi-legal activities and professions as well, smuggling, pirating, thieving, pimping, prostituting, and begging. Both by preference and by the imperatives of labor and survival, less privileged early Americans appear to have interracially mixed more freely than did other cohorts. Not all were destitute, nor were all destined to remain in the impoverished classes; many lived lives of social mobility, upward, downward, and lateral.[19] Shaped by waves of immigration and forced mobility, they were a transnational and multiethnic cohort. The many local forms of the low constituted a mobile, circumatlantic cohort; they worked the markets, waterfronts, and the ships that moved goods and bodies around the circumatlantic periphery. Many of them circulated involuntarily, arriving in America as conscripted, impressed, indentured, transported, or enslaved.

Stage underclasses dramatize the offstage transformations of labor and economic social organization affecting early American workers. As historians such as Sean Wilentz have shown, the paternalistic relationships that had enmeshed master, journeyman, and apprentice in supportive and manipulative personal and professional relationships gave way between the 1770s and the 1830s to new relationships among workers and employers.[20] The social experiences of cross-class contact changed perceptibly, shaping the representation of class on stage. Forms of political participation and representation also changed significantly during the period. The American Revolution, as historians have shown, energized new forms of collective action among plebeian classes, even if many of the period's emancipatory and egalitarian promises remained largely unfulfilled in the early national period.[21] The broadening of the white male franchise and the Jeffersonian "Republican Revolution" of 1800 brought increased visibility and perhaps real forms of participation for the lower sorts. Institutions such as New York City's Tammany Hall, for example, displayed working-class capacity to organize and participate in politics. Later, the early 1830s saw the emergence of new and often racialized working-class identities in the forms of Jacksonian democracy. These practices of low participation, representation, and self-representation inevitably shaped the stage representation and the offstage acts of America's lower sorts.

In the face of such everyday acts of affiliation and self-representation, observers tended to stereotypically associate poverty and dependence with criminality and immorality. Outsiders, including the stage, frequently seem intent on criminalizing poverty and alternately disciplining or rescuing the less fortunate from their own misfortune. Of course, early America's lower sorts did periodically transgress, following their own cultural trajectories. They performed acts of resistance and self-possession, misbehaving, running away, deserting, and even conspiring in rebellion. Breakout energies and breakaway mobility (the "excarceration" so important to Peter Linebaugh's work) remain important aspects of the experiences and theatre of the American low.[22] Rebellion and escape almost certainly represent exceptional acts. Frequently, they simply failed.

Those acts, however, make good theatre. As society criminalizes them, the stage ironically celebrates such acts, and the early American stage seems particularly fascinated with its problem underclasses. They transgress, do not submit easily to discipline, and perform alternatives to society's dominant forms of social organization. The constant interplay of scripts and bodies, persons and personalities, acts and actors makes it extremely difficult to cleanly sort "real" from "theatrical" types. The playhouse itself had a hand in this, providing a site where on- and offstage performances collide. Motley and often unruly lower sorts physically entered the early American playhouse. Servants, slaves, prostitutes, sailors, soldiers, yeomen, artisans, and mechanics attended the theatre in varying mixes. They bought tickets, snuck in, visited on furlough, held their masters' places, shifted scenery, and worked the balcony for their pimps. They seem to have enjoyed theatre as much as anyone did. Early American playhouses, with tiers separating (but never entirely insulating) parts of the audiences, hosted significant cross-class contact and mixing.

The recorded presence of lower sorts in the theatre betrays a bias toward the sensational and spectacular, and especially toward the violent and catastrophic. Mixing and misbehaving, theatre's underclasses seem frequently on the edge of riot. In the early 1800s, even a tame night at the playhouse could be a dangerous experience for the genteel playgoer. Washington Irving's theatrical persona "Jonathan Oldstyle," for example, famously describes New York's gallery residents, who traditionally used the upper reaches of the theatre as a site of rowdy self-expression and perhaps even inchoate class-based protest. Irving's well-known comic commentary broadly suggests the kinds of social performances occurring in the theatre: "Somehow or another, the anger of the gods seemed to be aroused all of a sudden, and they commenced a discharge of apples, nuts, and gingerbread, on the heads of the honest folks in the pit, who had no possibility of retreating from this new kind of thunderbolts." Audience members freely interact with each other and with the actors on stage. The audience liberally demonstrates its displeasure, inspired by particular actors, performance quality, or factors entirely unrelated to the performance. Although "a little irritated at being saluted aside of my head with a rotten pippin," Irving follows his fellow spectator's advice against remonstrating with the "gallery gods." Such displays of genteel resentment would only draw further unwanted attention and escalate playhouse rowdiness. Strategically placed constables regulate the "waggery and humor of the gallery," but that order seems rather tenuous.[23]

Irving narrates an audience on the edge of riot, the stage hosting the ultimate underclass performance of the ludic turned agonistic. A significant proportion of antebellum rioting, contemporary accounts and historians concur, probably originated with plebeian types such as sailors and laborers. It would be a mistake, however, to conclude that violence or disorder always originates with the low, or that such violence was simply socially disruptive or radical. Rioting and mobbing, as Paul Gilje notes, had achieved a "quasi legitimacy" in early American culture, and most Americans of early national period "viewed popular disturbances with less apprehension" than did elites.[24] Unruly theatre crowds, as the extensive 1809 Old Price riots at London's Covent Garden theatre showed, could act to restore durable and even conservative traditional social imperatives. Theatre rioters frequently coordinated protests and allowed nonparticipants, especially women, a chance to clear the playhouse.[25] The elaborate Anglo-Atlantic conventions of theatre rioting suggest that the playhouse provided a regulatory space in which audience members of various classes could act low, playing at class-inflected unruliness. As Irving's account suggests, theatre mobs often seem more entertaining than threatening, and they could certainly be profitable. Philadelphia manager James Fennell understood this, writing in his 1814 memoir that the "prospectus of a disturbance, or, as some call it, fun, is the most attractive bill that can be made out."[26] The unruly but appealing convergence of play and violence measures the extent to which rogue performances shape Atlantic and early American theatre culture.

Outside the playhouse, early Americans encountered the laboring, common and lower sorts perhaps most frequently in revels and self-conscious displays. Penitent prisoners and hanging thieves seem the most transgressive of the frequently misbehaving low. They exhibited criminality, insubordination, and penitence before audiences of juries, judges, churchgoers, and crowds at public executions. Captive sailors and soldiers enacted the possibilities of renegade conversion and treason, reinstating the boundaries of the national imaginary upon their return home. Americans watched militia musters, parades, effigy burnings, and other forms of low participation in public politics. Taverns and oyster cellars hosted various performances, including singing, toasts, and general misbehavior. On street corners and markets, laboring and indigent people hawked their wares or advertised their trades. Underclass types of all types engaged in recreational (and sometimes ritualistically competitive) performances such as dancing and singing, sometimes in secret sites, sometimes in hyper-publicized scenes, sometimes hidden in plain sight. The low negotiated new forms of self-expression, affiliation, and political participation in America's public performance spaces. Throughout, they appeared always already theatrical, on display, playing roles, entertaining themselves and others. Early American culture accordingly observed and constructed its underclasses as performers, dissemblers, and spectacular objects. The stage frequently scripted such types as active, unruly, even criminally threatening. Such low scenes show the potential for social friction, for chaos, even for resistance, while trading on their charisma and entertainment value.

* * *

This study centers on the rogue, the gang, and the mob, the same types that had emerged from the theatre to rescue Ann Brown. As generic elements, such types embody low forms of social organization on stage. The rogue, for example, consistently demonstrates singular and spectacular outlaw charisma. The type, popularized by *The Beggar's Opera*'s Macheath, reappears in the forms of housebreakers, thieves, deserters, traitors, Maroon bandits, pirates, and even rebellious Roman slaves. Macheath's heirs are predominantly male characters, but the rogue's masculinity verges on the dandiacal, enacting the values of style, fashion, and performance above mere action. The rogue also frequently becomes the object of female desire, and equally transgressive female characters (and actors) seek after and replay his performance. Macheath, for example, attracts a coterie of women of varying degrees of ill-repute in *The Beggar's Opera*, and its sequel *Polly* sends the breeches-wearing female adventurer in search of him in a stage parable of feminized theatricality and rogue desire. As Ann Brown demonstrates in 1776, the female character appears joined with rogues and outlaws, showing that such transgressive acts are not the exclusive privilege of male actors.

Behind the rogue, the stage frequently imagines underclass organization and disorganization in the shapes of gangs and mobs. The "ideology of the gang," as

Michael Denning terms it, poses the potential of the low to enact independent forms of socialization.[27] The gang frequently appears as a secretive, masked social form, its terms of membership mysterious and restrictive, its members initiated and disguised. At the same time, its members are also ironically exhibitionist, recognizable on stage by the public display of those secret signs. The gang inevitably performs, celebrating its occult social organization in play. Following its outlaw leader, the criminal gang can enact fears of criminal conspiracy and outlaw influence. Gangs articulate rogue theatre's ability to inspire mimicry, corrupting and inspiring its audiences. Such stagings of low collectivity speak variously to different audiences, and the gang can just as well figure the plebeian sociability occurring in early American taverns, workshops, and workingmen's organizations.

Mobs, for their part, figure the more chaotic and violent forms of underclass social organization. Early Americans knew the force of active and unruly mobs, and the stage rabble figures violence that some audience members feared could be overwhelming their streets. At the same time, many in the audience knew that the mob was not just a chaotically thrashing many-headed hydra, enacting structured expressions of collective will. The eighteenth-century neologism *mobile vulgus*, gradually shortened to "mob," codes deep suspicions of the low and their collective potential. From its beginnings, the term conveys a double sense of commotion and motion. We "might as easily translate the term as 'the movement,'" as Peter Linebaugh argues, and that movement can seem both chaotic and purposeful, riotous and revolutionary.[28] The term coded disdain and its own inversive potential; later commentators, as W. T. Lhamon observes, punned "mobility" against "nobility" to describe blackface minstrelsy's precocious and pretentious "Ethiopian Mobility."[29] Scenes of rogues, gangs, and mobs collectively stage the interplay of individual agency and subjectivity, staging the class-related problems of imitation, deference, agency, and independence that saturated early American culture.

The theatrical low cohort repeatedly emerges from and defines certain kinds of spaces. They frequently stand on the cart or the gallows, the elevated stages of punishment that historians have so evocatively recovered as spaces of contested performance.[30] Hidden spaces—taverns, grottos, below decks—also host such scenes. Still other acts occur in purgatorial and penitential spaces of confinement, the cells and brigs of an increasingly incarceral Atlantic culture. Offstage spaces shape stage performance, hosting dialectics of motion and stillness, speaking and silence, subjugation and freedom. Like its counterparts outside the playhouse, the boards bound, enclose, foreclose, and force performance. At the same time, such spaces also seem capable of protecting, nurturing, or allowing rogue performances. The stage, like the gallows, operates as a site of simultaneous display and discipline, hosting and shaping the forms of play and punishment. Occasionally, as Macheath and his heirs demonstrate, such theatrical spaces can even allow escape. Underclass acts stage breakouts, erupting into open public spaces, where they seem most dangerous and most in need of conceptual and literal containment.

If theatre frequently imagines the absent bodies and acts of the offstage low, it also (and just as importantly) stages the audience to themselves. In February of 1823, New York City's Park Theatre hung an enormously costly and heavy mirrored curtain that reflected the audience back to itself at the end of the performance.[31] The stunt reveals theatre's self-consciously constructed relationships to its audiences. Theatre reflects but also creates and disciplines its audiences. Although the Park Theatre may have meant its reflective curtain to evoke Hamlet's declaration that the stage holds "the mirror up to nature," the stage acts as more than just a mimetic space.[32] Instead, playhouses operate as socially constitutive and self-referential spaces, producing and glossing social relationships. Against theatre's onstage play and amidst the heterogeneous mixing of the playhouse itself, audiences work out their own forms of social organization, negotiating and contesting relationships in practices that ranged from tiered seating to theatre riots. The audience, after all, often seems a version of the mob, albeit one shaped by the tenuous discipline of ticket pricing, spatial arrangements, and conventions of spectatorship. The spaces of the theatre enable multiple social exchanges—between the play and its audience, among members of the audience, and among management, troupe, supporting staff, and patrons. In such settings, plays become live, embodied, symbolic transactions. Audiences do not passively receive performance, just as the script does not determine every aspect of a performance. Theatre reveals multiple parties participating in, around, and through its plays.

Early American theatre culture thus stages itself and its outsiders in the roles and scripts of rogue performances. Onstage acts rehearse the interactions of actors (and management), troupes, and audiences. Embodying, identifying with, and even imitating the low, the stage does not simply produce an absent and manipulated "other." Stage underclasses appear like the audience, even *of* the audience. Those disdained but powerful characters perform the cultural work of simultaneously imagining the outer limits of society and the ways in which the marginalized infiltrates and constitutes the community.[33]

EARLY AMERICAN THEATRE'S ROGUE PERFORMANCES

Rogue performances reenact the paradoxical mix of desire, danger, and discipline in early national performance culture. On its face, this claim revisits the old and ultimately unsatisfying argument that early American theatre operated as an outlaw institution against strict Puritan opposition. While that seems true in some instances, recent scholarship has shown early American theatre history to be at once more complicated and more mundane. Certainly, the colonial resistance to theatre seems a persistent relic of early modern English laws against vagabond strollers.[34] Actors and vagrants share professional techniques, and by that logic, onstage rogue acts mark the conflict, embodying theatre's rogue part in society.

However, the "structuralist dyad of Puritan versus Players," to borrow Odai Johnson's formulation, hardly explains the gradually materializing desire for performance in early American culture.[35] Theatre articulates vexed drives for status and cultural power. As Jeffrey Richards has written, "theatre will out in one form or another," and by the 1830s, the formal stage seems permanently established on the American landscape.[36] That sequence tracks more than a triumphal narrative of theatrical invasion and conquest, of outlaw performance overcoming prohibitions. Outlaw status does not represent a liability that theatre overcomes, but an asset that it actively preserves and reenacts. The stage versions of outlaws and outcasts sign theatre's continuing ambivalent social position. Their vexed scenes of simultaneous marginalization and empowerment offer meaningful resources for the stage. Theatre parlays onstage outlawry into networks of patronage, dependency, and privilege, the institutional forms that Heather Nathans examines in *Early American Theatre from the Revolution to Thomas Jefferson* (2003).[37] Likewise, early American dramatic texts construct troubled and often marginal forms of identity, as Jeffrey Richards's *Drama, Theatre, and Identity in the American New Republic* (2005) shows, but even their transgression follows established and conventional routes of transmission.[38]

Vexed instabilities of class and status shaped theatre's onstage acts. Its ostentatious displays offered playgoers the chance to act above their station in ways that could become transgressive. The communal luxury of a playhouse itself functioned in much the same way, as a public and material sign of desires for status. Plays often internalized this logic, posing actors as the Atlantic world's elites. At the same time, theatres hosted scenes of downward social mobility. The stage authorized imaginative slumming, offering access to otherwise secret scenes of low life, and observers frequently expressed satisfaction at the chance to infiltrate and observe the low, and even to mimic their moves. Rogue performances consistently drew audiences to the playhouse, offering vicarious and ambivalent experiences of action, agency, and even rebelliousness. Audiences could both applaud and fear low revolts, both mourn and celebrate their defeats. Even in its desire for respectability and status, theatre seems compelled to rehearse its own disdained and contentious histories, incorporating itself through the performance of lowness. The urge to rise ironically produces scenes of class falling. The habitual return to such themes suggests that early American theatre desired the cachet of dangerous and profitable renegade lowness as much as or more than respectability.

Those vexed vertical desires shaped early American theatre's social and geographical movements. Indeed, colonial troupes seem as vagrant as the characters they embodied in plays such as *The Beggar's Opera*. In the first half of the eighteenth century, players traveled circumatlantic routes that extended in the Americas from Nova Scotia to Jamaica. Walter Murray and Thomas Kean's troupe, for example, traveled those routes from 1749 to 1752. Soon after, Lewis Hallam Sr. organized the London Company, working American towns, as its name suggests, into an extended provincial circuit of the London theatre. Hallam's longer seasons brought theatre into the population centers and daily routines of early American cultural life. The troupe spent the middle of the 1750s in New York, Philadelphia, and Charleston, finally sailing for Jamaica, where Hallam died. In 1758, the troupe returned to the North American colonies under the management of David Douglass, who established early America's first permanent (if not permanently occupied) playhouses. The move, as Odai Johnson argues, established theatrical troupes "not as strollers raking the frugal resources out of the small economies in a one-time raid, but as seasonal residents engaged in a sustainable relationship with the community."[39] In the mid-1760s in Charleston, Douglass renamed the troupe the American Company, displaying early outward markers that performers traveling the most extended English provincial circuit had begun to identify with their territory.

The Continental Congress's 1774 ban on theatre and other sumptuary displays sent the troupe in retreat along established routes to Jamaica for the duration of the war. As Jason Shaffer and Jared Brown have shown, the conflict and the interdiction of theatre did not put a complete stop to performance, but the turmoil did disrupt organized professional theatre's regular schedule.[40] After the Revolution, troupes reestablished their circuits, continuing regular but still itinerant seasons. Within a few years, cultural elites reenergized theatre in the northern colonies, as Heather Nathans has shown. Against those attempts to claim cultural hegemony, the lower constituencies of artisans and mechanics also shaped their own spaces and practices. Philadelphia, continuing and even augmenting its theatrical bans until 1789, waited until 1794 for a new playhouse, the Chestnut Street Theatre. With the organization of the elite Tontine Society, Boston's elite produced the means to build the Federal Street Theatre in 1794 as well. In New York City, Hallam's John Street Theatre reopened soon after the war, although the Park Theatre replaced it in 1798, patronized by the city's merchant elites. While not the only sites of early American performance, the New York, Boston, and Philadelphia theatres constituted "a powerful—if competitive—triumvirate" that competed and shared circulating actors, managers, and scripts.[41] Those cities were among the most active in constructing playhouses and bringing Atlantic scripts and actors to American audiences.

Even as theatre established its permanent material presence in America's more populous cities, it continued performing marginalized mobility. Troupes followed the old circuits that continued to offer profits. Baltimore audiences, for example, could attend performances by Philadelphia actors; Boston players performed in Newport and other nearby towns during their low seasons. In the 1790s, other competing venues sprang up within cities, usually offering less formal entertainments to wider audiences. A circus and then a theatre opened on Philadelphia's Walnut Street, for example; Boston's middling artisans at the Haymarket Theatre competed with the Federal Street Theatre, and in New York City, the Greenwich Street Theatre hosted visiting acts in the 1790s. All along, the metropolitan circuits shared personnel and dramatic material with performers in Virginia and the Carolinas.[42]

Even after decades performing in early America towns, theatre's struggle for legitimacy continued to affect materially the spaces available for performance. Colonial theatres such as the Murray-Kean playhouse often occupied spaces on the social and political margins of the city.[43] The location materially substantiates theatre's literally edgy position within the social spaces of early American communities. Even after its gradual move to the centers of colonial life, theatre continued to work through forms of resistance. Philadelphia theatregoers, for example, could see an entire season of plays in disguise as "moral lectures" as late as 1788.[44] The thinly veiled subterfuge seems a joke more than an earnest evasion. The history of illegalized strollers repeats as farce the second time around. Despite the farcical qualities of such episodes of proscription, actors remained itinerant and vagrant, sometimes inadvertently and sometimes deliberately chosen. Even the antebellum achievement of apparent good community standing, appearing most famously at P. T. Barnum's family museum, hardly represented the unqualified whitewashing of theatre. The upwardly mobile desires of antebellum theatre created new cultural underbellies, as Lawrence Levine, David S. Reynolds, and others have shown.[45]

Early American theatre was in some ways an imaginatively inclusive place. Its repertoire of roguery circulated among a variety of Restoration and eighteenth-century

British offerings that ranged from low farce to popular revisions of Shakespeare. Anglo-Atlantic theatre, of course, interacted with and incorporated other cultural traditions. Ballad opera's vernacular musical forms, for example, emerged from century-long competition with Italian opera in London.[46] Anglo-Irish playwrights exercised a strong influence on late-eighteenth-century drama and comedic acting styles. Jewish American playwrights such as Mordecai M. Noah contributed significant dramas to the American stage as well. William Dunlap, one of the early American culture industry's most energetic professionals, also translated and staged an astonishing number of German playwright August von Kotzebue's melodramas.[47] That variety appears in tireless repetition, and the most successful early American plays appeared in multiple venues and encored in subsequent seasons. Generic repetition also reiterated forms within a variegated mix of circulating, repeating, and interacting cultural traditions. Theatre relied on standard offerings to draw playgoers, suggesting that audiences did not particularly demand scripted innovation or literary originality from their drama. Even dramatic novelties (as were a few of the plays in this study) recycled elements of convention, borrowing from other plays. Often they drew from familiar or vernacular offstage performance conventions or parlayed London success into provincial popularity. Live theatre, appearing amidst a rich blend of cultural influences, produced complex and ever-changing repertoires from conventional scripts.

If institutional theatre had routinized certain aspects of performance, the 1820s saw important sea changes in the quantities and qualities of theatrical and paratheatrical entertainments. Circus acts, which had formerly toured rural areas and occupied venues in the off-season, began to construct permanent urban arenas, establishing themselves within theatrical topographies. Pleasure gardens, especially visible in New York City, set up shop and began offering musical and dancing entertainment along with their refreshments. At the same time, perhaps not coincidentally, the older generation of management began retiring or expiring. The position of the theatrical professional in early American culture had changed radically in the previous decades. Theatrical memoirs appearing in the 1820s and 1830s suggest that the clubby, personal world of managers and actors who had produced the post-revolutionary generation of theatre had begun to disappear.[48] In its place, several major institutional changes affected theatre personnel, audiences, and the plays they presented. Proliferating playhouses provided niche spaces for specific clientele, sites for constructing and demonstrating group identities. Early American theatres, as Nathans writes, had consistently functioned within the early American public sphere, defining "those in the center and those on the periphery of political and economic power."[49] Such practices continued and even accelerated in the nineteenth century. New York theatre spaces, for instance, would diverge by the 1840s, elites fleeing to the Astor Place Opera House and rowdier working classes claiming the Chatham and the Bowery Theatres. Moving into the 1830s, this account finishes amidst the increasingly active public displays that Rosemarie Bank describes in *Theatre Culture in America* (1997).[50]

This account also ends with the rising influence of the southern and western U.S. circuits. The unstable economics and competition of the 1820s sent managers such as James H. Caldwell, Sol Smith, and Noah Ludlow southward and westward, developing theatres along the Ohio and Mississippi Rivers and the Gulf Coast. Serviced by touring companies, those venues saw the apprenticeships of aspiring actors such as Thomas Dartmouth Rice and Edwin Forrest, who would become the twinned stars of minstrelsy and melodrama. Increased competition and a public appetite for novelty meant that repertoires changed more frequently. Urban scenes in New Orleans and the provincial laboratories of blackface minstrelsy in the 1830s suggest that southern and western American venues played crucial parts in Atlantic theatre culture. Indeed, the increasing success of "American-authored" plays such as *The Gladiator* might be an effect of such economic and institutional instability. Advances in transportation and communication meant that a few actors could position themselves to take advantage of multiple markets, and a star system emerged to promote and transport headlining actors between cities in irregular and higher-paying engagements. Stock actors and local troupes suffered accordingly as the stars increased in salary, range, and celebrity.

Even in earlier periods, Early American theatre was certainly not limited to the northeast. The playhouses of Boston, New York City, and Philadelphia frequently served as waypoints in the wider circulations of theatre. Caribbean culture and theatre remained intimately connected to the northern colonies well after the colonial period, and plays such as *Three-Finger'd Jack* and *Polly* recognized and restaged those connections.[51] Although formal playhouses appeared later and saw shorter seasons in rural, inland, and southwestern areas, careful study of those venues and their plays would surely show that they hosted as much activity and cultural complexity as northeastern urban venues. Many (if not all) of the plays in this study appeared in those smaller and more itinerant locations, and those acts of recycling, revision, and revival remain important to this study. Although they have left fewer traces, the on- and offstage theatricals occurring away from the northeastern theatres remains an important part of this study. At the same time, the densely populated cities, economic activity, and especially transatlantic connections of the eastern seaboard made institutionalized theatre more active, more visible, and (unfortunately, but perhaps most importantly) more copiously archived. In the end, my emphasis on circulating performances does not imply closed narratives or completed circuits. Instead, it offers an invitation and a challenge to continue examining the theatricals that the traditional archive has neglected or less comprehensively preserved.

STAGES OF UNDERCLASS PERFORMANCE

Rogue Performances examines a selection of plays alongside the offstage acts with which they interacted. Proceeding in roughly chronological order, these chapters show the multiple interchanges among vernacular and formal theatrical forms. Transmissions and survivals, substitutions and surrogations show underclasses inhabiting and shaping early American theatre culture. Although premieres and periods of popularity crucially show how a play initially meets its public, theatre history of course does not end with first nights—or even last nights. It hardly ever ends at the edge of the stage or the proscenium arch dividing audience from actors. Instead, the roles, scenes, gestures, and scripted actions of the low circulated quickly to other forms, and this study takes as one of its objects the tracing of such transmissions.

Chapter two, "Gallows Performance, Excarceration, and *The Beggar's Opera*," examines the effects of a century's worth of *The Beggar's Opera* on an Atlantic culture of theatrical criminality. The play, premiering in London but one of the most popular in colonial and early national America, stages a spectacular version of offstage roguishness in the highwayman Macheath and his gang. The ballad opera perpetuates the residual cultural memory of criminal cultures and rehearses the public spectatorship of offstage criminals in an emergent American culture of theatrical crime. It provides interpretive models for the condemned and their audiences at early American executions. Performing escape from the hangman's noose, *The Beggar's Opera* also enacts a wishful, excarceral alternative plotline for the condemned and his sympathizers. The influence of *The Beggar's Opera*'s appears in print as well as performance. Broadsides, confessional texts, and other representations of the rogue struggle over representation of the criminal classes. The textualized acts of outlaw cohorts shape the eighteenth century's carnivalesque print culture.[52] Ultimately, even the texts shaped and authorized by moral and legal power reveal traces of Macheath's breakaway acts, revealing rogue theatricality at play.

The next two chapters turn from transported acts of English culture to examine two plays more conventionally identified with emergent "native" American authorship, Susanna Haswell Rowson's 1795 *Slaves in Algiers* and William Dunlap's 1803 *The Glory of Columbia—Her Yeomanry!* In their attempts to imagine national identity, both plays rely on low-comic cohorts linked to older forms of roguery. Chapter three, "Algerians, Renegades, and Transnational Rogues in *Slaves in Algiers*," examines Rowson's *Slaves in Algiers*, which is perhaps most known for its brief and not terribly popular attempt to profit from public concern over American sailors held hostage in Algeria. Although noteworthy as an early feminist critique of the limits of early American political participation, her play also stages the convergence of playing captive, Algerian, and rogue, foregrounding the slippery transformative potential of the early American underclasses. *Slaves in Algiers* gathers and condenses public displays of American captives and exotic orientalist exhibitions and embodies acts disseminated in a variety of textual Algerian performances. The play ultimately sorts out its ethnic and class categories, triumphantly resolving the American struggle for freedom, but not before it has continually embodied the threats and rewards of becoming ethnically and culturally alien.

The fourth chapter, "Treason and Popular Patriotism in *The Glory of Columbia*," examines the forms assumed by stage underclasses in the service of patriotic performance. The 1803 *The Glory of Columbia* revises a much less popular tragedy, Dunlap's 1798 *André*, by overwriting the death of Benedict Arnold's coconspirator Major John André with low comedy. The revision, appearing after the "Republican Revolution" of 1800, reveals shifts in the roles and scripts of lowness. As *Slaves in Algiers* does, Dunlap's play recruits the lower sorts to constitute a new national imaginary. The strategy works in part by producing formal theatre from popular forms of patriotic display and public performance. It stylizes the collective organization and chaos of the plebeian parades and processions in which many early Americans participated. It also stages satirical or mocking scenes of treason and loyalty, articulating the possibilities of low political participation. Dunlap's revision reveals not the failure of dramatic form, but the re-articulation of complex relationships among on- and offstage acts in early American theatre culture.

Even as patriotic themes conscript the low, the breakout acts of Macheath and his gang continue to influence Atlantic theatre. They become new forms of stagey circumatlantic roguery, taking race as a defining form. Chapter five, "Pantomime and Blackface Banditry in *Three-Finger'd Jack*," examines circumatlantic blackness appearing in the late eighteenth century. *Three-Finger'd Jack* transforms banditry into blackface rebellion and invents new Atlantic stage outlaws. It also continues the cultural work begun by *Polly*, the sequel to *The Beggar's Opera*. The long-dormant *Polly* (first performed in 1777) stages perhaps the first self-conscious blackface performance of the formal Atlantic stages; *Three-Finger'd Jack* in turn produces rogue blackness as "pantomimical drama." Outcast badness becomes blackness, and slave rebelliousness and resistance in the Americas becomes a defining site of Atlantic roguery. The blackface masking of *Polly* and *Three-Finger'd Jack* also signals an increasingly important theme in early American culture—the possibilities of interracial mutuality that appears most threatening amongst servants and laborers in urban areas.

Such acts work hard to imagine the troubling undersides of early American society. The final two chapters examine the stage rehearsal of new forms of blackface minstrelsy and melodrama in the 1820s and 1830s. Chapter six, "Class, Patronage, and Urban Scenes in *Tom and Jerry*," examines urban scenes as emergent sites of American class imagination. The play's scenes of "Life in London" seem an alien export to American stages, but they also reveal the Atlantic processes of making theatre. Amidst the gradual and uneven transitions in American urban life, stagings of W. T. Moncrieff's *Tom and Jerry* vault the underground urban low to prominence in the 1820s. Interpreted as the return of *The Beggar's Opera*, *Tom and Jerry* reshapes underclass performance through the infiltrating, patronizing, and imitating acts of elite flâneurs. *Tom and Jerry* also troubles imagined binaries of mimicry and authenticity by imagining elite patronage and reenactment. Offstage, the play energized emerging sporting cultures in which privileged urban youths gained slumming access to low scenes. The play eventually becomes a vehicle for performing local American versions of the performing Atlantic low, transforming London scenes into American minstrelsy in the 1830s.

The seventh chapter, "Slave Revolt and Classical Blackness in *The Gladiator*," examines a play that enacts slave revolt before the same urban audiences that *Tom and Jerry* conjures into existence in the 1820s. Unlike the intimate interactions with urban blackness in Moncrieff's play, Robert Montgomery Bird's *The Gladiator* turns to the remote classical past to imagine the stage heroics of low characters. In the embodied, physical forms of Edwin Forrest's stage celebrity, Bird's tragedy presents the low in a double disguise, in classical garb that itself covers blackness on American stages of the 1830s. Forrest, trained in the provincial laboratories of popular blackface performance, presents a counterperformance to minstrelsy's emergent tricksters, embodying slavery in the form of classically whitened, heroic, and muscular working classes. Forrest's performances also provide a model around which white working class audiences seem to have coalesced, sharing forms derived from the cultures of low interracial kinship but intent on upward mobility and affiliation with power.

The epilogue turns to the 1839 melodrama *Jack Sheppard*, examining American theatre's returns to the scenes of rogue performance that effectively set these plays in motion. The play resurrects Jack Sheppard, the housebreaker and escape artist that had energized modern rogue performances in the 1720s, even inspiring Macheath in *The Beggar's Opera*. In 1839, outlaw acts return to their Atlantic roots for renewed energy, and theatre continues to recycle and reenact performances. The play presents some of the most salient themes of the stage low, including their transgressive theatricality and excarceral urges. The themes of costuming and extravagant display appear in breeches performance, as female actors play the youthful male thief. The play also resurrects scenes of mob violence, allowing us to reexamine the ways in which underclass performances refuse containment, transmitting unruliness through and against disciplinary forms.

* * *

These plays reimagine class as matters of performance and stage aesthetics. Ballad opera, for example, turns outlaws into singers. Such performed charisma frees Macheath, the theatrically gifted outlaw, and performance itself becomes roguish and potentially resistant. Patriotic celebrations likewise mingle with transgressive acts of masking and cross-dressing to produce plots of national affiliation in *Slaves in Algiers*. The underclasses provide the supernumerary displays that flesh out the spectacle of early American patriotism, although they recede at play's end, excluded from the scene of dramatic triumph. Dunlap's *The Glory of Columbia*, however, troubles such conventional notions of lowness by thrusting yeomen to the fore, presenting a series of plebeian stage displays that later become holiday spectacles. In *Three-Finger'd Jack*, pantomime embodies black rebellion as voiceless. If *Polly*'s self-conscious blackface masking imagines outlaw abilities to enact transracial performance, *Three-Finger'd Jack* naturalizes such forms into rebellious blackness. Acting black also dooms such characters. In the 1820s, onstage forms of flâneurie and patronage reimagine the sites of the low in American cities, literally producing scenes for privileged spectators. By the time Forrest stars in *The Gladiator* in 1830, neoclassical forms and tragic melodrama shape the dilemmas of the underclass hero. Forrest's muscular aesthetic also articulates the heroic slave rebel, drawing upon the physical presentation of America's laboring and even enslaved classes for definition.

The stage underclasses simultaneously shape and respond to the genres of stage lowness. Audiences recognized ballad opera, pantomime, blackface, ethnic masking, spectacle, and melodrama as the popular or vernacular genres of Atlantic theatre, alternatives to elite dramatic forms such as tragedy. Drawing from those varied forms, rogue performance becomes a sort of genre itself. It improvises on and revises earlier patterns, and relatively stable patterns of usage emerge by the early nineteenth century. Songs and dances recycle from play to play; characteristic catch phrases reappear. Conventional scenes with reused props materialize night after night before slightly altered backdrops. Stabilized plots continued to resurrect the very character types they seem so intent on executing and reforming. The material realities of theatre—costuming, makeup, prop closets, even stock actors—create some of this repetition and stability, but such conventionality is also an effect of the need for comprehensibility and imaginative coherence. In effect, the stage rehearsed lowness into a theatrical genre.

Rogue performances intimately intertwine acts of low badness with forms of stage blackness. The contributions of low and popular performance to the shaping of stage blackness seems one of its most important innovations, and one of my goals is to begin accounting for early American theatre's vigorous flirtation with blackness. Theatrical roguery shapes stage blackness in key ways, giving it an early and habitual insouciance that survived minstrelsy's disdain and dilution. Early outlaw performances stage troubled hierarchies, and later acts replay those scenes with blackened characters. Rogue performances coded class affiliations first, but (as Gay's *Polly* shows) they relied upon stage versions of racial difference from a very early date. As Dale Cockrell has compellingly argued, blackface stylizes unruly preindustrial noise; racial motifs layer over preexisting forms solidified through the experiences of Atlantic capitalism and industrialization.[53] The plays in this study variously participate in such processes, sometimes overtly, sometimes in masked acts. Early minstrelsy's unruly and outlaw types stage alternatives to *Othello* and *Oroonoko*'s noble characters. Those masked productions figure mutualities of class working through racial differences. As W. T. Lhamon Jr. argues, early minstrelsy embodies the "lateral sufficiency" of low and mobile cohorts of the Atlantic world.[54] While Eric Lott has identified an "unstable or indeed contradictory power" in blackface's racial mimicry, his recognition that such force "issues from the weak, the uncanny, the outside" seems essentially an understanding that the racial politics of blackface require and reinforce hierarchical structures recognizable as class.[55]

Stage practices produce interlinked hierarchies of venues, audiences, performances, and character types. For example, institutional and political pressures, like English royal monopolies, limited the spoken drama to London's patent theatres and produced other forms as "illegitimate," oppositional, and even outlaw.[56] Those illegitimate forms of spectacle, pantomime, and musical theatre crossed the Atlantic as both low and popular. Associating certain modes of performance, such as circuses, with diverse and popular audiences led some observers to imagine those forms as popular and hence low. The inversive mocking of clowns, Harlequins, and eventually Jim Crow characters linked such roles to traditions of lowness. Although colored by long-standing associations, none of those performance elements exists as inherently low, and this study asks how certain forms and characters produced each other as low. Rather than taking for granted the timeless existence of such qualities, this project questions the processes that stage the weakness, uncanniness, and outsider status that feeds into blackface performance. Audiences and producers collectively and unevenly negotiate such meanings over time, both on and off the stage. Abjection also produces its opposite, and those traits become compelling and masked assets. New publics affiliate around underclass characters, from the traditional rogue to related types such as the frontiersman, the Yankee, the Bowery B'hoy, and Jim Crow. Early ballad opera and pantomime feed into antebellum patriotic spectacles, Yankee plays, urban voyeurism, minstrelsy, and melodrama. Such acts repeatedly rehearse the assumptions and contradictions of class, training audiences to discern and respond, consciously or not, to orders of cultural power. Even acts that seem to treat race as a detachable cultural category still rely upon conventional forms of lowness.

Blackface and its theatrical relatives honed preexisting forms of self-conscious mimicry and self-mimicry. By the middle of the eighteenth century, such imitation seems one of Atlantic theatre's well-established conventions. Before the sacralization of culture had begun to shape disciplined spectatorship and naturalistic representation, theatre constantly and openly renegotiated its own social standing in prologues and epilogues, self-consciously posing before the audience in efforts to curry favor and gate profits.[57] Such acts produce radically destabilized performance experiences, and the early American stage seems self-referential to the point of metatheatricality. Such self-conscious theatricality, Janelle Reinelt explains, emphasizes "theatrical processes instead of their contents, their indeterminate possibilities rather than their fixed cultural meanings."[58] For contemporary audiences as well as later observers, such acts valuably code the processes of theatre itself. Self-conscious spectatorship and display offer a primary frame of reference for American audiences, even for unscripted and offstage performances. Moreover, the self-aware theatricality shapes perceptions outside the playhouse as well. Scripting, casting, rehearsing, improvising, and prompting provide key forms by which eighteenth- and nineteenth-century people understood their own actions. Likewise, the modes of producing theatre—costuming, makeup, props, scene-setting—provide compelling techniques for imagining the world in general. The metaphorics of theatre, the idea that "all the world's a stage," also supply the defining forms of experience, as Jeffrey Richards has shown.[59]

These chapters hardly exhaust the topic, pursuing limited arcs and overlapping storylines in the transmissions and adaptations of popular culture. This study resists stories of linear progressive development from one kind of performance to another, although it does trace certain broad changes. *The Beggar's Opera*, though remaining popular, gives way to plays shaped by novel structures of celebrity, larger venues, and new performance styles. Transforming under such pressures, underclass performances reveal a consistent cultural traffic in forms of low badness. That traffic produces connections, although energetic and not entirely predictable or traceable. Theatre reuses and redeploys the roles, scenes, plots, and gestures of rogues and other low characters, turning them to ever-shifting purposes. At the same time, the lowness of such characters produces apparently unstoppable cultural energies. Rogue actors circulate on- and offstage and pass from hand to hand, embodied by actors of various motivations, backgrounds, and degrees of skill.

Although I limit this study to a few of the more popular and more striking characters, the early American stage exceeds Marx's lumpenproletariat catalog in diversity. The nautical melodramas of the 1830s, for example, present sailors, mutineers, pirates, and the forecastle fraternity. Operatic Italian banditti and romanticized gypsies both enjoy their heyday on Atlantic stages. Servant comedies such as *High Life Below Stairs*, made famous in the 1760s by David Garrick, satirically present class difference and low unruliness. I pass by some of these cohorts somewhat arbitrarily, but also because many represent simplified or conventional attitudes toward class hierarchies. The sentimentalized stage types, the pitiable inventions of compassionate reformers and disdainful elites, seem to perform different urges. While they remain important indicators of theatre culture's imagined undersides, such abject characters interest me less to the extent that they do not share the forceful charisma that drives rogue performances. This study works to overcome certain limitations established by rogue thematics. Independent and outlaw characters frequently appear as singular individuals. Such rogues (along with their audiences and institutions) also seem most frequently male and male-dominated. The masculine individuality seems both a dominant mode of those acts and an illusion created by rogue charisma. Two important correctives, however, appear on stage alongside the singular outlaw: the conscripted cohort of stage extras and the gendered counter-performances of female characters. Underclass stagings frequently call up the crowd of extras embodying the rabble to produce rogue performances. As Bruce McConachie's *Melodramatic Formations* (1992) has shown, the crowd of supernumeraries enables both the operation of the playhouse and the staging of the theatrical low.[60] The crowding audience also represents a significant counterpoint to individual celebrity, as many rogue scenes show. In similar fashion, the stage playhouse is hardly a males-only endeavor, as scholars such as Faye Dudden and Leslie Ferris have shown.[61] Female characters (often in some form of cross-dress) frequently mirror, mimic, and reinvent the masculine sociability central to rogue performances. Although often centered on the spectacular starring male figure, underclass acts stage much more complex dynamics of individual and collective identity. The mob and the female, although frequently in supporting roles, play crucially constitutive parts.

In the end, I use a label like theatrical underclasses carefully and with real reservations. Lumping the various low types together in one category, as Marx did, runs the risk of erasing the human diversity and imaginative variety of low characters both on and off the stage. For that reason, I do not attempt a methodical taxonomy of theatre's low types. Regimenting lists of totalizing categories can hardly do explanatory justice to the numerous underclass characters constantly revised and improvised in playhouses and on the streets. At worst, they reproduce the acts of exclusion that so often render such types invisible. At any rate, "class" itself is best understood here as a discursive construct, as Gareth Stedman Jones has argued—and, as theatre shows, part of ongoing rehearsals, repetitions, and improvisations.[62] Lowness and class positioning, while related to material and economic conditions, frequently become negotiated and transacted traits available under differing conditions to various parties. Such negotiated ideas of class appear subject to reconstruction and relocation, too; the figures of the theatrical low can shift meaning precipitously and unpredictably. The characters appearing on stage and in this account do not stabilize into a fixed set or a steady collective. The very act of naming the low is useful only insofar as it shows the stage, like Marx, imaginatively trying to identify and recruit troubling cohorts. This study, of course, hopes to avoid replicating those disciplinary acts, instead attending to the diverse ways in which theatre trafficked in uncontrollable, charismatic, and fascinating versions of the low.

TRANSMISSIONS OF PRINT-PERFORMANCE CULTURE

Performance, as Peggy Phelan has described, occurs on and against its own erasure.[63] Paradoxically, its disappearing acts frequently generate documentation, tenuous material, and textual records. Theatre history's own spectatorship frequently relies upon the textual tools used to produce performance—scripts, promptbooks, playbills, and newspaper advertisements, for example, all point toward the performance that we suppose happens. Indeed, such sources can yield a fairly reliable reconstruction of certain aspects of the playhouse experience. Such documents, however, prescribe rather than describe performance, and they point toward an ideal and nonexistent theatre. We conventionally turn to other documents for evidence of what may actually happen after "what ought to happen" meets the real world. Diaries, letters, newspaper reviews, and memoirs allow glimpses of performance in the real world. They require careful cross-referencing to parse perspectives, rhetorics, biases, blind spots, and audiences, intended and actual. Even so, textual evidence poses significant limits to recreating performance history. Documents are most obviously not performances, only the "residue and traces of ghosted performance," as Odai Johnson writes.[64] There always remain absences toward which the archival record can only gesture. Read carefully, however, those gaps themselves can provide a valuable record. The practices of theatre history do not so much reconstruct events as read the tensions and the cracks within an archive. Fissures among prescriptive and descriptive texts, stage and street acts, archives and repertoires all provide traction for reading a performance event.

The interplay of text and performance plays a central part in this study. Performance comes to inhabit, travel through, and shape the material archive of print culture, which in turn affects the performance it bears and produces. Early American theatre built powerful alternate public spheres on modes of participation and spectatorship distinct from those characterizing literature, visual art, and other cultural forms.[65] The "scene of reading" that Nancy Armstrong has examined, a middle-class performance of interiority and imagined discipline, certainly shaped a textually literate early American culture.[66] At the same time, those practices were more public than they appear. Early American readers performed a variety of relationships to the word, leaving traces of their active engagements with the text. They scribbled marginalia, dismembered texts, and incorporated them into chapbooks and letters. The word performed its own cultural work, as Cathy Davidson and Jane Tompkins have both shown.[67] The varied uses and practices of literacy converged with visual, musical, and even theatrical sensibilities. Understanding that internalized and disciplined reading practices coexisted with externalized, public, and unruly modes of spectatorship, Julia Stern has usefully argued that early American fictions imagine the "connection between vision and emotion, spectacle and sympathy."[68]

Even more than literary texts do, early American theatre's scripts instigate active and participatory imaginative experiences. Theatre does not provide audiences with the solitary experience of a self-contained playworld. Rather than textually working toward the abnegation of embodiment that Michael Warner has detected in early America's republican print discourse, the texts of the stage presuppose, rely upon, and reproduce the staged presence of the subject.[69] Moreover, the experience of theatre reinforces the importance of embodiment in representative acts. The presence of the actors, the materiality of the playhouse, props, costuming, and makeup all call attention to theatre's self-presentation. The documents of the stage reveal these interactions, and live performance leaves its traces on these documents. Theatrical texts contain, replicate, and transmit performance in the material forms of print culture. Ultimately, although they accomplish distinct kinds of cultural work, "print culture" and "performance culture" interact across porous boundaries, and the archive and the repertoire frequently appear intertwined.

As Diana Taylor argues, performance survives in both the privileged written archives and the fugitive repertoires of vernacular performance.[70] Records of offstage acts preserve traces of performance repertoires (perhaps ironically) in written archive form, and the archives invoke and depend upon the embodied presence of actors. This study takes as a guiding principle the notion that text and performance, speech and writing, do not exist in simple binary relationships, whether as supplements, opposites, or causes and effects. Instead, those modes operate in mutually constitutive and interacting processes of communication and transmission. Print culture's various texts can generate, produce, reproduce, provoke, respond to, riff on, revise, and even reenact performances; performance in turn demonstrates a similar array of active relationships with print culture. At the very least, it seems a two-way, mutually constitutive street.[71] This study thus hopes to remain aware of interactions of print and performance, and the ways in which our understanding of theatre relies upon textual evidence.

Print culture and the institutional stage also maintain symbiotic relationships with offstage performance. Vernacular acts lead secret lives within the formalized conventions of the playhouse. The stage lives of the Atlantic underclasses draw from the variegated offstage performances of outsider and outcast. In this account, hangings, black Caribbean dance, public demonstrations, effigy burnings, tarring and feathering, militia musters, street busking, and the practices of naming and advertising ethnic and racial identity all actively traffic between streets and stages. If we look harder, more forms will inevitably burst into view. The stage institutionalizes such acts, operating as an active and tensioned intermediary between cultural categories in the Atlantic world. Offstage forms in turn repeat and revise onstage acts. Early American theatre thus links sites of oral and literate, local and cosmopolitan, elite and plebeian cultures. Articulating diverse and mongrel constituencies, such handoffs link various kinds of groups and identity categories, frequently revealing the problems and opportunities of cultural production and transmission across boundaries of class, race, and ethnicity.

Low, vernacular, and popular forms maintain irregular and fugitive presences within theatre culture. To borrow Raymond Williams's term, such vernacular acts code complex and subtle "structures of feeling," rather than clearly articulated philosophies. They represent offstage versions of the "theatrical formations" that Bruce McConachie sees as subtly conditioning actor, act, and audience.[72] The vernacular formations circulate in the transient but solidly embodied forms of music, dances, and gestures. They move through instinctive interactions and remembered scenarios. They embody the interplay of emergent industrial and residual preindustrial cultural forms. The cultural forms that we variously identify as "folk" and "popular" coexist, overlap, and share with "mass" and "commercial" forms of culture. Theatre, as Victor Turner has influentially argued, indeed operates as and alongside ritual.[73] This study crucially does not imagine theatre as the "more sophisticated" form of vernacular customs. The anthropological work of James Clifford and Clifford

Geertz has shown that the performances we variously identify as folk, popular, or vernacular are not merely the naïve, un-conscious acts of pre-theatrical people, and this study tries to internalize that understanding.[74] If anything, such acts seem frequently already theatricalized, appearing alongside, repeating, and revising onstage forms. The stage demonstrates various relationships to its offstage contexts, drawing from, continuing, revising, and contributing back to offstage acts, which in turn act in a similarly various ways. Theatre certainly draws upon informal acts, occasionally even publicizing such pilferings to bolster its own authority. In contrast, popular acts do not always make such citations, nor do they always keep formally archived records of their sources. That the folk so often seems the raw material of formal theatre says more about the ways the culturally empowered appropriate and divulge their borrowings, producing newly useful pasts by erasing the less valued elements.

It seems to me that underclass performances occupy special positions within print-performance culture, mediating among the differently documented lives of the powerless and those who live within and through print cultures. Real, material conditions of the eighteenth century defined and even produced rogues, gangs, and mobs, and the material, lived experiences of the less privileged were certainly important to early American theatre. Stage personnel had contact with the low; many came from the ranks of the working and impoverished classes. The less privileged worked behind the scenes and even appeared on formal stages, shaping theatre in significant ways. The stage took up the low, reshaped them through stage techniques, and trafficked in their experiences. Public re-performances of vernacular, folk, popular, and mass culture further circulate such acts, revealing the workings of the Atlantic entertainment industries.

Early American theatre's inventions and conventions produce entertainment out of the mobile and disorderly low. Certainly, theatre has committed the crimes of misrepresenting the low and screening audiences from the often gritty, unpleasant, and even abject realities of their lived. Just as bad, propping such characters up on stage inevitably transforms them into the objects of privileged spectatorship. At times, it warps their acts and persons beyond recognition. Even so, rogue performances do not merely manipulate pliant and abject underclasses. Riotous acts and extravagant spectacles openly display resistance to the ordering practices of Atlantic modernity. That cagey theatricality and the consistent desire for it center this study and its own fascination with the outlaw characters. Rather than trying to peer behind duplicitous masks or recover real lives from outsider misperceptions, this account seeks to reconstruct some of the ways in which the "acting low" and the "reenacted low" intertwine. Outlaw acts embody conflicting urges to rise and fall; they trouble hierarchies. They play with the possibility that the low can act independently. Such potential seems covert, furtive, but nonetheless real and potent. Rogue performances represent considerable resources for cultural production and transmission, and this study means to recover their power.

2. Gallows Performance, Excarceration, and *The Beggar's Opera*

On October 21, 1773, Levi Ames, a twenty-two-year-old petty thief, died by hanging on Boston's gallows. His ascent to the scaffold made him for a short time a celebrity of sorts, the focus of a spectacular performance of punitive power. His arrest, trial, and execution aroused significant public interest, and a sizeable audience turned out to see him "turned off" the scaffold. Although for only a few brief moments, Ames played a complexly scripted role in the Atlantic theatre of criminality and punishment. The language of performance and celebrity does not just describe Ames's act casually or figuratively—it was indeed theatre, or perhaps a paratheatrical event. Ames, an amateur in his first and last performance of the kind, ended his life on a temporary stage specifically constructed for punitive display. Documentary evidence reveals the theatrical conventions that shape the public performance of a hanging. The execution itself presented scripted, conventionalized acts calculated to draw a crowd of spectators. Those performances started before Ames's September 7, 1773, trial, with the confessions, evidence, and juridical proceedings included in some of the accounts. His September 10 sentencing sealed his fate, and on the Sabbath prior to the execution, Boston pastors Samuel Stillman and Andrew Eliot both preached sermons commissioned and supposedly attended by Ames. At the Thursday lecture, on the day of Ames's execution, Samuel Mather preached another sermon, "Christ sent to heal the broken hearted," which also claimed Ames's attendance and bore his supposedly autobiographical life story when printed. At various times before the execution, auditors recorded versions of Ames's "Last Words and Dying Speech," which they then reprinted as warnings to other would-be criminals. At his execution, Ames supposedly delivered an "affecting speech," recorded as by a "dying penitent."[1]

As Ames stood on the gallows, printed versions of his dying theatrics circulated through the audience. Broadside poetry bid farewell to Ames and even protested against the injustice of theft as a capital offense. Others created more imaginative and affective performances, purporting to reproduce Ames's last words and even his dying groans. Many of these poetic effusions must have been composed before the execution date in order to appear in print on the street on the fateful day. Someone, perhaps the Elhanan Winchester who claims the songwriting credit, sang an "Execution Hymn" to Ames two days before his death.[2] The prisoner allegedly requested an encore of the song for his execution day. As Stillman's printed sermon advertised on its title page, Ames conducted a conversation with the pastor "as he walked with him from the prison to the gallows."[3] The staged act of conversing and walking to the execution site enacted the thief's penitence and the minister's spiritual authority.

Ames's stagey execution also supplied the raw materials for further performances. Illustrated broadsides (some purporting to represent Ames's acts on the scaffold and others standing as independent performances) were produced and sold near the site of the execution. The spectacle of Levi Ames's theatrical death provoked texts, which in turn can provide a window into the theatrical construction of those moments. Such performances, although informal and vernacular, responded to and reshaped preexisting print and stage practices. The event's theatricality becomes visible alongside the practices of institutional theatre—the spaces, scripts, actors, and audiences of the formal stage offer compelling insights into the street theatre of rogue actors and punitive performances. Although Ames's execution enacts the power of the state, as Foucault would argue, it seems far from a decisive demonstration of uncontested authority.[4] Ames's gallows performance shows complex and sophisticated forms of theatricality, even play, emerging from the disciplinary show. Ames's rogue performance reveals scripted qualities as well as a frequently stagy awareness of display, exhibition, and the various roles enacted by its participants and audiences.

Before he even stepped up onto the gallows, Ames appeared already theatricalized—hyper-mediated, thoroughly scripted, always already performing within the stage of public execution. The execution points toward the workings of print-performance culture, as disembodied and disseminated texts conjure Ames's corporeal presence. Those documents offer a sense of the possibilities and threats posed by the criminal body—the ways, perhaps, in which outlaw actors such as Ames both submit to and trouble the templates. In key ways, the formal stage supplied those scripts. The century's most popular performance of low roguery, *The Beggar's Opera*, staged imaginary alternatives to Ames's execution. The *Opera*'s potential counterperformances remain submerged and only partially articulated in scenes of punishment and discipline.

THE BEGGAR'S OPERA AND ATLANTIC CRIMINAL CULTURE

Ames's execution—a relatively rare occurrence in the 1770s—seems at once unusual and conventional, an extraordinary event happening in the usual way. The spectacle emerges from the playhouse's durable and flexible forms of staging criminal performance. As various prohibitions foreclosed certain formal kinds of theatre, especially in Ames's late colonial New England, paratheatricals played an important role in the public life of Atlantic performance. The Atlantic forms of criminal performance had developed collectively and casually in the streets. Even when officially suppressed, however, the formal stage exercised considerable influence. John Gay's 1728 *The Beggar's Opera*, one of the most popular plays in the eighteenth-century Anglo-Atlantic world, performed precisely the sorts of uncontrolled, uncontrollable criminal acts that hover in the background of Ames's theatrical execution. Gay's ballad opera popularized the acts of the highwayman Macheath and his gang, making them specifically theatrical, and making that theatricality suggestively resistant and perhaps even emancipatory. *The Beggar's Opera*, famously staging the low as disdained, outcast, and outlaw, represents a center of gravity both for this study and for its eighteenth-century Atlantic audiences.

John Gay wrote *The Beggar's Opera* in 1728 as a social and political satire on Robert Walpole's Whig administration and contemporary elites. It introduced a radically new and popular stage form, the ballad opera, which restructured the found materials of English music to mock the elite genre of Italian high opera.[5] Appearing an unprecedented sixty-two nights in its first season at Lincoln's Inns Fields, and almost as many the second, the opera became one of the Atlantic world's first commercialized popular entertainment phenomena, inspiring decades of celebrity biographies and promotional merchandising. It quickly produced new forms of theatrical celebrity, shaping an emergent eighteenth-century Anglo-Atlantic theatre culture.[6] The logic of this theatre culture was popular (though not precisely democratic) and commercial; audience demand for popular performance and celebrity displaced elite standards of taste and decorum.

In their celebrated scenes, Macheath and his gang restaged the eighteenth-century Anglo-Atlantic public crisis of property, criminality, and punishment. The play capitalized, for example, upon "Newgate Calendars," the popular accounts of criminals whose careers inevitably ended on the Tyburn gallows. Such texts accompanied the rise of hanging itself as a spectacle, reproducing the conventions of its repeated enactment. *The Beggar's Opera* restages most immediately the celebrity escapes and final punishment of the London thief Jack Sheppard, whose history (some versions of which have been attributed to Daniel Defoe) thrilled audiences in 1724.[7] As the entertainment industry's response to the emergent theatricalization of crime, Gay's ballad opera vaults the rogue and his cohort into newly modern forms of celebrity. Diane Dugaw has shown that Gay's use of popular and plebeian materials (ballads, catches, mummer's plays, country dances, fables, and so on) not only ensured the *Opera*'s enduring favor with audiences, but also allowed his work to comment on the emergent forms of modernity and the eighteenth-century constructions of social hierarchies.[8] *The Beggar's Opera* shaped the rogue and his relationship to Atlantic modernity, mediating class formations and plebeian expressions in performances into the nineteenth century.

The Beggar's Opera centers on the character of the rogue, embodying its outlaw underclasses in sumptuous costuming, insouciant attitudes, and catchy vocal performances. As Michael Denning has argued, Gay's underclass ideology fuels an inversive and ironic critique of English class structures and their legal underpinnings.[9] The pervasive irony of the *Opera* accuses elites of roguish behavior while charging rogues with the greed, disloyalty, and duplicity of modern society.[10] Rogues such as Macheath do not precisely function as "social bandits," to use Eric Hobsbawm's debated formation—Macheath, in fact, serves as Hobsbawm's counterexample.[11] At the same time, he performs a popularized social and political critique of the elites, and ultimately Macheath seems to have become a heroic embodiment of underclass agency. Most alarmingly, the *Opera* makes entertainment out of such class critiques. The play grants the highwayman the social authority, even cachet, to become the

Anglo-Atlantic world's "social critic par excellence," as Andrea McKenzie argues.[12] Such meaning, of course, can vary in performance, and later performances may have dulled Gay's edgy irony, which prevents the *Opera* from sentimentalizing or romanticizing low criminality. Moreover, in the hands of impromptu and popular re-enactors—provincial troupes, theatre audiences, or even condemned criminals—Macheath seems to have provided a malleable form for the heroic imaginings of charismatic lowness and roguish individualism.

Macheath was not the first highwayman on stage, but he seems the most popular and compelling rogue of his kind. As Gillian Spraggs has observed, Macheath is no ordinary criminal; his status as "gentleman thief" makes him an inversive and charismatic figure, troubling to the social hierarchies that he pilfered.[13] His offstage thieving and multiple escapes even become heroic, and the play trades self-consciously in Macheath's status. Audiences can agree with Mrs. Peachum, Macheath's unwitting mother-in-law, who exclaims, "Sure there is not a finer gentleman upon the road than the captain!"[14] Even her husband, the fence and thief-taker Peachum, begrudgingly notes Macheath's "Personal Bravery, his fine Stratagem."[15] The attribution of gentlemanliness gauges the role's ironic and inversive social critique, but it also shows the importance of "style" to the rogue. Such performances occurred offstage as well; Gay's play loosely follows the trajectory of the historical housebreaker Jack Sheppard, who, together with the play, helped inspire an eighteenth-century genre of rogue street style.[16] Tyburn's condemned criminals enacted theatrical exhibitions of style, staging what Peter Linebaugh describes as a "flaunting, ostentatious display of opposition to the severities of the law."[17] Turned into an iconic character of popular culture, Macheath arguably dominates the *Opera*'s rogue performances. On stage, Macheath carouses in taverns with numerous women in larger-than-life demonstrations of his sexual prowess and charismatic appeal. Such scenes shape rogue performance—Macheath, for example, leads his gang in blustering and audacious boasts about the rights of highwaymen to pursue their vocation with panache.[18] Likewise, songs such as "Youth's the Season" (which Macheath sings during Act 2, surrounded by admiring women in a tavern) articulate the play's outlaw style.[19] Macheath's appeal came from his theatricality—his roguish style, self-conscious posing, and ultimately, the ability to perform.

The role's charisma exceeded the confines of the playhouse. Some accounts tell of Thomas Walker, the first Macheath, taking the part from its intended actor when Gay overheard him humming the tunes in a lively manner. Much later in the century, accounts remember Walker for having "thrown an easy and dissolute air into the character."[20] There is no way to tell whether Walker lent his style to the part, or if the story represents a retroactive myth that developed in the wake of Macheath's popularity—and that may be precisely the point. *The Beggar's Opera* makes an actor's theatrical style and his character's charismatic roguishness interchangeable qualities. Rogue style also notably crossed and mediated class boundaries offstage. The young Scottish aristocrat James Boswell, for example, seems to have been particularly taken with roguish styles, writing in his journal of an encounter with two young prostitutes in 1763 in which he "toyed with them and drank about and sung *Youth's the Season* and thought myself Captain Macheath."[21] More dangerously, impressionable youths apparently took inspiration from Macheath, going on crime sprees after viewing the

Opera. Condemned criminals, as Andrea McKenzie has noted, seem particularly inclined to reenact the role of Macheath. They read the play in their cells for consolation, as Isaac Darkin reportedly did in the early 1760s, sang songs from the ballad opera, and died resolutely, performing their own versions of Macheath's roguishness.[22] The reports of imitation, of course, may represent rumors rather than actual acts of mimicry, formless fears of the outlaw imitation that could produce gangs of aspiring rogues. Even if only paranoia, such reports more importantly reveal the rogue's conventional allure. Paranoia and rumor, reputation and infamy point precisely to my quarry—the diffusions and transmissions of roles and character types away from their supposedly "authentic" sources and into wider arenas. That diffusion scripts Macheath as a part available for offstage imitation.

John Gay's supporting cast of petty thieves also models potential audience behavior, following Macheath's roguish lead. Peachum (based on the historical Jonathan Wild, a duplicitous receiver of stolen goods and thief-taker) peruses the "Register of the Gang," detailing a rogue's gallery of characters, including such colorfully named characters as Crook-finger'd Jack, the gang's most productive thief and Wat Dreary, "alias Brown Will, an irregular Dog." Contemptible drunkards such as Tom Tipple contrast with Mat of the Mint, a "promising sturdy Fellow" valuable for his thieving prowess.[23] Like Peachum himself, who operates in a self-consciously proclaimed "double capacity, both against rogues and for 'em," the *Opera* simultaneously celebrates and censures criminal actions.[24] Ultimately, Gay's gang outperforms Macheath's roguishness by betraying him to authorities. Its rogues become, in the ironic logic of the opera and in the economy of theatrical celebrity, at once despicable and admirable. In the end, they remain empowered by their ability to conjure popularity over and against dominant moral and legal practices.

The condemned criminal remained a popular spectacle outside of playhouses well into the nineteenth century. Reenactments of the role reappeared, for example, as the late-eighteenth-century Sixteen-String Jack, a fancy-dressing highwayman known for his elaborate laces and genteel manners. Macheath appeared again in the 1830s with the revival of Jack Sheppard, the housebreaker on whom Gay had originally modeled the highwayman. William Harrison Ainsworth's 1839 novel *Jack Sheppard* and successful stage adaptations (in London by John Baldwin Buckstone, in Philadelphia by Jonas Phillips) further developed the Atlantic thief's legend. The more sensational forms of the *Newgate Calendar* and the *Malefactor's Register* eventually gave way to the purported disciplinary functions of the mid-nineteenth-century *Police Gazette*. Even so, the proliferation of cheap popular editions of such texts suggests that their policing functions (while not entirely incidental) took a back seat to their value as popular, "vulgar" entertainment, as appealing representations of charismatic criminality.[25] A similar doubling of vulgar entertainment and moralizing exhortation shapes the theatrical execution of Levi Ames and other Atlantic outlaws.

THE BEGGAR'S OPERA IN AMERICAN THEATRE

In America, Ames and others like him acted within a culture of theatrical criminality shaped by the *Opera*. The playhouse, unrestrained by the prerogatives of penitence and power, staged alternatives to the street theatre of public executions. The stage rogue, unlike Ames, did not have to die. In what seems a wishful fantasy of gallows performance, Macheath parlays his theatrical prowess into freedom at the end of *The Beggar's Opera*. The criminal celebrity of the jailbreak artist Macheath and his thieving gang provide a model that persisted and circulated through the Anglo-Atlantic world during the course of the century, remaining popular well beyond the eighteenth century. In an age that scholars have described as thoroughly theatrical, the stage provided discourses, forms, and cultural logics for understanding criminal behavior, state authority, and even resistance to and revision of those forms—so much so that as Ames goes to the gallows, theatre does not simply provide a metaphor for the rogue; it actually constitutes his criminality and discipline.[26]

The American rogue represents the logical continuation of Gay's London stage gang, who were bound for the Americas all along. *The Beggar's Opera* implicitly sentences the gang to transportation, condemning them ironically to further mobility and criminality. The audience, the script instructs, "must have supposed they were all either hanged or transported."[27] Gay's sequel *Polly* (published in 1729 but not performed until the late 1770s) confirms this supposition when the gang, breaking out of their London underworld, terrorizes the West Indies. The gang's onstage sentence of transportation emphasizes theatre's own more profitable transportation of such characters. Eighteenth-century theatre culture exported the characters, and Ames's 1773 audience would have been familiar with them. The *Opera* first appeared in the English colonies in 1733, only five years after its metropolitan premiere. It eventually became the fourth most popular play and the single most popular ballad opera in America. With thirty-two known colonial performances, it trailed the most frequently performed play (*Romeo and Juliet*) by a mere three performances over the entire colonial period. Gay's ballad opera seems in a virtual dead heat for the prize of "most frequently performed," if not "most popular." Those numbers reflect the incomplete evidence of early American theatre; quantitative measurements, moreover, do not reveal qualities of reception or interpretation. They do, however, suggest that American travelling troupes presented the play as a regular offering, and that local theatregoing audiences had a good chance of attending a performance of *The Beggar's Opera*.[28]

Fifty years after its American premiere, the cultural work of *The Beggar's Opera* had shifted and diversified. Its topical political satire had little force by the end of the century—it no longer found a target in Walpole's administration as it had in 1728. The *Opera* in America also no longer exercised the power to make instant celebrities out of its actors, although it gained status as a test of theatrical skill in American theatres. It had produced a new Atlantic genre; ballad opera had become one of the Atlantic theatre world's most frequently performed modes. Ballad operas such as Isaac Bickerstaff's 1762 *Love in a Village* dominate lists of frequently performed American plays, perpetuating the plots and performance techniques of *The Beggar's Opera*.[29] The charisma of Gay's highwaymen also persisted in Anglo-Atlantic culture. Acting criminal, Macheath reminded audiences, meant overtly self-conscious performance and celebrity. Street theatre such as Levi Ames's execution reinvents such associations. Although Ames and his ilk rarely perform their way to a reprieve, they still convert the gallows into a stage, a place of appealing and charismatic underclass performance.

Theatre troupes traveled the routes implied onstage by the threat of transportation, staging rogues and gangs in nearly every major port of call in the circumatlantic world. The continued American productions of *The Beggar's Opera*—local performances before local spectators—creolize the play and its circumatlantic cohort. The first known New World staging of *The Beggar's Opera* occurred in 1733 in Jamaica. William Rufus Chetwood's 1749 *History of the Stage* retells the story of the play's inauspicious American premiere:

> [A] company in the year 1733 came there and cleared a large sum of money.... They received three hundred and seventy pistoles the first night, to *The Beggar's Opera*; but within the space of two months they buried their third Polly and two of their men. The gentlemen of the island for some time took their turns upon the stage, to keep up the diversion; but this did not hold long, for in two months more there were but one old man, a boy, and a woman of the company left. The rest died either with the country distemper or the common beverage of the place, the noble spirit of rum punch which is generally fatal to new-comers. The shattered remains, with upwards of two thousand pistoles in bank, embarked for Carolina to join another company in Charlestown, but were cast away in the voyage.[30]

The account suggests the dangers of acting low in the Caribbean, rehearsing the associated dangers of morally suspect activities, tropical climates, and circumatlantic mobility. The report construes actors as rogues, concluding that had "the company been more blessed with the virtue of sobriety they might perhaps have lived to carry home the liberality of those generous islanders."[31] The theatre troupe remains unnamed, effectively becoming their roles in Chetwood's account, even reenacting the low dissipation and inebriety that supposedly separates newcomers from the island's "gentlemen." The "diversions" of theatre, in the typical veiled logic of antitheatricality, only lead to disease and dissipation. The actors withdraw to another part of the Americas, eventually ending their retreat in shipwreck. Victims of their own mobile, marginalized, and ultimately castaway status, the troupe lived out the possibilities of Macheath and his gang.

Nevertheless, Gay's criminal antiheroes became a staple of American theatre as companies invariably attempted the ballad opera in the late eighteenth century. The first New York City performance of *The Beggar's Opera* occurred on December 3, 1750, at the Nassau Street Theatre operated by Thomas Kean and Walter Murray's theatrical company, one of the first to present regular dramatic seasons in the American colonies.[32] This popularity persisted; according to William Eben Schultz, "[n]o decade of the century, beginning with 1750, finds Gay's piece absent."[33] American theatre historian George Overcash Seilhamer notes that "for almost half that period [between 1728 and 1769] there was no American company so 'mean and contemptible' as not to sing or attempt to sing" the *Beggar's Opera*.[34] Gay's theatrical underclasses generate cultural cachet from the otherwise disdained low, and the *Opera* reiterates Atlantic culture's conversion of "mean and contemptible" characters into charismatic and appealing, even prestige-generating, theatrical roles. Macheath and his gang propagated a sort of carnivalesque inversion on an epic circumatlantic scale.

A sampling of playbills in the Harvard Theatre Collection confirms that companies staged the *Opera* up and down the eastern seaboard of North America, from

Charleston, South Carolina to Portsmouth, New Hampshire. Theatregoers in the northern communities as well as southern enjoyed the play. Its second part, *Polly*, does not seem to appear on the American stage until much later, but even Boston, with its more consistent opposition to institutional theatre, saw one-man performances of *The Beggar's Opera* in the 1760s and 1770s presented as lectures interspersed with songs.[35] The *Opera*'s consistent popularity suggests that American audiences found something compelling in staging such rebellious low characters. Gay's *Opera*, performed frequently enough to be familiar with most playgoers, offers an alternative to the serious tragedy and genteel comedies of manners that otherwise dominated American stages. It saw more performances, for example, than George Lillo's 1731 *George Barnwell; or, the London Merchant*, a contemporary of the *Opera* that redeems its wayward underclass apprentice with a "staunch advocacy of Christian temperance and a stringent work ethic."[36]

In outperforming other underclass acts, the *Opera* helped construct a New World culture in which underclass characters enacted a roguish social critique shaped by the forms of publicity and theatrical celebrity. Its American performances, like their counterparts in London, habitually skirted the bounds of propriety, but *The Beggar's Opera* seems to have gone underground with its social critique by the 1770s. By that time, London performances of the *Opera* frequently appeared as cross-dressing spectacle, offering popular titillation rather than serious social or political critique. Contemporary audiences and critics alike have generally received these productions as little more than amusing diversions, but they also reveal a significant trend in the way that the late-eighteenth-century stage rewrote and rerouted the cultural work of Gay's rogues—the *Opera* had begun to operate behind a mask of frivolous humor. The evidence suggests that *The Beggar's Opera*'s criminal characters frequently appeared in a mix of social peril and entertainment appeal. In Williamsburg, Virginia, a 1768 performance by the Virginia Company of Comedians (a troupe that may have shared some performers as well as its repertoire with the Lewis Hallam's American Company in New York) presented the *Opera* in partial cross-dress. A Mrs. Parker played Polly for her benefit, while a Mrs. Osborne played Macheath. Similarly, at least one of two 1769 Annapolis performances by the New American Company featured some cross-dressing antics. A cast list printed in the *Maryland Gazette* indicates that while Mrs. Parker played Polly to William Verling's Macheath, a Mr. Walker doubled up in the roles of Crook-Finger'd Jack and Moll Brazen.[37] These particular performances do not seem to have inspired much opposition, but a travestied *Beggar's Opera* might still have been an unusual event, linking underclass and homosocial spectacles.

Such performances generated opposition to *The Beggar's Opera*, subjecting it to suspicion and even censorship from time to time. A number of altered or censored versions of the play appeared. Performances in Charleston, South Carolina, in 1793, 1794, and 1795 had "exceptionable passages obliterated"; in 1799, the ballad opera appeared in a "deodorized version" in Philadelphia, according to William Eben Shultz. Shultz calls the censorship evidence of a "countercurrent to the general popularity of *The Beggar's Opera* in America," but such official intolerance seems to indicate just the opposite, responding to the play's popularity.[38] No details of the censorship remain, but one might argue that official rewrites attempted to manage the dangerous appeal of an onstage lumpenproletariat. That theatres occasionally watered down these performances suggests that troupes performed the *Opera* despite and through opposition, adapting it so that it could appear, even if in altered form. Performances of Macheath's charismatic gang combined with the always-precarious social and legal positions of theatres and actors to produce uncomfortably popular stagings of mobility, of shifty disorder, of dangerously lionized criminality. Indeed, after Josiah Quincy attended a performance of the *Opera* in 1774, he reveals the conflicted relationship of some early Americans to the stage. "I am still farther satisfied that the stage is the nursery of...vice, & diseminates [*sic*] its seeds far & wide with an amazing & baneful success," he wrote in his journal. Ironically, he also demonstrates familiarity with the *Opera*'s characters and appreciates the striking "elegancies [of] gesture voice & action" and the "powers of eloquence" he witnessed in the playhouse.[39] Quincy, attending the theatre repeatedly during his stay in London, seems just as taken with the stage's appealing eloquence as fearful of its potentially corrupting influence.

The Beggar's Opera perhaps seemed most dangerous around underclass audiences, who seem already inclined to imitate its scenes—Gwenda Morgan and Peter Rushton, for example, observe a "certain irony" in the *Opera*'s performance in Maryland, "the colony with the highest proportion of transported convicts in its population."[40] At times, Gay's characters and scenarios certainly could have numbered America's lower sorts among its audiences. Its acts would have resonated openly with the actual lived experiences of its American audiences. Such possibilities, however, probably represent exceptions to the usual circumstances of encountering the *Opera*. Many of early America's lowest sorts could not afford tickets, as Odai Johnson argues, limiting the possibility that *The Beggar's Opera* performed its transgressive acts before low audiences.[41] Gay's play may never have presented genuinely popular class-leveling entertainments. At the same time, such obstacles to underclass spectatorship may have increased the chances that servants or slaves saw the play illicitly, sneaking in or remaining in a seat intended for one's master. Underclass spectatorship of *The Beggar's Opera* may have itself been a transgressive act, albeit as exception rather than rule.

More historical evidence attests to *The Beggar's Opera*'s cross-class cachet. The rogue's appeal mediates social positions, and a great number of its audience would have encountered its underclass theatrics as a sort of imagined slumming. Alexander Graydon, a Pennsylvania lawyer, Revolutionary officer, and politician of the early republic, reveals a thorough familiarity with the *Opera* in his evocative, if not completely historically reliable, memoirs. "Mr. L," a local minister, also turns out to be surprisingly familiar with its tunes:

> One day, as I was strumming a tune from the Beggar's opera, upon a fiddle I had purchased, with a view of becoming a performer upon it, he ["Mr. L"] entered my apartment. What, says he, you play upon the violin, and are at the airs of the Beggar's opera! He immediately began to hum the tune I had before me, from which, turning over the leaves of the note-book, he passed on to others, which he sung as he went along, and evinced an acquaintance with the piece, much too intimate to have been acquired, by any thing short of assiduous attendance on the theatre.[42]

Not only does the parson's familiarity imply his slightly scandalous "assiduous attendance" of the theatre, but it also suggests repeated stagings of *The Beggar's Opera*. Its audiences, as well might expect, recruit the *Opera* for different purposes. Graydon ostensibly used it to polish his genteel musical skills; somewhat more shockingly, the minister (who could pass theatrically "from grave to gay, from lively to severe" as circumstances required) reenacts a version of rogue theatre.[43] Minister L, as Graydon notes, also plagiarizes sermons and makes his talent as a jockey a point of pride, and the ballad opera marks his participation in rakish sporting culture. Graydon imagines the two connecting over Macheath's airs despite these apparently superficial differences, using the criminal subculture of *The Beggar's Opera* as both excuse and pattern for privileged male socialization in early American culture.

If eighteenth-century productions seem to have habituated audiences to the *Opera*, the routine always seemed capable of becoming unusual spectacle. The *American Monthly Magazine and Critical Review* in 1817, for example, reports the hissing of the *Opera* at New York City's Park Theatre. Rioting occurred, and watchmen arrived to keep the peace; arrests and legal action followed. Although the play had become an everyday occurrence, the volatility of the mob continued to punctuate rogue performances. The account blames the play itself, pointing a finger at the "disgust produced by the representation of this vulgar and licentious *burletta*."[44] That prim reaction, however, may actually camouflage the real reason for the disturbance. Notices in the *American Monthly Magazine* leading up to the disturbance suggest that the riot may have responded primarily to poor acting by its English performer, Charles Incledon, who replaced a local favourite. Incledon's performances had disappointed audiences, and an "American patriotic song of British manufacture" hardly made up for the absence of "those powers which the company had assembled to admire."[45] The reviewer's moralizing focus on the play's vulgarity seems opportunistic. If, as E. P. Thompson has argued, English (and arguably Anglo-American) rioting frequently defended traditional prerogatives, the disturbance might demonstrate that audiences, in fact, had enough exposure and experience to expect certain qualities in performances of *The Beggar's Opera*. They were prepared to preserve their perceived theatrical rights violently as well, acting like the stage low in order to defend the cultural representation of such types. In such scenes, the rogue seems consistently capable of producing his audience as the mob, if only to riot for the traditional dramatic prerogatives of early American theatregoers.

ROGUE ACTS AND THEATRICAL EXCARCERATION IN *THE BEGGAR'S OPERA*

Theatres at the end of the eighteenth century staged *The Beggar's Opera* in suggestive juxtaposition with an early American criminal crisis. Especially after the French and Indian wars, New England experienced an upsurge in the same kinds of property crime that had created a London crisis of legal authority and helped inspire *The Beggar's Opera*. Americans began to perceive a problem class that Daniel Cohen describes as "neither insiders, nor outsiders, but undersiders, participants in a flourishing quasi-criminal subculture."[46] The undersiders were a development of the eighteenth century, distinct from the "sinful but integral members of local communities" and the social outsiders such as pirates that appear most frequently on earlier gallows.[47] Mobile and displaced cohorts such as decommissioned military personnel, transported convicts, immigrant laborers had trouble achieving or restoring integrated positions in early American communities. They resembled neither the servant classes nor the alien outsiders on whom early American communities had traditionally blamed criminal activity, and they seemed to be recruiting. Certainly, the evidence suggests that crime waves did occur; at the same time, such events remain inevitably a function of perception and representation, dependent upon the practices of legal detection and print communication. The historical "upsurge" appears in the form of newspaper reports, alarmed clergymen and official paranoia, new and revised laws, and increased numbers of arrests, convictions, and executions.[48] Such evidence shows the difficulties not only in controlling the low, but also in understanding them, capturing the fugitive realities of criminal culture. The crisis occurred in the realm of perception as much as in sociological reality.

It is tempting to see early American performances of *The Beggar's Opera* during this period as a sign of the times, an expression of the criminal crisis. However, if the *Opera* reflects the statistical upsurge in crime and capital punishment, it also represents an extravagant, wishful response to those offstage contexts. Theatre does not precisely concern itself with statistical realities, anyway. Spectacular exceptions make entertaining plays, and theatre remains fascinated with the possibilities of evasion, escape, and even resistance. Indeed, the popularity of Macheath and his gang seems a result of the threat of punishment that shadows them throughout the play. As Mrs. Peachum comments, audiences (notably women) are "so partial to the Brave that they think every Man handsome who is going to the Camp or the Gallows."[49] Criminal display, in Mrs. Peachum's ironic logic, creates the illusory performance of bravery, which colors the subject as "handsome," which in turn augments the display, reinforcing the circular logic of theatrical criminality. The cohort repeatedly asserts a bold, charismatic indifference to punishment. Even when imprisoned at the play's end, Macheath's spirits hardly seem dampened; "Since I must swing," he asserts, "I scorn, I scorn, to wince or whine."[50] Macheath's performances of charismatic indifference and even bold defiance eventually short-circuit the processes of discipline that loom over Macheath and his crew. Macheath's escapes, his last produced by his ability to perform, eventually come to define the power of *The Beggar's Opera*'s outlaws. While increasingly imitated on early American streets, such acts rarely had the same spectacular and liberating result.

Breakout potential remains the signature of the theatrical rogue, and the *Opera*'s excarceral acts make its hero Macheath an enduring model, the epitome of stagey transatlantic criminals. As Peter Linebaugh has convincingly demonstrated, the restructuring of Atlantic commercial, legal, and penal systems accompanied the broad reorganization of criminal and underclass cultures as well. Linebaugh's "Tyburnography" in *The London Hanged* has shown that court sessions at the infamous Old Bailey prison (in what must have been an unintended consequence of efforts to suppress crime) produced performances and authorized the archival preservation of the criminal classes' stories. The newly visible bodies, gestures, and words of the criminal underclasses share at some root level the cultural forms of Macheath and his gang. Most strikingly, they exhibit "excarceration"—the resistance to and even escape from the grips of punitive authority "played out," Linebaugh writes, "in escapes, flights, desertions, migrations, and refusals."[51] Escape seems a real possibility in the eighteenth century, one to which criminal confessions attest with their rich catalogues of previous crimes and jailbreaks. Moreover, the enforcement of justice was as uneven in America as in London, and reprieve or escape a real possibility.[52]

The ballad opera's third act best displays the magical excarceral potential of performance. Macheath, betrayed by Jenny Diver, languishes in a prison cell, condemned to die before he can escape again. As the plot's tension heightens, the play's criminalized characters begin to perform, singing and dancing as if in response to punishment. The performers, Lucy Lockit reveals, are those "whose trials are put off till next session," now noisily "diverting themselves" in the "condemned hold."[53] In Newgate Prison's condemned hold in the 1720s, prisoners had used such disorderly performances as cover for escape attempts. Linebaugh describes a mass breakout attempt that took place as Jack Sheppard (the "original" Macheath) evaded authorities. Prisoners persisted in "Huzzaing, and making Noises...to prevent the workmen being heard"; they almost succeeded in transforming performance into excarceration.[54] In *The Beggar's Opera*, deferral of capital punishment provides the incarcerated underclasses with reason enough to celebrate. The prisoners' dancing procession—at once stage display and prison fun—foregrounds the popular entertainment that such character types came to provide on Atlantic stages. Criminal (though not necessarily criminalized) acts become opera, and the *Opera*'s ending performs a sort of cultural cause-and-effect relationship between official incarceration and underclass performance: imprisonment generates theatre.

Macheath has better reason to exult than the other prisoners do in the ensuing scenes. The fettered hero sings a series of airs, emphasizing his onstage function as simultaneously escape artist and vocal performer. Each role complements the other, and both form his identity as a heroic representative of the stagey criminal class. Excarceration and theatrical performance merge in the scene. Macheath sings a musical inner monologue, making entertainment a matter of his oscillating moods. The airs constitute ballad opera's distinct form, culturally constructing lowness as entertainment. Macheath's vocal performances are not gratuitous, nor are they an incidental matter of form, separable from the play's content. Interdependent theatrical form and rogue content create a mutually reinforcing wave pattern in the opera's ending. Macheath's theatricality makes him more than a representative of underclass attributes. Instead, as John Richardson argues, Macheath is primarily "an invented stage figure who challenges our notions" of both class and the stage by "being so insistently a stage figure."[55]

Performance, as Macheath's staginess confirms, defines the low. To be criminal, Macheath confirms, is to act, and performance focus the charisma and the threat of the criminal in the moment of stagey self-representation. Moreover, *The Beggar's Opera*'s performance of criminality enables Macheath's jailbreak. In the play's final scenes, the condemned highwayman launches an energetic critique of upper-class immunity from the law ("gold from law can take out the sting"), voicing the outcast low's jab at the powerful.[56] He performs theatrical protest, an outspoken inversive act claiming the right to judge authorized by his criminal bravura. The songs set in motion a cause-and-effect sequence of self-conscious theatricality. As Macheath's end draws near, the Player and the Beggar, low figures who introduce and regulate the opera from the framing locale of London's St. Giles slum, interrupt the ending. The interruption makes sense only from offstage, from the position of the audience member. Macheath's performances and the appearance of the regulating underclass formally perforate the play's diegetic boundaries. They subject the rogue not to the judgment of the criminal justice system, but to that of the Atlantic theatre-world, ultimately bringing Macheath a reprieve.

The "catastrophe" of Macheath's execution, the Player insists, is "manifestly wrong, for an opera must end happily"; form dictates content.[57] The Beggar acquiesces to Macheath's reprieve, observing that "in this kind of drama 'tis no matter how absurdly things are brought about."[58] The lines satirize the conventionalized opera forms that held the elite stage into the nineteenth century, but they also link entertainment forms to the exercise of class power. Before Macheath reappears on stage, the *Opera* reminds its audience that the rogue owes his last escape to audience appreciation—the Beggar directs the "Rabble" to "cry a Reprieve," all in order to "comply with the taste of the town."[59] Theatrical celebrity and popular taste supplant the interlined logics of justice, propriety, logic, and even aesthetic value as the spectatorial mob in effect springs Macheath. In the iconic concluding scene, Macheath (surrounded by an adoring audience of his multiple wives) dances and sings, performing within the performance, his theatricality punctuating the *Opera*'s ending. Macheath's final act ritually reenacts the theatricality of such slippery outlaw characters that enables their excarceral breakouts.

THOMAS MOUNT'S *CONFESSION*: REENACTING THE ATLANTIC GANG

The imagined potential for criminal self-staging makes *The Beggar's Opera* at once popular and proscribed, charismatic and dangerous. Such rogue self-representation appeared in other forms as well. Criminal confessions, for example, although their collective and cooperative authorship could minimize the role of the condemned in producing a text, still reveal negotiations over modes of representation and self-expression. The criminal confession represents a textual counterperformance poised against the excarceral acts of *The Beggar's Opera*.

The narrative of Thomas Mount reveals the complex possibilities scripted into rogue performances. Before his hanging in Little Rest, Rhode Island, in 1791, Mount admitted to a series of thefts in a formulaic confession. Like Levi Ames, he participated in the wave of eighteenth-century petty property crime, paying with his life for his numerous thefts. As in the execution spectacle of Ames, his text does not—perhaps *cannot*—appear without the sanction of religious right, legal power, and cultural authority. At the same time, it reveals the subtly contested scripts of rogue performance. Mount had supposedly engaged in an epic series of burglaries, beginning during his military service during the Revolutionary War and continuing through his personal trajectory of marginalization, displacement, and dispossession. He had thieved his way nimbly through the New England countryside, sometimes in the company of accomplices, frequently working alone. When caught, local authorities seem to have relied on traditional paternalistic responses to his transgressions. Mount replaced stolen goods or submitted to halfhearted corporal punishments and even a few brief incarcerations. Just as often, he contrived escapes, sometimes by betraying or abandoning his companions.

Mount's *Confession*, like many other documents of its kind, produces and documents a conventional performance of penitence.[60] As Dan Cohen explains, the generic practices of textualizing crime underwent broad shifts in the eighteenth century, most noticeably changing from religious to secular framing. While earlier execution accounts produced "confessions," later ones produced "speeches" and autobiographies. The document no longer renders public proof of an essentially private confession; instead, dying statements emerge as always-already public performances, as speeches and public rhetorical gestures. While Mount's title retains the older style of characterization, his account is very much of the later secular type, cataloguing his misdeeds and detailing his criminal career. These broad changes emerge from new modes of cultural authority enacted in the practices of print culture—printers, for example, supplanted ministers in managing the production and dissemination of execution documents.[61]

The genre also underwent aesthetic and social repositioning. By the end of the century, authors such as Royall Tyler would observe the public's declining taste for the "dying speeches of Bryan Sheehan, and Levi Ames, and some dreary somebody's Day of Doom" as older popular textual forms gave way to bookshelves full of more sentimental and cosmopolitan material.[62] In addition, prominent opponents of execution such as Benjamin Rush linked sentiment to reformed state authority as public attitudes toward executions changed at the end of the century.[63] Popular criminal accounts seem to have become a residual, old-fashioned part of American culture by the end of the century, but Mount and Ames demonstrate that even in the 1790s, a hanging could still produce an enthusiastic array of printed accounts. Only later in the nineteenth century would *The Beggar's Opera* transition from a performance frequently labeled as "low" and "vulgar" to one of aestheticized value and musical virtuosity. The separate trajectories of printed execution texts and ballad operas suggest a transformation in the entertainment value of criminal performance. The relative scarcity of institutionalized theatre in New England alongside the prominence of execution accounts suggests not just that New England executed more criminals than some other regions (which was to some extent true), but that the execution offered a stage play of criminal performance that was otherwise less available to the common observer.

In content, the criminal confession of Thomas Mount produces penance even as it outlines the adventurous career of a theatrical thief in the mold of Macheath. In form, his 1791 *Confession* elaborates remarkably on *The Beggar's Opera*—its criminal account moves beyond the performance of penitence and social control, recruiting audiences to reenact and participate in its rogue performances. Mount's *Confession* displays performances built on and around the body of the condemned criminal. An editorial address "To the Public" prefacing Mount's confession operates as a sort of theatrical prologue, explicitly guiding and regulating its audience. Making the text a public document, the preface imagined Mount's execution as a matter of public safety. Smith implicitly resurrects the distant threat of Macheath and his gang, asserting the alarming presence of a "company of foot-pads and highway-men" in New England. "Almost all the persons who have been hanged of late in North America," he claims, "have belonged to this company."[64] The text alleges that London gang members had infested Connecticut, and that Mount's confederate, James Williams, had met the criminal underworld in the form of printed books in London. The better to bolster its alarmist assertions, the text imagines the connections between American thieves and the London flash companies not merely as long-range echoes and imitations, but as traceable cultural transmissions—American thieves have direct links to the London criminal underworld represented by Macheath's gang.

The company possesses its own language and secret customs, making it what Smith calls a "Flash Company." Rumors of such secret societies, groups of dangerously self-contained, self-defined, and secretive rogues, had long persisted in Anglo-Atlantic culture.[65] Mount's confession includes a dictionary, a list of terms reproduced "to gratify the public," an oath administered to initiates in the criminal underworld, and finally "several flash songs."[66] These textual elements have more than one function, as the text's own editorial commentary suggests. On the one hand, they expose disguised criminal culture to the powers of public detection. On the other, the songs and the oath satisfy audience curiosity, providing entertainment as much as they reinforce legal discipline. The flash dictionary, for example, provides simple transliterations of terms, offering titillating glimpses into the criminal underworld; a "man," for example, is a "cove," a "house" is a "ken," a son is a "young cove of the ken."[67] After demonstrating the compelling ways that criminal culture can transform the terms of everyday life, it moves quickly to more sensational terms, presenting phrases such as "ringing blue bit" (passing bad money).[68] Another section translates more complex phrases of criminal-speak, introducing readers not only to the language, but also to criminal concepts such as "queering the quod" (breaking prison).[69] Enacting a sort of textual criminal apprenticeship, the reader vicariously moves deeper into the complexities of the criminal underworld. A section of reprinted "flash songs" also stage criminal culture as entertainment, inviting audiences to enjoy the adventures of "roving scamping blades," as one lyric proclaims.[70] The songs, like the ones Macheath and his gang had repeatedly encored throughout the Atlantic world, celebrate the exploits as well as the performances (the "roaring song"[71]) of highwaymen and footpads.

Imagining the criminal underworld as a place of secret languages, performances, and rituals produces the outlaw as a revealed and detected subject. However, Mount's text also transforms criminal acts into entertainment, even participatory theatre. Taken together, the dictionary, oath, and songs script a self-consciously theatrical experience; they work through the conventions and forms of the stage to shape the reader's subject position. Suggestively positioning the mundane before the secret terminology of the criminal underworld, the dictionary assists readers in translating their own words and concepts into criminal language. Its function is not detection at all (which might be better served by a list of criminal terms translated into non-specialized language), but scripted play, converting the reader into a reenacting subject. The dictionaries and songs of Mount's narrative allow the reader to inhabit the persona of the novice newly exposed to criminal jargon. Even the more complicated phrases (logically following the first section's simple term building) use the first and third persons, presupposing identification with and reenactment of the criminal role. The first-person couching of phrases such as "I am spotted" ("*I am disappointed, somebody saw me*") and "let us sterry" ("*let us make our escape*") position the reader vicariously among Mount's gang of thieves.[72]

The dictionary subtly scripts theatrical transformation. In the implied narrative logic of its sequence, people become thieves and property becomes stolen goods. Its collection of everyday actors and props, people and household items, also gradually transforms into the more complicated criminal practices of stealing, escaping, counterfeiting, and jailbreaking. The logic of this transformation applies to the spectator as well; exposure creates familiarity and even produces imitation. After introducing its audience to the world of criminal practices, the text readies the reader for initiation. An oath cements the reader's role by detailing the ritual admission of "flats" into the "Flash Society."[73] The first-person text of the oath ("I will be true—never divulge their secrets, nor turn evidence against any one of them") presumes the reader's imaginative participation as the act of reading becomes the ritual process of taking the oath.[74] Immediately following the oath the text prints a flash song, producing a figurative eruption of musical performance. Rollicking flash songs celebrate the exploits made possible by full (performative) membership in the Flash Society. Although the text does not include musical notations, the lyrics conjure reperformance, if only in the reader's imagination. In a very direct way, the text thus constitutes a reenactment of *The Beggar's Opera*, which defined and celebrated its criminal classes through their ability to burst into songs. Mount's flash songs, like those in the opera, appear as vernacular criminal productions; their language marks them as specialized subculture performances.

The catalogue of Mount's crimes represents another such doubled criminal performance. It operates ostensibly as detection and confession technology, listing his illegal acts with meticulous detail, the better to condemn and punish him. In its urge to enumerate Mount's crimes in exhaustive detail, it also produces a sort of penal overkill. Mount emerges in his criminal excess as almost heroic, enacting a scandalously real version of Macheath's legendary rogue prowess. Mount's document is not alone in this; Levi Ames's "Last Words and Dying Speech" includes a similar fine-print catalog of his criminal acts, the specificity and sheer number of which suggest an improbably extravagant rogue performance. The script of penitence, followed all too well, can redeploy confession against the authorities to enhance criminal reputation. Mount's confession potentially transforms the script of penitence into mimicry of Macheath's virtuoso criminality. The performance may have obstructed justice as well. Mount admits to at least one other case in which he took "all the blame upon myself and cleared the other two"; he received a whipping and imprisonment, from which he easily escaped.[75] His final confession could just as easily be a false confession. Such textual performances are not precisely the weak outsmarting the powerful; in his introduction, Smith voices suspicions that Mount turns blame into credit for deeds he may not have committed. Nevertheless, Mount's official amanuensis allows the performance to continue. Playing the role of Macheath seems permissible, so long as the guilty still ultimately submit to punishment. Clearly, the various actors have competing but not mutually exclusive stakes in playing out the scenario. Both the condemned and the authorities can manipulate Macheath's model of theatrical criminality; the stage offers flexible scripts for enacting criminality, punishment, submission, and resistance.

Most importantly, Mount's text reveals authoritative acts of detection producing counter-performances. Mount's confessions, while perhaps aimed at staging the abject penitent, might also produce the opposite effect, enacting Macheath's virtuoso criminality. His act might even mask and protect other guilty parties. Ultimately, Mount's performance can also produce antiauthoritarian identification with the criminal. This sort of text wields performative power, not simply to produce and represent the acting criminal body, but also to reenact outlaw acts. It presents a textual version of *The Beggar's Opera* that celebrates the outlaw and even transforms spectators themselves into rogues.

LEVI AMES AND PUBLICLY MEDIATED GALLOWS PERFORMANCES

Mount's performative narrative reveals the complexities of scripting outlaw acts, but it ultimately presents relatively faint textual echoes of live performance. To recover a more elaborate (albeit still a textually mediated) sense of offstage rogue performance, we must return to the execution of Levi Ames. Surviving evidence shows at least fifteen distinct broadside, pamphlet, octavo, and quarto publications appearing directly before and after Ames's execution.[76] The veritable explosion of print artifacts provides one of the richest extant archives of gallows performance. The texts include four different sermons by three ministers who preached sermons over Ames's penitent attendance, took his confession, and accompanied the prisoner to the gallows. Besides the sermons, at least eleven broadsides deal with Ames's life and death. Most produced evidence of penitence and warnings to onlookers in danger of going the way of Levi Ames. Broadsides ventriloquize Ames's "Exhortation to Young and Old" and subsequent "Solemn Farewell," simultaneously constructing the performing criminal and his attentive audience.[77] The range of production values and genres implies the rogue's various imagined audiences. As did Macheath, Ames presents a performance that seems capable of engrossing readers of sermons, illiterate viewers of broadside images, singers and auditors, poets and poetasters. Some of the broadsides take socially critical and aesthetically radical approaches, demonstrating the ends to which spectators could turn the raw material an execution. A broadside poem entitled *Theft and Murder!* stands out as an "insurgent" text, as Dan Cohen argues, using the execution as a chance to protest the unfair harshness of making thieving a capital crime.[78] The fantastic *Speech of Death to Levi Ames* supplements rational exhortation with a vision of skeletal Death dropping by to comment on Ames's criminal career and sad end.[79]

The collection of broadsides creates a dizzying swirl of circulating performative texts, a remarkably thick print archive of a popular performance. Although it is impossible to tell how many of these items spectators would have actually encountered at the hanging, most claim the execution as their immediate context. If indeed they passed from hand to hand at the execution, the documents helped produce an event that its spectators could experience as a richly multivalent, multilayered performance. Circulating throughout the city of Boston before and after the hanging, the printed matter almost certainly maintained and diffused the theatre of Ames's death. Ultimately, the material but disappearing performing body remains at the center of this complex matrix of print-performance documents. The sermons and broadside construct a theatrical body as abject and compliant, but they also position Ames as the authorizing agent for the acts—not merely a follower of a script, but a potentially empowered acting body.

Elhanan Winchester's broadside *Execution Hymn* reveals the processes of conjuring and authorizing the presence of the criminal Ames on the public stage. Winchester composed the hymn "on Levi Ames" (in the broadside's perhaps telling phrase, leaning as it does on the body of the condemned) at least two days before the date of the execution itself. The document, taking the date of the execution as the present tense, but anticipating the actual hanging, literally tracks the processes of producing performances. The slippage between present and future tenses suggests that the performance of execution became a textual effect, generated literally and literarily in narrative. Performance becomes a textual effect, and "original" acts become inseparable from the documents that repeatedly conjure up the act. Audience members, even if not present at the prison on October 19, could find the broadside either at the shop of E. Russell, "next the cornfield, Union-Street," or presumably near the gallows. An unknown performer, perhaps Winchester, supposedly sang the execution hymn to Ames "and a considerable audience, assembled at the prison." The song was entertaining or affecting enough that the prisoner reputedly desired an encore at the execution itself. Although no corroborating accounts of the hymn's performance survive, its diffusion via print culture produces readers as arrayed spectators of the act. Ames functions simultaneously, in this account, as criminal body on display, subject and audience of performance, and authorizing agent of its repetition.

Ames participates in problematic but crucial ways in the staging of his own execution. Whether his acts are imagined or real, and however motivated, they play a central role in the events. Although the evidence does not conclusively reveal him as an independent actor, neither does he become simply a passive, manipulated subject of the theatre of punishment and social control. The broadsides and newspaper accounts generally agree that Ames participated actively by delivering his last words in a conventional and penitent dying speech. Winchester's execution hymn, for example, also includes a "dying soliloquy" purportedly spoken by Ames. Likewise, the *Last Words and Dying Speech* claim to be "Taken from his own mouth, and published at his desire."[80] One broadside poem even purports to have been "sent to him for his improvement" and then published at his request.[81] Ames's supposed participation and authorization appears repeatedly in these texts, suggesting that they rely upon the collective fiction—the conventional performance—of Ames's agency, if not his actual willing participation. Even if he did not actually speak and authorize as the documents claim, such participation produces a compelling plot, so deeply embedded in the performance conventions that its absence would produce an entirely different, and perhaps even unthinkable, scenario.

The broadsides perform Ames's compliant and rational authorization; at the same time, their print-mediated emotional appeals rely upon the condemned's physical person. The "affecting speech of Levi Ames, taken from his own mouth, as delivered by him at the goal in Boston the morning of his execution" takes as its main title *The Dying Penitent*.[82] In a slippery rhetorical substitution, the speech act recedes behind the display of a perishing penitent body. Taking a slightly different and more striking tack, the broadside *Dying Groans* of Ames signals the suffering body's presence though its groans, its inarticulate effusions.[83] Both popular broadsides and sermons rely upon the produced presence of the acting, theatrical criminal. They display Ames's disciplined body before a receptive audience, with or without articulate words. These acts perform the central fact of the execution—the sometimes speaking, sometimes groaning, but ever dying and performing body of Levi Ames.

Even as the broadsides extract and produce theatre from Ames's body, it remains a mystery how (or even if) Ames might have exercised any real agency. Ames could have voluntarily performed such penitent groans, enacting rational authority over his contrite performance. We ought to take seriously the possibility that Ames is not simply passive—that he willingly participates in the performances of his execution. At the same time, we ought to retain enough skepticism of authority's claims not to simply dismiss that participation as meaningless docility or painless compliance with the performance of power. The authorization extracted from or given by Ames, for example, presupposes some agency. The scripts seem to allow, even to require, Ames to retain some control over his texts. Perhaps this helps produce the effect of Ames's independence and moral agency that underwrites judgments and claims of moral superiority. Alternatively, Ames could simply have desired, and received, some degree of control over his own story for various reasons. He could have acted out of desire to curry favor with legal and spiritual authorities, from a hope of attempting to manipulate audience sympathies, or even of seeing some monetary profit from publication. The broadsides' repeated claims to affecting acts offered the prisoner a preexisting format for making an emotional appeal to the crowd. Such acts were not impossible, as Kristin Boudreau has argued.[84] Undeniably, an actor in Ames's position has the capacity for self-expression, even if conventions and practices allow the exercise of that agency only in certain ways—editorial authorization, penitent speech, or affecting groans. The practices of print-performance culture do not entirely foreclose or consistently compel the self-representation of the criminal low. The performative texts, however, never give Ames a real opportunity for radical self-expression. He does not act in an openly rebellious scene, unlike the impenitent pirate William Fly, who had fifty years before confronted Cotton Mather's sermonizing performance with obstinate refusal to act.[85] Even so, the texts tantalize with the possibility of the performing criminal body's capacity for eloquent self-expression and silence.

Although the broadsides only tentatively explore the possibility that Ames could be seizing the performance, other texts suggest that bystanders knew well the transgressive possibilities of rogue performance. The danger of rogue performance, they suggest, was not in the rogue, but in the crowd. Andrew Eliot's sermon, for example, rigorously produces Ames as a penitent subject; it turns abruptly near the end to address the possibilities of performance. "You are to die," Eliot intoned, "before a numerous multitude of spectators"; even as he speaks to Ames, Eliot's attentions seems riveted on the watching crowd.[86] The multitude surely represents a more dangerous entity than does the bound and penitent criminal sitting in the church pew. In part, this is because spectacular display produces not simply penitential self-abasement, but theatrical exposure. The gallows produce publicity and celebrity. The "infamy, which will attend your death," Eliot protests, "is a circumstance of little importance"[87] Eliot's address continues poking at its own tender spots. His spiritual guidance, he insists, represents "the only way in which he can be saved."[88] The anxiety of his claims (is there another way to be saved, to literally escape the gallows?) becomes increasingly palpable as he turns to his audience:

> Let me exhort and intreat [sic] all who may attend the execution of this poor condemned criminal, to lay to heart such an affecting sight, and to behave with decency and seriousness on such a solemn occasion. And may the awful spectacle be a means of instruction and amendment to sinners! May they find their hearts suitably impressed; and sincerely resolve that wherein thy have done iniquity, they will do so no more![89]

Eliot's anxieties, his need to remind his audience of the necessity of displaying "decency and seriousness" at the "solemn occasion," the need to point out the opportunity for edification, point to the obvious but unstated alternative ways of construing Ames's execution.

Eliot hints at a performance in which the audience takes the main role, transforming it from an exercise in penitence and authority into an "indecent" and "unserious" spectacle, recreation rather than edification. The image heading Ames's broadside *Exhortation* dramatizes just such a scene (figure 2.1).[90] The image is likely a stock woodcut; printers resurrected its visual elements repeatedly for such broadsides. The basic generic components of scaffold, elevated criminal, guards, and crowd appear, for example, in the slightly different stock image decorating the broadside *Solemn Farewell to Levi Ames*. That scene, however, pans out to depict the stage of execution within a wider scene of community life. A windmill and buildings, for example, overshadow the scaffold, people mill about in the foreground, and a flock of birds fills the sky above the gallows.

The more focused scene of the broadside *Exhortation* visualizes theatrical criminality, the spectacle of punishment, and the audience for such theatricals. The image counters its own promised moral exhortation with criminal celebrity. Ames perches on top of the cross-pole of the gallows, legs swinging in what could seem casual composure. Indeed, the conventional criminal might have real reason to seem casual: the scene quite possibly does not picture execution at all, but instead, a non-event—or more precisely, the rehearsal of execution. A broadside memorializing the punishment of Elizabeth Smith and John Sennet in 1773 glosses an analogous scene as an act of deterrence: each of the convicted is "sentenced to set upon the gallows for the space of one hour, with a rope round their necks."[91] Despite Ames's sentence of death, the conventions of broadside publication actually avert the scene of capital punishment. Despite his impending execution, Ames appears on the broadside at the last possible moment of freedom, the final scene in which he can still play the role of Macheath. Rope around neck, he sports a stylishly long

An Exhortation to young and old to be cautious of small Crimes, left they become habitual, and lead them before they are aware into those of the most heinous Nature. Occasioned by the unhappy Case of *Levi Ames*, Executed on *Boston*-Neck, October 21st, 1773, for the Crime of Burglary.

I.

BEWARE, young People, look at me,
   Before it be too late,
And see Sin's End is Misery :
   Oh ! shun poor *Ames*'s Fate.

II.

I warn you all (beware betimes)
   With my now dying Breath,
To shun Theft, Burglaries, heinous Crimes ;
   They bring untimely Death.

III.

Shun vain and idle Company ;
   They'll lead you soon astray ;
From ill-fam'd Houses ever flee,
   And keep yourselves away.

IV.

With honest Labor earn your Bread,
   While in your youthful Prime ;
Nor come you near the Harlot's Bed,
   Nor idly waste your Time.

V.

Nor meddle with another's Wealth,
   In a defrauding Way :
A Curse is with what's got by stealth,
   Which makes your Life a Prey.

VI.

Shun Things that seem but little Sins,
   For they lead on to great ;
From Sporting many Times begins
   Ill Blood, and poisonous Hate.

VII.

The Sabbath-Day do not prophane,
   By wickedness and Plays ;
By needless Walking Streets or Lanes
   Upon such Holy days.

VIII.

To you that have the care of Youth,
   Parents and Masters too,
Teach them betimes to know the Truth,
   And Righteousness to do.

IX.

The dreadful Deed for which I die,
   Arose from small Beginning ;
My Idleness brought poverty,
   And so I took to Stealing.

X.

Thus I went on in sinning fast,
   And tho' I'm young 'tis true,
I'm old in Sin, but catcht at last,
   And here receive my due.

XI.

Alas for my unhappy Fall,
   The Rigs that I have run !
Justice aloud for vengeance calls,
   Hang him for what he's done.

XII.

O may it have some good Effect,
   And warn each wicked one,
That they God's righteous Laws respect,
   And Sinful Courses Shun.

Figure 2.1 Exhortation to Young and Old […] Occasioned by the Unhappy Case of Levi Ames, Executed on Boston-Neck, October 21st, 1773, for the Crime of Burglary (1773), broadside; Historical Society of Pennsylvania.

coat and a three-cornered hat. Like generations of eighteenth-century rogues following Macheath's lead, he dies game, displaying his extravagant sartorial resistance to the severity of the law. Below the image, the stanzas of his exhortation, though moralistic in content, formally convert the gallows scene into entertainment. The printed verses construe Ames as a latter-day Macheath, singing via the broadside for popularity, perhaps even acting for a reprieve. In the end, even the broadside's conventional warning ("And see Sin's End is Misery") assumes Ames's charismatic theatricality, which can inspire a dangerous imitation.

Despite his sentence, Ames has not yet undergone punishment. The scene does not yet reveal the abjected end-result of official punishment. It is almost certainly a stock (and thus not literally representative) image, but its very conventionality foregrounds the suspenseful moment before execution. It even hints at the escape possible in Ames's execution ritual. Generic slippage, meaningful if inadvertent misapplication of convention, highlights the last moment of uncertainty. The broadside presents an entirely different set of excarceral possibilities than does, for example, *Theft and Murder!*, with its subdued and lifeless hanged body. Above his poetic *Exhortation*, Ames is still alive, and audience members gaze on.

The mob surrounding the scaffold—perhaps including the "young People" to whom the broadside addresses itself—provides the most compelling, and perhaps most neglected, part of the image's drama. Images of the audience appear in other broadsides, constituting a stable element of the genre. The *Solemn Farewell*'s crowd, for example, puts the gallows scene in wider perspective. The broadside *Theft and Murder!* shows an even larger crowd, although its evenly regimented figures seem disciplined; moreover, they watch the consummation of discipline, the criminal's dead body about to be lowered from the gallows. *Exhortation*'s crowd, in contrast, seems a lively, motley bunch. At least two dozen spectators occupy the space around the gallows. They do not appear passive watchers, either. Among the group of spectators, some appear to brandish staffs or cudgels at the condemned. Opposite them, a man appears tied to one of the gallows poles while a nearby figure raises a multiple-tailed whip in what appears public corporal punishment. Two smaller figures—children, perhaps—and a small animal tilt kinetically in the scene's foreground. Executions provided occasions for play and perhaps traditional forms of community policing as well as punishment. Eliot's sermonizing attempts to regulate spectatorship surely responded to these broadside scenes of active and unruly spectatorship. The visual disorder of the crowd is framed by four riders (one on the left, and a tight group of three on the right), who appear armed with swords and whips. Although they might contain the mob, the regimented riders hardly discipline the crowd's energy.

If Ames acts as a Macheath at the gallows, the press of the crowd below the scaffold pictures the audience whose low taste frees the rogue repeatedly in more than a century's worth of *The Beggar's Opera* performances. Such spectatorship, as Kristin Boudreau has argued, posed problems for early American authorities, who understood that "the sympathies of the crowds who watched were always volatile and unpredictable, and could easily pour out against the agents of execution."[92] Even when policed by armed riders, the mob's presence and power at hangings seems persistently problematic. The scene presents a telling acknowledgment of the presence and even the power of the crowd. They are, after all, the public for whom multiple broadsides have been produced, to whom three different ministers address their warnings, and who threaten to become imitative converts to flash gangs and a short life of petty thievery.

The crowd also suggests the multiple audiences, some of whom would have been less literate or less interested in the sermons and literary confessions produced to supplement the drama of hanging. To those audiences—who neither read nor produced literary responses to the performance—broadsides, speeches, and poems may have ultimately been more notable as image, spectacle, and entertainment, rather than for their moralizing texts. Viewed alongside conventions of theatrical and popular performance, the textual relics of Ames's crimes and punishment reveal elaborately constructed, embodied, and deeply theatrical scenes. Criminal performance staged the interplay among the intertwined and even indistinguishable forms of oral and literate, folk and commercialized culture. As Andrea McKenzie has argued of earlier English rogues, highwaymen "reflected and facilitated a dynamic interchange between patrician and plebeian culture," operating as "media personalities whose currency transcended class lines."[93] The theatrical execution of Ames seems to have generated such scenes. While the act may certainly have reinforced class lines even in trafficking across them, the sheer number and variety of documents produced suggest that his appeal reached outward to a broad audience. His audience certainly included the common observer on the street, but also cultural producers, those with the means to compose, print, and distribute broadsides, as well as deliver sermons and print elaborate confessional texts. Ames's case and its celebrity value seem to have appealed in some way to a broader constituency, even if he only crossed class lines as a financial opportunity. Those documents achieve complex meaning in public performance, negotiating differently motivated scripts performed by variously empowered actors. The complexity seems a function of the execution's theatricality, an effect of the fact that these documents specifically emerge from and represent live performance.

Ultimately, the early American theatre of criminality presents scenes in which various parties, including the audience and the criminal himself, attempt to produce and manage performance around the condemned body. The very presence of the outlaw body represents both opportunity and quandary for the stage managers. Such negotiations may not have generated any radical underclass expression on Boston's gallows and in the streets. Nevertheless, the possibility that preachers and confessional texts try to ward off appears repeatedly in early American culture. As Ames stood on the gallows, *The Beggar's Opera* entertained early American audiences; together, they performed the dangerous appeal of rogue performance and mob spectatorship.

CIRCUS ESCAPES

Fascination with criminal and punitive acts reappeared in a shocking and fascinating performance in the early 1820s. Stoker, a "rope-vaulter," briefly appeared at Joe Cowell's Philadelphia circus. Cowell remembers Stoker in his theatrical memoirs:

> he, among a variety of liberties he took with himself, used to hang by the neck, not till he was dead, but just long enough to give his audience reason to believe that he might be;

and this faithful imitation of the last agonies of a malefactor, in a spangled jacket, drew together, nightly, quite as large a crowd as a public execution always does.[94]

Semiconsciously tapping into the street theatre of Levi Ames's execution and the stage drama of *The Beggar's Opera*, the act improvises outrageously on the themes of rogue performance and excarceration. Stoker's "faithful imitation" of a punished criminal displays the stage accoutrements of theatrical roguishness. His "spangled" jacket, for example, recalls Macheath's customary style, his gaudy display of fashionable criminality. Stoker's spectacular reenactment of imminent punishment and last-minute escape transforms dramatic plot of *The Beggar's Opera* into rope dancing and acrobatics. Theatrical roguery and popular reprieve become circus. Stoker apparently always escaped at the last minute, but his shocking act packed the circus with curious audiences. An observer, ironically naming himself "Tim Spectre," describes the act and his own voyeuristic delight in a letter published in the Philadelphia *Aurora*.[95] "I was all eyes," he writes, as Stoker ascended to begin his act. The acrobat began swinging, tantalizing the audience with a death-threatening dive arrested at the last moment by a noose around his ankle. After "various feats of surprising dexterity," the observer recounts, "he arrived at the climax of the exhibition, which was to hang himself by the neck.—I could hardly contain myself for joy." At the height of his swing, "Spectre" writes, "he slipt from his seat, and to my unspeakable delight, there he was, *sus per Col*; hanging dingle dangle." After a calculated moment of suspense, Stoker ascended the rope again. The shrieks subsided and the "awful pause which pervaded the theatre was succeeded by a busy hum from all quarters."[96]

The act generated its own publicity mechanisms, as Cowell remembers; "Fortunately for the management, several ladies fainted the first night he appeared."[97] Once the threat of accidental death and audience alarm was publicized, "the boxes were always filled with the fair sex whenever the feat was advertised."[98] The female spectatorship ironically recalls Mrs. Peachum's commentary on the condemned criminal's appeal to women, who "think every Man handsome who is going to the Camp or the Gallows." In 1824, the transgressive circus spectacle of averted death fills the stands with female spectators, although they presumably watch for different reasons than those of Macheath's admirers. Spectre's audience member enthusiastically declares the circus "the place for me," purchasing a season ticket and attending in hopes of an "appalling accident."[99] When no actual deaths prove forthcoming at the circus, Spectre turns instead to a collection of Spanish Inquisition torture devices to satisfy his desire for macabre entertainment. His satire concludes with the jaded spectator examining the criminal dockets and funeral listings for further morbid pleasures.

Stoker's act radically transforms Macheath's gallows performance into the physical feats of circus entertainment. The threat of hanging becomes a function of physical comedy and acrobatic skill. Stoker's performance, heir to the itinerant clowning of earlier eras, reveals the theatrical legacy of stylized vagabondage. By the 1820s, the conventions of the entertainment industries had reshaped that stylized lowness. The plot of the *Beggar's Opera*, a story of charming, theatrical, and comic outlaws reprieved from the gallows, recedes behind a new one that foregrounds the intermixed anticipations, horrors, and pleasures of a deeply conflicted audience. As it had done in 1728, the self-conscious taste of the town still provides a—perhaps *the*—central part of the performance. The act's satire no longer targets state power; instead, Spectre takes aim at vulgar, low, and popular entertainments. His commentary, while showing intimate familiarity with such forms, mocks the offensive pleasures and the naïve expectations of popular entertainments in the 1820s.

Spectre's circus attendance follows a survey of the vulgar commercialized entertainments available to contemporary audiences. He visits "all the waxworks and puppet shows in the city," enjoying Marie Antoinette's beheading, the Baron Trenck wasting away in prison, and even a restaging of Othello smothering Desdemona, before those acts "finally lost their pungency."[100] After Stoker's circus act loses its own appeal, the narrator turns to funerals and criminal dockets in hopes of finding real entertainment. Stoker's macabre circus act coexists with a variety of popular entertainments, which in turn appear alongside the formal theatre of scripted dramas throughout the late eighteenth and early nineteenth centuries. Although such forms continued to coexist, the account suggests a shift in the character of early American entertainments. Macabre displays of hanging, once the province of popular spectacle and informal street theatre, were becoming commercially available in forms (including waxworks and displays of torture devices) that Spectre and others dismissed as low and tasteless. In a sense, the commentator satirically descends into the increasingly sensational ethos that helped shape melodrama in the coming decades. Notably, *The Beggar's Opera* does not appear in his survey of morbid entertainments, despite the possibility that its popularization of condemned criminality had arguably whetted appetites for the very forms Spectre derides. By the 1820s, as Karen Ahlquist suggests, the *Opera* seems safely ensconced as "high entertainment," increasingly presented under the auspices of the opera house and the legitimate dramatic stage rather than the circus, the street, or the arena of commercial exchange.[101]

Ultimately, Stoker's act shows the versatile shape-shifting persistence of rogue performances on American stages. Transforming outlawry into dexterous circus clowning, Stoker's skill resurrects not just the rogue, but also a persistent kernel of Atlantic cultural memory. The performance reenacts the entertainment value of the hanged criminal and his narrow escape from capital punishment. Stoker's later act revives the execution scenes, such as those of Ames and Mount, which reveal the shaping influences of similar structures of theatricality and spectatorship. Those stagings use the forms of the playhouse in different ways, and with varying levels of self-consciousness, but such roguery remains evidently theatrical. Such acts suggest the interconnected and mutually constitutive emergence of various acts and actors, occurring both on and off formal stages. The lives and deaths of characters such as Levi Ames and Thomas Mount were real, embodied, and lived. At the same time, they seem, like Macheath, media effects. Rogue performances become visible and legible primarily through disciplinary rituals and through the forms of print mediation that continue circulating them. Those scenes render the low, the outcast, and the disenfranchised in outlaw accents. The rogue's gallery of underclass characters also finds ways of manipulating the scripts, with varying effects and degrees of success, reconstructing the theatrical forms even as the forms construct them.

On the stage, *The Beggar's Opera* rehearses those characters as charismatic, entertaining, and even resistant. Macheath's self-assured carousing, his singing and dancing, and his final defiant stand finally release him. That escape helps make the theatrical criminal a durably resistant figure on early American stages. Responding to theatrical celebrity, the taste of the town rewrites dramas of criminal transgression and punishment. Gay's ballad opera condenses scenarios and gestures, amplifying them in performance and reinjecting them into popular culture. From there, they emerge again in various artifacts of print and performance culture. Disciplinary displays become entertainment, and Gay's ballad opera suggests the potential of offstage rogues and audiences to act in ways that the enforcers can never fully control. Levi Ames's scaffold speech and Mount's confessional account reveal various impulses, not all in line with the state's punitive intentions. Likewise, *The Beggar's Opera* variously represented low satire of elite pretensions, a vulgar performance of lowness, a conventional musical test-piece, or a bit of comic relief. Stoker's circus act could be both rogue performance and a demonstration of acrobatic skill, both evidence of sadly declining popular tastes and a popular display. Ultimately, the theatre of criminality—whether on the gallows, in the Park Theatre, or in a circus—reveals scenes in which no one party fully controls the performance. Perhaps because of such multivalent possibilities, rogue performances draws spectators well into the nineteenth century and beyond, and the stage continues to renegotiate acts of roguishness, punishment, and escape.

3. Algerians, Renegades, and Transnational Rogues in *Slaves in Algiers*

Even as America's "social undersiders" stood on the gallows, they also appeared in early national dramas of citizenship and national identity. In 1794, Susanna Haswell Rowson's *Slaves in Algiers; or, A Struggle for Freedom* presented a spectacle of Americans enslaved in Algerian captivity. The play premiered at Philadelphia's New Theatre (later the Chestnut Street theatre) on June 30, 1794, the proceeds benefiting Rowson and her husband. Rowson wrote and acted in the play, and her growing fame as the author of *Charlotte Temple* probably helped attract spectators as well. *Slaves in Algiers* hardly achieved overwhelming success, but it saw occasional performances in Baltimore, New York City, and possibly in Boston (where Rowson would perform and reside after 1796) before the end of the decade.[1] The play's scenes of Algerian captivity had some durability. In 1816, for example, at the end of another American conflict with the Barbary pirates, the play reappeared in Boston.[2] *Slaves in Algiers* participated in a broader theatre culture of "acting Algerian," showing how offstage performances infuse early American theatre.

Algerian captivity, drawing on other acts of Anglo-Atlantic roguery, invokes the underclasses that had begun to trouble American communities in the eighteenth century. The play condenses early American culture's tendencies to depict freedom and bondage, to perform ethnic masking, and to reenact the power of the lower sorts. In staging its national struggle, *Slaves in Algiers* differentiates its low into three cohorts: the exotic, despotic Algerian rogues, the troubling transnational cohort of rebellious slaves, and redeemable plebeian characters. From among these types, Rowson's play imagines unruly underclasses that it can turn to the productive ends of patriotic stagings. Even as they help shore up the racial and religious underpinnings of national identity, those characters also reveals a deep ambivalence about the outcasts and outsiders coming to the fore at play's end. *Slaves in Algiers* improvises on the impulses that kept *The Beggar's Opera* popular, and that would reemerge as other patriotic performances, such as William Dunlap's *The Glory of Columbia*. Such "homegrown" products still rely upon the circulating and conventional rogue performances of the Atlantic world. *Slaves in Algiers* demonstrates American theatre's creation of topical, local, and patriotic performances "cobbled together," as Jeffrey Richards argues, "from stock elements inherited from British drama."[3]

Slaves in Algiers, Rowson's only surviving play script, follows an Anglo-Atlantic family taken captive and held separately by Algerian captors. Hassan, an Anglo-Jewish convert to Islam, holds the matronly Rebecca, while her son Augustus and daughter Olivia (played by Rowson) endure separate captivity. They presume that their long-lost husband and father Constant had died in English military service, but the twists of imperial fate have deposited him into Algerian captivity as well. The circumatlantic routes of commerce and conflict have, unbeknownst to the captives, brought them together in the unlikely locale of North Africa's Barbary Coast. In the play's final scene, the proto-American family reunites in the extravagant court of Muley Moloc, the Dey of Algiers. Reunion, however, does not entirely solve the play's problems. To avert the threat of her father's execution, Olivia must enter the Dey's harem, trading sexual slavery for her family's lives. She virtuously plans to commit suicide before the Dey can consummate their forced relationship.

At the last possible moment, however, freed European captives crash into the room, forcefully wresting control from their Islamic captors. Muley reacts to the ensuing confusion with futile commands: "Vile abject slave obey me and be silent—what have I power over these Christian dogs, and shall I not exert it. Dispatch I say—(*huzzah and clash of swords without.*)—Why am I not obeyed?—(*Clash again—Confused noise—several Hazza's,—*)."[4] The slaves' plot, signaled textually by the intrusion of a confused clamor, has matured in the course of the play, and their presence breaks the deadlock of virtuous obligation that stalemates the emergence of the proto-American family in formation. As despotic Algerian rule collapses, the Anglo-American family unit emerges reunited and triumphant. Arguably, however, the most proactive and self-empowered—not to mention entertaining and attractive—characters in Rowson's play were not the decorous ideal American citizens who stand triumphant at play's end, but the transnational cohort of low characters who eschew patriotic eloquence and simply fight their way to freedom. Opposing the piracy of the Algerians, the roguish captives free themselves, deposing an abjectly reformed Dey in the last act. Those freed slaves represent significant (and vexed) emancipatory energies in a play dedicated to sorting out the conditions of early American and Atlantic inclusion and enfranchisement. If the triumphal American family emblematizes national identity at play's end, it can do so only because Rowson's script has recruited and deployed the underclasses to achieve success in the collective "struggle for freedom."

Slaves in Algiers, then, demonstrates the complications of rehearsing American identity and the complexities that remain even after achieving resolution. In the end, it stages a scenario of early American constituency in which an ethnically, religiously, and economically compact grouping of (blood-related) characters emerges as the ideal American subject. Those ideal subjects, however, skirt dangerously close to outsider identities throughout the play, threatening conversion, as Olivia does in offering herself to the Dey. In the end, the emergent American family exists only through the exertions of the more diverse group of low characters, and only because American actors—the onstage surrogates of real American captives—embody the same Algerians and rogue underclasses that the play eventually displaces from the (final) American scene.

PLOTTING ALGERIAN SLAVERY

Rowson's play reenacts public scenes of Americans enslaved in North Africa. As Paul Baepler notes, the genre of Barbary captivity narratives had emerged with the early modern English involvement in the Mediterranean, but it had become a particularly pressing aspect of America's early national experience in the 1790s.[5] Newly independent and lacking the protection of the British navy in the Mediterranean (perhaps even actively opposed by the British), Americans found themselves in ongoing conflicts with the Barbary States of North Africa. Between 1785 and 1793, Algerian pirates had captured over one hundred American sailors, precipitating what some historians have called America's first hostage crisis. The U.S. government controversially negotiated for the slaves' freedom rather than overextend their military forces. By 1797, the United States had paid ransoms to Tunis and Algiers totaling $1.25 million, or one-fifth of the government's annual budget.[6] This tribute would only temporarily solve the international friction, and the United States would continue diplomatic and military conflicts with North African states until 1815.[7] The experience accentuated America's tenuous position in international affairs.

Algerians and their American captives remained a public matter, appearing repeatedly in newspapers periodicals, broadsides, and on the stage. Alongside Rowson's play, over one hundred other Barbary captivity texts, including plays, autobiographical accounts, and novels appeared in publication or performance in the early national and antebellum periods. Those texts and performances repeatedly employed figures of Algerian captivity to work out scenes and plots of American national identities. As Elizabeth Maddock Dillon has argued, Rowson's play represents the early national imaginary operating "within a set of global relations" (primarily trade and commercial) and depending "upon peoples beyond the enclosure it seeks to make immanent."[8] Its exotic "struggle for freedom" seems crucially a matter of domestic politics as *Slaves in Algiers* explores American identity through the externalized figure of the Algerian captor and his slaves. Frank Lambert recognizes the Barbary conflicts as an "extension of America's War of Independence," an economic rather than a religious struggle or a clash of civilizations, and *Slaves in Algiers* seems to support this thesis on the cultural home front.[9]

Ultimately, Rowson's version of Algerian captivity repeats jingoistic proclamations of national triumph. As Olivia, Rowson delivers the play's closing lines: "Long, long, may that prosperity continue—may Freedom spread her benign influence through every nation, till the bright Eagle, united with the dove and olive-branch, waves high, the acknowledged standard of the world."[10] The conclusion seems an open endorsement of Federalist policies and social values, despite the fact that William Cobbett, a prominent Federalist and one of Rowson's most merciless critics, called *Slaves in Algiers* "a most excellent emetic" while famously deriding female participation in the political public sphere.[11] Cobbett's biting elitist satire (he would also refer to his Democratic Republican political opponents as "butchers, tinkers, broken hucksters, and transatlantic traitors!") perhaps responds to the class associations of Rowson's chosen form and venue.[12] Federalists dominated the organization and financial support of the Philadelphia and Boston's theatres, at times limiting the modes of spectatorship, as Heather Nathans has shown, but they could not entirely foreclose the playhouse's heterogeneity. Ultimately, the theatre functioned as, in Marion Rust's phrase, a "forum for urban diversity" bringing together spectators of varying socioeconomic and ethnic backgrounds.[13] The playhouse of the 1790s housed a conflicted public space, frequently hosting competing expressions of partisan loyalty, and Rowson's play asserted itself troublingly in this arena. The heterogeneity and popularity of theatre overwhelms the essentially conservative, and even Federalist, content of *Slaves in Algiers*.

Cobbett's attack also articulates exclusionary attitudes toward female participation in the public sphere. Rowson's play scripts the Algerian captivity crisis as a domestic and gendered business, defined by family relationships and gendered critiques of slavery. The play's epilogue reveals its investment in a feminized national imaginary, as Rowson facetiously proclaimed the purpose of men to "adore, be silent, and obey."[14] Although the lines immediately following temper the statement, it points to the play's reliance on contested notions of tenuously empowered femininity. The play itself centers on the freedom and captivity of women, who morally guide the play's struggle for freedom. Pairings of American and Algerian female characters imagine transcultural feminized homosociality and rehearse the play's final reconstruction of the American family. The play's identity revelations and family reunion establishes national identity as a matter of "filiopiety," as Elizabeth Dillon argues, in contrast to the "prodigal acts" of other national fictions, including Rowson's own *Charlotte Temple*.[15]

Recent feminist readings of *Slaves in Algiers* have astutely recognized Rowson's theatrical brand of gender politics, understanding the relationships and actions of the female characters as key to the home-front politics figured in Algerian captivity. Cross-dressing, potential conversion, and self-sacrifice give the play's women perhaps the most complex roles in the play. Their freedom of action, however, seems limited to the concerns of early American gender and political participation. Critics, for example, generally concur that, although audiences could easily have made the connection, the play fails to deploy (or audiences simply did not perceive) white slavery as an overt critique of American slavery.[16] Moreover, the play's Federalist and proto-feminist agenda establishes itself on the passive, manipulated bodies of the comic low.[17] Essentially, Rowson's play allocates power carefully among the characters in a zero-sum cultural calculus, variously enabling the captive subjects as agents of revolution. The female American captives, for example, although unable to revolt, find literary consolation in Addison's *Cato*. The male captives take direct action and seize their freedom, in the process becoming a mob. Some characters have it both ways, but at the cost of being expelled from the racial or ethnic consensus—the comic Anglo-Jewish woman Fetnah, for example, cross-dresses and joins the revolt, but must remain with her father, an Islamic convert, in the end.

Those comic low also act as the play's rebelling mob, a cohort that modulates magically from comic to heroic in the last act to enforce the achievement of American freedom. While the more genteel characters appear incapable of acting effectively or resolutely, the lower characters emancipate themselves, collect arms and supplies in a grotto, and eventually emerge in the third act as the play's proactive liberators. The play introduces its underclass revolutionary plot through Frederic, a "wild young Christian" who had managed to ransom himself before the play began.[18] Frederic negotiates with Hassan to "assist a parcel of poor devils to obtain their liberty" by purchasing a vessel and supply the slave's escape.[19] The characters of the low plot (like Sebastian, the son of a Spanish barber, or Frederic, the hapless American adventurer) produce humor out of the time-honored faults of good-natured indulgence in alcohol and confusion over social hierarchies. As the plebeian characters bumble their way through their plot, their misidentifications and undesirable unions provide foils for the deadly serious business of recognizing and reuniting the American family. The spectacle of a cross-dressed female adventurer and a Spanish rube making love to her father in petticoats, for example, precedes the anticipated protonational family reunion.

The violently self-emancipated captives constitute a compelling and troubling element, to which the playhouse's gesturing bodies, spectacular display, and potentially promiscuous class mixing lend immediacy. The play's rebellious cohort resonates with the periodic revolts and widespread popular insurgency in early America. In the decade before *Slaves in Algiers*, Shays' Rebellion in western Massachusetts and the Whiskey Rebellion in Pennsylvania provided some of the most spectacular and public examples of the unruly low's potential for revolt.[20] Those uprisings also offered firsthand referents for literary and dramatic depictions of the lower sorts. A decade before publishing his 1797 novel *The Algerine Captive* (which I discuss later in this chapter), Royall Tyler played a role in suppressing Shays' Rebellion. Closer to home, American cities experienced a resurgent plebeian political sphere after the Revolution, which made demonstrative and sometimes disorderly acts visible in the streets.[21] Substituting Algerian slaves for the rural insurgents and urban mobs of the early republic, *Slaves in Algiers* reimagines the role of underclass revolt in the national imaginary.

STAGING CHARACTER

Algerian captivity in the 1790s provoked complex performances of the problems of American identity. Both inside and out of the playhouse, such captivity appears not just as subjugation, but also as a theatrical act, threatening the possibility of conversion or transformation. As Heather Nathans has argued, *Slaves in Algiers* "offered only two points of view: that of the tyrant and that of the American patriot/audience member," and the play seems to overtly reinforce such dualities.[22] At the same time, the playhouse experience complicates simple dichotomies by presenting Americans actors transformed into exotic Algerian tyrants and their captives. The theatre, like Algerian captivity itself, operates as a space of transformation and conversion. Its forms—acting, role-playing, masking, and transformation—clear a space for audiences to experimentally "try on" roles, as Marion Rust argues. Watching *Slaves in Algiers*, audiences could try on the modes of public participation made possible by low and outcast characters.[23]

Stage Algerians and their American captives enact struggles over the constituencies of the young nation. Their negotiations centered on the representation of character. Rowson herself, performing literary humility, points this out when she credits the efforts of the actors who "so ably supported their several characters."[24] Character and role—who one is on stage—function crucially in Rowson's staging of the dilemmas and solutions of American and alien identity. Rather than trading in verbal eloquence, musical virtuosity, or spectacular displays, *Slaves in Algiers* seems to rely upon easily identifiable characters that fall into well-defined categories. The play's effectiveness depends in part upon quick audience recognition of straightforward types—whether American, Algerian, Christian, Islamic, genteel, or underclass. Moreover, the play requires ready detection and evaluation of the relationships among those characters. The play only makes sense within a complex preexisting network of well-defined types and conventional relationships.

On the face of things, *Slaves* differentiates among polarized groups of Algerians and Americans; at the same time, it produces those performances of difference by literally masking American ethnic and national identity behind Algerian. The play's "Dramatis Personae," a commonplace part of the play's textual presentation, suggests the ways in which the play categorizes characters in categories of ethnic, gendered, and national identity (figure 3.1).[25] The cast list functions as what Gérard Genette's has called a paratext, preceding and conditioning the meaning of the text.[26] Precisely because it seems such a conventional, predictable, and utterly unremarkable element of the text, it signals crucial but unstated assumptions about the performance and its textual rendering.

The Dramatis Personae organizes the play's presentation of character, listing the play's most prominent roles and the players who embodied those parts. The text, however, communicates more through form than content. The listing reenacts in print the play's mechanisms for delineating character and stage type, pointing out the relationships among imagined characters and real performers. It generates meaning by organizing its content, producing articulations and correspondences, rendering visible on the page lines of division and connection. Quite simply, the parallel columns (roles beside actors) mark articulations of actor and role, and the groupings literally delineate the various social cohorts of the play (men, women, and extras). Moreover, the cast list subdivides those categories, creating multiple groupings at once. The textual and visual divisions suggest the play's reliance on complex sets of co-affiliations and divisions amongst its characters.

The list's most noticeable categorization is its deliberate and clean separation of cast members by gender. The listing suggests an institutional context deeply defined by gender—a category that inflects all other considerations of character. In a play whose patriotic message depends ultimately upon gender politics, this division can hardly be incidental, of course. The form marks conceptual divisions between the functions of male and female characters, as well as corresponding divisions among the men and women who enacted those roles. This seems perhaps a commonsense observation, but one that defines the performance of "Islamic other" and "American captive." Men, the cast list signals, perform a certain set of roles, and women another. Male actors embody a diverse range of character types, including captives, slaves, Renegadoes, and Algerian officials. Women, in contrast, perform a narrower range of functions in the play, labeled more simply by their gender and an ethnicity. The subdivisions of each gender, signaled by brackets, suggest certain roles available on the stage—options for character types constructed as what we might identify today as "ethnic." The oddly incongruent binary categorization of female characters into "American" and "Moriscan" becomes much more complex among the men. Male characters can appear as western or eastern. If western, they can take the form of American or European. European (but not American) men can also appear as slaves or captives, an

DRAMATIS PERSONÆ.

MEN.

MULEY MOLOC, *(Dey of Algiers,)*	Mr. Green.
MUSTAPHA,	Mr. Darley, jun.
BEN HASSAN, *(a Renegado,)*	Mr. Francis.
SEBASTIAN, *(a Spanish Slave,)*	Mr. Bates.
AUGUSTUS, ⎫	Master T. Warrell.
FREDERIC, ⎬ *(American*	Mr. Moreton.
HENRY, ⎨ *Captives,)*	Mr. Cleveland.
CONSTANT, ⎭	Mr. Whitlock.
SADI,	Master Warrell.
SELIM,	Mr. Blissett.

WOMEN.

ZORIANA, ⎫	Mrs. Warrell.
FETNAH, ⎬ *(Moriscan Women,)*	Mrs. Marshall.
SELIMA, ⎭	Mrs. Cleveland.
REBECCA, ⎫ *(American Women,)*	Mrs. Whitlock.
OLIVIA. ⎭	Mrs. Rowson.

Slaves.—Guards, &c.

Figure 3.1 "Dramatis Personae," Susanna Haswell Rowson, *Slaves in Algiers; or, a Struggle for Freedom* (1794); Historical Society of Pennsylvania.

important distinction to the play. In a replay of sentiments expressed during America's revolutionary period, the play avoids applying the label of "slave" to an American citizen—Algerians, according to the Dramatic Personae, merely hold Americans "captive," while the condition transforms other nationalities into "slaves." Likewise, the cast listing's refusal to name women as slaves suppresses the threat of sexual slavery, perhaps the preeminent fear accompanying western imaginative forays into the orientalist seraglio.[27] The cast list suggests that while Rowson's characters first appear easily divisible into clear categories, the plot enacts rather complex and overlapping networks of affiliation and differentiation. Despite the textual attempt to sort and demarcate ethnic boundaries, *Slaves in Algiers* presents a cast of characters who repeatedly shift or threaten to shift categories, ultimately conceiving of identity as performance.

The "dramatis personae" does not simply organize imaginary characters. It bridges on- and offstage roles, tracking the correspondences between actor and role. Audiences used the text (and similar notices printed in newspapers and playbills) to connect the performing bodies onstage with the characters acting in the plot. Although this seems a commonplace and pragmatic function, the listing represents a textual technology for relating the "representational" to the "real," stage personae to personal identities. In that sense, the document presents a more complex negotiation of the performance of identity, simultaneously sorting real from "imaginary" by separating them into columns, yet implicitly connecting the two. The actor simultaneously is and is not the character. The form of the list recapitulates that juxtaposition, constructing and unraveling binary schemata—American actors are of course not Algerian, and the play represents an imaginative foray into ethnic, religious, and national differentiation. At the same time, the play turns centrally on the doubling act, the ability of Americans to play Algerian. As audiences might have understood, *Slaves in Algiers* is only partly "about" ethnic conflict and difference, about Americans opposing Islamic others. Just as crucially, such performances always seem about the playhouse experience of Americans acting as ethnically and racially defined others. Early American audiences did not attend the theatre to witness Algerians; they paid to view Americans playing Algerian. That doubled performance, as the cast listing suggests, shapes the characterization and the plot possibilities of *Slaves in Algiers*. The play makes "acting Algerian" its central theatrical technique, both the problem and the chief pleasure of the play.

The doubling act tracks alongside another hierarchical ordering. As the listing proceeds downward, the overlapping depictions of role and actor govern its order. The order of the characters indexes both the characters' position in the plot and the actors' professional standing, tracing the practices of theatrical celebrity and publicity. Eighteenth-century Atlantic theatre relied upon "lines of business," standardized but flexible conventions that associated actors with certain kinds of roles, communicating character types efficiently and effectively. American theatre inherited that tradition; as James C. Burge writes, "the stock company format which the Hallam troupe brought with them was a classic example of provincial joint stock organization."[28] Managers cast parts, at least in some measure, based on actors' suitability for various kinds of parts; conversely, playwrights often wrote parts to suit established lines of business. The cast listing's order thus simultaneously inscribes the prominence of the role and the celebrity of the actor.

The neat descending sequence of characters and lines of business collects, finally, at the bottom of the page in one final segment. An underclass, a literal subcategory, of "Slaves—Guards, &c." appears at the bottom of the page, demoted beneath the notice of publisher, reader, and ultimately the play's audience. Listed collectively, the supernumeraries—the less celebrated among the newly recruited "company of performers, consisting of fifty-six men, women and children" whose 1793 arrival James Fennell describes—receive no individual notice.[29] Excluded and segregated from the troupe's premier actors, the extras at the bottom of the cast list faintly signal the play's reliance on conventional notions of class and status. Behind the liberated Americans, the dramatis personae shows, stands a motley collection of other characters conscripted by Rowson's plot. Although the roles themselves played a crucial function in the play, they occupy the space of disdain or disregard on the list as in the plot. The extras signal the commitment of the script (and, through its adherence to convention, of Atlantic theatre culture) to notions of class, agency, and individual identity. They have no individual identity; a mob of undifferentiated characters without speaking lines, they would exercise little apparent agency within either the play or the playhouse. Their collection at the bottom of the page perhaps marks the play's most fundamental act of social distinction, the textual manifestation of a generally unspoken theatrical "habitus."[30] The most significant division among the characters onstage, the cast list suggests, is not one of ethnic, religious, or racial difference, but the hierarchical distinction between individual agents and a low collective of extras.

FEAR OF TURNING TURK

The play's low cohort emerges from behind a set of remarkably destabilized characters whose captivity threatens conversion and transformation. *Slaves in Algiers* participates in broader cultural and theatrical strategies of representing American captivity in Algiers. Rowson's play shows the importance of a wide array of politicized, but also trans-partisan acts of Algerians and enslaved Americans; those performances stage troubled standards for inclusion before broad American audiences. They also enact fears of Algerian slavery's potential to turn insiders into outsiders—and perhaps, to reveal those insiders as actually alien all along.

North African captivity had consistently figured such possibilities in Anglo-Atlantic culture. The long-standing conventional fear of becoming a "renegado" or of "turning Turk," as Dan Vitkus has noted, appeared amidst early modern Anglo-Mediterranean commerce and cultural traffic. It shaped everyday acts and performances as iconic as Shakespeare's *Othello*.[31] Rowson's early national audiences understood the problems of cultural crossing, conversion, and cultural contamination posed by Algerian captivity. By the eighteenth century, North African narratives shared forms with Native American captivity tales, both exploring "generic and cultural changes, divisions, and differences occasioned by the captives' cultural crossings," as Christopher Castiglia argues.[32] American slavery in Algiers subjected captives to the threat of various kinds of transformations. On the surface, slavery under Islam threatened—or tempted—religious conversion. More broadly, captivity posed a cultural or ethnic threat as the experience transformed the language, clothing, and appearances of early American captives. Finally, and perhaps most elusively, captivity posed the imaginative possibility of figurative racial transformation, as freeborn Americans found themselves subject to the same oppressions suffered by Africans enslaved in the Americas. The threat of Algerian captivity thus represents the triple threat of religious, national, and ethnic conversion into the amalgamation of black, North African, and Islamic identities. Such potential transformations trouble Rowson's play, which simultaneously embodies and vigorously resists such conversions.

Representations of Algerian captivity tend to play on such fears of religious and cultural conversion. The playwright and novelist Royall Tyler's 1797 *The Algerine Captive*, for example, provides a contemporary literary representation of such conversions. The novel turns on an overtly theatrical chapter in which characters performatively blur lines between key identity markers such as religious and ethnic affiliation. The novel's protagonist, Doctor Updike Underhill, engages in a religious debate with a Mollah; although ultimately fruitless in his efforts, the Islamic proselytizer certainly presents his case persuasively. Conversion poses an attractive alternative to Underhill's captivity, since Islamic Algerians reputedly did not enslave coreligionists. The conversion process alone offers a respite from work, and the pressure to go renegado seems almost irresistible. The Mollah himself, an English convert to Islam, embodies the possibility of conversion that Underhill fears and resists. The possibility of conversion imagines religious, ethnic, and national identity as roles that one can discard or assume, as the Mollah already has and as he encourages Underhill to do. Underhill remains unable to justify his own religious tenets but unwilling to convert, and ultimately, he simply abandons the dialogue. Notably, Tyler depicts his protagonist's resistance to conversion as a matter of simple stubbornness, and not the result of any innate religious superiority of Christianity. The Mollah's rhetorical victory raised some contemporary eyebrows, and Tyler had to defend his novel against accusations that it "savored of infidelity."[33] The scene presents the threat of conversion, but also the menace of embodying and allowing an Islamic character to speak for himself. Even more dangerously, the Mollah as renegado suggests that the self-representing Algerian could actually be American.

Slaves in Algiers rehearses variations on the theme of religious conversion, although the play finally forecloses such dangerous possibilities. Olivia, for example, offers herself as a sacrifice in order to save her fellow American men. Perhaps predictably, Muley commands her to "renounce your faith—consent to be my wife."[34] The plot of threatened feminine virtue pressurizes the drama of threatened religious conversion, and Olivia plans to comply with the Dey's demand but then to commit suicide. Although it avoids the perils of Christian apostasy, the play more willingly contemplates Muslim conversion to Christianity. Zoriana, the daughter of the Dey, for example, identifies herself as a "Christian in my heart"; in love with the American Henry, she plans to run away and abandon her Islamic faith.[35] Notably, her figurative conversion changes the ethnic or racial markers used to describe her—in the next scene, she becomes a "fair Moriscan."[36] Throughout the play, Zoriana continues enacting the role of convert, displaying misplaced but admirable filial loyalty; when Henry, the object of her love, turns out to be already engaged to Olivia, she demonstrates commendable Christian restraint. Although technically still Muslim, Zoriana declares, "I will, tho' my heart breaks, perform a Christian's duty."[37] Her potential conversion offers a logical and, ironically, perhaps more sympathetic counterpart to Olivia's resistant superficial conversion. Both cases, however, follow the logic that internal and hence more "genuine" motives count more than external and superficial performances. The incomplete conversions thus actually provide antitheatrical inoculations against captivity's threatened transformations.

Cultural conversion (as opposed to willful religious conversion) poses trickier problems, and offstage captivity accounts suggest looming fears of inadvertent and performative cultural conversion. In a letter to his mother published in his hometown newspaper in 1795, American captive John Foss, for example, relates the experience of conforming by degrees to his captors' culture. Foss narrates his assumption of Algerian costume and eventual adoption of Algerian language and customs.[38] Algerian "habits," in the dual sense of costume and custom, partially convert the American captive. Foss's experience suggests adaptation more than transformation or conversion, but other captives reveal the more striking effects of cross-cultural contact. The 1817 autobiographical narrative of Robert Adams, who experienced captivity in North Africa in the first decade of the nineteenth century, narrates a remarkable transformation.[39] Unlike the fictional recruitment of Tyler's *Algerine Captive*, Adams's transformation seems to have been entirely unintentional; it also seems much more thorough than Foss's short-term experience. A letter by Robert Dupuis, "the British Vice-Consul at Mogadore" (now Essaouira in Morocco), prefaces Adams's narrative, recapitulating the startling changes wrought by North African captivity:

> The appearance, features and dress of this man upon his arrival at Mogadore, so perfectly resembled those of an Arab, or rather of a Shilluh, his head being shaved, and his beard scanty and black, that I had difficulty at first believing him to be a Christian. When I spoke to him in English he answered me in a mixture of Arabick [*sic*] and broken English, and sometimes in Arabick only. At this early period I could not help remarking, that his pronunciation of Arabick resembled that of a Negro, but concluded that it was occasioned by his intercourse with Negro slaves.[40]

Captivity provokes a threatening and theatrical transformation. In costuming and facial appearance, Adams seems more Arabic than American; his language, too, reveals traces of conversion. As the account suggests, he had played Arabic so long that he has (at least temporarily) become Arabic. The former captive's appearance and comportment suggest a much more complex and messy performative process of acculturation, transformation, and adaptation than the neatly dichotomous choice of conversion faced by Tyler's Underhill. Adams recovers his American habits and appearance, but his conversion contaminates him with a suspicious theatricality; antitheatricality provides conversion's logical antidote. Dupuis, for example, relates Adams's "involuntary exultation at the sight of the American flag," an instinctual and unscripted reaction that proves his national identity.[41]

Dupuis's account, grasping at forms through which to understand the returned captive, presents Adams as a problem of interpretation. Speaking a mixture of broken

English and Arabic, his pronunciation inflected by "Negro" intonations, Adams simultaneously enacts the performative markers of Arab, sub-Saharan African, and "Shilluh," an Arabic label applied dismissively and indiscriminately by westerners to Berber "vagabonds" and other exotic low characters.[42] Appearing as both slave and vagabond, Adams experiences a class falling associated with racial and ethnic crossing. Dupuis explains (or perhaps exacerbates) his case with rumors of miscegenation, recounting a rumor that Adams's "mother was a Mulatto, which circumstances his features and complexion seemed to confirm."[43] Dupuis's confusion at Adams's conversion amplifies when he realizes that the converted captive might actually be American. Ironically, his labeling of Adams as mixed-race converts the captive again even as it attempts to stabilize the performance of identity.

Adams's performative conversion to black/Berber/Arabic makes him in a sense the low-comic conscripted offspring of race mixing, the logical extension of Othello's threatening cross-racial appeal. The stage figure of Othello represented for some viewers the dangers of cross-racial desire. By that logic, desire and racial mixing produce conversions of sorts, the generational transition from white to nonwhite. Such dangers, voiced most famously by John Quincy Adams, emerge later in the celebratory chaos of T. D. Rice's satirical 1844 *Otello*, which embodied the possibility of conversion in the biracial child of Othello and Desdemona.[44] Reputedly already one-quarter black, Adams embodies the results of cross-racial intimacy. Unlike Othello, he does not openly threaten to produce mixed-race offspring or tragically kill his lover. Instead, Adams performs his own inclination to turn Turk, re-scripting Othello's early modern "drama of conversion."[45] The captive sailor also falls in standing, reversing the noble Moor's upward trajectory toward European respectability. Adams's story, like the broader genre of Islamic captivity narratives, revolves around the threat of such performative transformations. The captivity experience gains force in proportion to the imminent threat of conversion and pollution. For its part, Dupuis's commentary both resists and repeats those turns in its attempt to reinstate Adams's "authentic" identity.

HASSAN'S RENEGADO CONVERSION

Slaves in Algiers wards off the implications of Adams's and Foss's cultural transformations by presenting its only explicit Renegado character as laughable and ultimately despicable. Hassan, an Anglo-Jewish convert to Islam, provides the play's most scathing (and ultimately, a deeply anti-Semitic) embodiment of transcultural conversion. Ben Hassan is the most genealogically complex character in the play, posing problems of national, ethnic, and racial social identity that the play must resolve in order to reinstate the republican Anglo-American polity by the finale. As a stage Jew, Hassan represents an evolving and variously deployed type in Anglo-American theatre. After the Revolution, early Americans encountered a "wider number of possibilities for reading the Jewish character both on and off the stage," according to Heather Nathans.[46] Innovations in Atlantic acting techniques as well as changing conceptions of Jewish characters combined to open up space for the representation of Jewish stage types. Enlightenment thinking and the observation-based approaches of actors such as David Garrick had transformed the stage Jew from a clown to a villain in eighteenth-century performances of plays such as *The Merchant of Venice*. Hassan demonstrates that both strains can coexist in the same character.[47] In 1794, Philadelphia partisans had debated Jewish political participation, but their arguments, associating Jewish people with other excluded minorities, reveal Jewish identity as a convenient figure for broader issues of constituency and participation in the early republic.[48] In *Slaves in Algiers*, Mr. Francis, a low comedian specializing in comic old men, played the role. His character appears roguish, avaricious, and untrustworthy, but also diminutive and ludicrously accented. As Nathans argues, Francis's Hassan represents an "unthreatening image of Jewish masculinity, as well as an exoticized one."[49] Ultimately, the play subjects him to humiliating cross-dress and the amorous advances of the oblivious and equally clownish Sebastian. Ultimately, Hassan's stage-Jewish identity seems more notably empty and unstable, rather than simply stereotypical. The category of "Jewish," while making him the butt of ethnic humor, also provides a convenient vehicle for his more complex cultural traversals.

Hassan also stands in for the transnational and transcultural Atlantic underclasses gone bad, as he demonstrates in a song in the play's second scene. His London education in the school of hard knocks has trained him first to profit from selling wares on the street, making him the sort of character that popular publications such as the *Cries of London* frequently displayed for Atlantic audiences at the end of the eighteenth century.[50] Hassan quickly takes his entrepreneurship to its illicit ends, learning to forge and fake his goods. He soon transitions to lending at exorbitant rates, a stereotypically Jewish offence, and eventually forges bank drafts. This crime, as he narrates, forced him to run to Algiers:

> So, having cheated the Gentiles, as Moses commanded,
> Oh! I began to tremble at every gibbet that I saw;
> But I got on board a ship, and here was safely landed,
> In spite of the judges, counsellors, attorneys, and law.[51]

Hassan's song stages his character as not only Jewish but criminal, performing the slippery slide from social outcast to outlaw. As William Pencak observes, anti-Semitic Federalist rhetoric imagined Jews as part of a "motley crew of Irishmen, 'Jacobins,' Frenchmen, African Americans, and the poor striving to wrest the government from a virtuous elite."[52] Perhaps unconsciously, Rowson's Hassan reprises such character typing. Criminalized and turning international fugitive, Hassan completes the process ("to complete the whole," as Frederic comments) by converting to Islam.[53] In the play's moral economy, Hassan's conversion to Islam simply proves the instability and untrustworthiness of his stage Jewishness. Hassan performs religious conversion alongside the more threatening possibility that such conversions are transgressive acts of the criminal underclasses. As if in Dantean exasperation with this theatrical excess, Rowson's plot finally forces Hassan into comic cross-dress, transforming him one last time into an undignified old woman in order to escape the Dey's punishment.

Going renegade, Hassan shows, constitutes a rogue act as much as it does a religious conversion; Hassan's criminality both produces his conversion and renders it outlaw. In broader contexts, his transgressive transformation has roots in the commercialized intercultures of the early modern Mediterranean world, in which English culture experienced a crisis of real and imagined conversion to Islam. At times, such conversions fueled spectacular roguishness as English subjects turned Turk to participate in Mediterranean piracy, as Dan Vitkus observes.[54] Hassan's act thus transmits forward and reenacts a kernel of residual cultural memories, the recollected Mediterranean experiences of the English, who (like Americans in the 1790s) had the will to power—the discourses and the poses of supremacy—but not the means. Americans, as Frank Lambert notes, frequently understood Algerians as pirates and rogues, "sinister characters" from rogue states whose cruelty and tyranny contrasted with the freedom-loving and peaceful free trade of American merchants.[55] Mathew Carey's 1794 *Short Account of Algiers*, published to explain and capitalize upon the American captivity crisis in Algiers, falls back on such common explanations of Algerian's outlaw origins and status. In a revised and expanded edition of the *Short Account* (which included a "copious appendix" with letters from American captives), Carey adds a passage explaining that the label "Arab" is "still applied to a race of tawny and independent barbarians, who wander in gangs about the country, and unite the double possessions of a shepherd and a robber."[56] The account's political and cultural geography of North Africa surveys a rogue's gallery of middle-eastern bandit types, "the sweepings of all nations blended together."[57] It shows palpable disdain for Mediterranean cultures, extending even to the "vagabond who founded Rome."[58] More importantly, those structures of feeling place ethnic others in existing class structures that recognize their mobility and trans- (or pre-) nationality as not only different, but as outlaw and outcast—to be dealt with accordingly. Outlawing alterity, such moves position rogues beyond the pale of Anglo-Atlantic civilization. *Slaves in Algiers* participates in those conventional Atlantic practices, articulating outlaws in exotic accents and redrawing the boundaries and constituencies of early American publics.

THE PLEASURES OF PLAYING ALGERIAN

Slaves in Algiers, especially in the demeaned Hassan, represses and regulates difference in the service of American national identity. At the same time, such scenes require—or even better, allow—early Americans to play Algerian. They literally become Algerian on stage, if only in play. Such acts require on stage the very move they appear to resist: the literal transformation of Americans, the masking and costuming of Americans as Algerians. Despite the dangers of conversion and transformation, playing Algerian seems to have provided a real resource to early American culture, offering significant pleasures. *Slaves in Algiers* thus registers the allure of becoming Algerian even as its plot counters such transformations. Such gratifications have political costs and benefits, and Rowson's patriotic ending ultimately reveals the politics of its Algerian pleasures.

Offstage, Americans frequently played Algerian, and imaginative renegado acts suggest that transformation did more than cathartically purge fears of difference. Numerous early American texts reveal the pleasures and payoffs of playing Algerian.

The 1787 fiction *The Algerine Spy in Pennsylvania*, for example, by Philadelphia dramatist and satirist Peter Markoe, produces an orientalist fantasy of American authorship and textuality.[59] The text develops an elaborate backstory of absent Algerian author, found papers, and translated text in order to present the author as a covert Algerian. The undercover-outsider role allows him to comment with humorous insight on current domestic events, including Shays' Rebellion and the debates surrounding adoption of the Constitution.[60] In 1790, Benjamin Franklin's last published letter textually performs a similar role, inventing a fictional Algerian slaveholder to satirize southern proponents of slavery.[61] Clearly, the imaginative and performative conversion of American author or actor into Islamic North African outsider promised specific rewards and opportunities for early American cultural producers. An even more striking example of early American playing Algerian appears in the November 24, 1794, *Federal Orrery*, in an ad offering "Forty Zequins Reward" for the return of an American slave, an "incorrigible infidel" who had run away from his master "Ibrahim Ali Bey" in Algiers. Proceeding in an apparent orthographic attempt to render Middle Eastern stage dialect, the advertisement describes the "ungrateful villain" as bearing the "evident marks of having worn an iron collar," and his back "well scored with the lash, and the soals [sic] of his feet with the bastinado, for his stubbornness." Ibrahim expresses outrage that his slave would respond to his "salutary punishment" and proselytizing attempts by absconding. The fictional piece, as we might expect, does not target Algerian captivity, but sectional and political opponents, implicitly comparing Algerian captivity with American slavery. The heading identifies it as a "most admirable satire on the *democratic professions*, and *despotic practices* of our ranting southern demagogues."[62] On the heels of Rowson's premiere, and alongside numerous depictions of Americans enslaved by Algerian captors, the advertisement plays politics by playing Algerian.

Such textual acts rely upon the theatricality of a variety of onstage performances featuring Americans posing as Algerians. Although *Slaves in Algiers* avoids extravagantly spectacular orientalism, it still displays embodied stage Algerians.[63] The play's most striking Algerian characters are Muley Moloc, the Dey of Algiers, played by a Mr. Green, who had also successfully played heroic bandit parts. (John Durang, for example, remembers Green in the part of "Abelino," probably the lead role in a popular Italian-style bandit opera.) Muley stands until his defeat in the play's final scenes as an extravagantly outlandish and powerful figure. The Dey's henchman Mustapha was played by the comic actor and singer John Darley Jr., who famously sang Rowson's comic song "America, Commerce, and Freedom" in *The Sailor's Landlady*. Darley also played the singing thief Robin of Bagshot in *The Beggar's Opera* in 1795.[64] Playing the Dey's lackey might seem out of Darley's usual line, but his roguish comedy as Macheath's gang member may have influenced his embodiment of exotic but comic subjection in Mustapha.

The available accounts do not specifically indicate the makeup or costuming techniques used to present the Algerians to early American audiences, but the actors quite possibly donned the burnt cork of later blackface or darker makeup to represent a "tawny" complexion, as they would for a more famous North African role, Othello.[65] The play script reveals few material details of the staging, but its dialogue can offer insights into the presentation of Algerian characters. Fetnah, for example, describes the Dey in the first scene. He is "old and ugly," she says, and "he wears such tremendous whiskers; and when he makes love, he looks so grave and stately, that I declare, if it was not for fear of his huge seymetar, I should burst out laughing in his face."[66] Preceding the appearance of the Dey, Fetnah's reaction conditions the audience reaction, envisioning him as an outlandish, powerful, and phallically gifted character, even if ultimately laughable. Fetnah describes Mustapha, likewise, as a "great, ugly thing" who bows until the "tip of his long, hooked nose almost touched the toe of his slipper."[67] Such lines cue audiences to recognize Muley and Mustapha as visibly different from American characters in costuming, complexion, and facial features. Although scripted descriptions do not definitively indicate staging, makeup (perhaps even facial prosthetics) and costuming might follow such spoken cues, if for no other reason than to ensure that audiences could easily recognize the characters.

The costuming of such characters probably consisted of elaborate renditions of oriental garb. English playwright George Colman's orientalist "dramatic romance" *Blue Beard*, which appeared in American theatres in various versions in the late 1790s, was perhaps the single most frequently performed middle-eastern spectacle.[68] An 1802 review of a New York production of *Blue Beard* describes the kinds of costuming audiences expected in plays such as *Slaves in Algiers*. The character of Abomelique, the review assessed, was "dressed gorgeously and properly"; likewise, the character played by John Hodgkinson wore "a large turban ornamented with a crescent sparkling with diamonds, a full Devonshire-brown satin robe, covering the person almost to the feet, satin trowsers of the same, reaching to the calf and fastened at the knee, with silk stockings and morocco slippers."[69] Describing performances occurring only a few years after *Slaves in Algiers*'s short run, the review gives an approximate sense of early American costuming of Islamic characters. The atypically lavish description suggests an unusual performance. At the same time, it also reveals that at least by 1802 (and perhaps before), American theatres possessed the costuming to create elaborate stage shows of Islamic dress. The New York playhouse perhaps recycled or adapted *Blue Beard*'s costuming from its earlier productions of *Slaves in Algiers*. The same review claims that the performance came off with "incomparably more splendor than in Philadelphia, and nearly vies with the *Blue Beard* in London." Although puffing reviews might exaggerate such claims, it seems less likely that they are patently untrue. Of course, some stages boasted better supplies than others, and it remains possible that American theatres even rivaled London's orientalist stage displays.

PATRONIZING AND PRODUCING ALGERIAN SPECTACLES

The early American stage frequently produced exotic Islamic and North African spectacles, many centered specifically on Algerian scenes. Just two years after *Slaves in Algiers*, the Philadelphia theatre staged a pantomime version of *Blue Beard*, "composed by Mr. Byrne" for his benefit.[70] In the next few years, Philadelphia theatres produced further runs of the play, capitalizing on the combined popularity of Colman's pantomime form and exotic depictions of the Middle East. The play was popular (and elaborate) enough to be performed repeatedly in Philadelphia in 1799, 1800, and 1801, where advertisements announced an exotic collection of "Spahis" and "Janizaries," many played by the same actors who starred in *Slaves in Algiers*.[71] Appearing simultaneously in Boston, the Federal Street Theatre's version of *Blue Beard* featured scenery depicting "A Turkish Village, with a distant Camp at Day Break" and a "Grand Procession of Abomeliques troops, descending from the Mountains," as well as "Blue Beard's Castle, and the Garden of the Seraglio."[72] It would go on to attract "many crowded houses" in Boston, traveling to nearby Salem and Providence, Rhode Island, as well.[73] The play also appeared as far afield as Baltimore, Maryland, and Charleston, South Carolina, and William Dunlap gave it a new lease on life when he revised it for the New York theatres in 1802.[74] The production, promising a "display of splendid pageantry such as has not been seen in this country," announced in playbills its processions of middle-eastern masters and their slaves reminiscent of the scenarios enacted in Rowson's *Slaves*.[75] John Brown's *Barbarossa, the Usurper of Algiers* (appearing in 1794 in Boston and Newport, Rhode Island in a "benefit of the unfortunate Americans, now held in bondage, in Algiers") also featured a cast of middle-eastern despots and their slaves.[76]

The extravagance of such displays could also boil over into the extravagant variety of popular spectacles. Five years after the performances of Rowson's *Slaves*, the Philadelphia Theatre produced as an afterpiece a "New Pantomimic Olio" entitled *The Arabs of the Desert; or, Harlequin's flight from Egypt*. The play improvised upon pantomime's traditional doubled characters—"Mustapho Lightfootero" became Harlequin Turk; Mahmoud transformed into the Father of Columbine; and Queronibus and Zobeide transformed into Pero and Columbine "à la Turque."[77] As each "Arab" character transforms into a performer of a traditional Anglo-American theatrical form, the spectacle of doubled identity restages the transformation at the theatrical heart of such exotic acts.

Such acts rely upon sumptuous scenic display, and Royall Tyler's novel *The Algerine Captive* offers a textual fantasy of Algerian exhibitionism that comments tellingly on the early American stage. After being "suffered to view the city, and the immense rabble," Updike Underhill's captors took him and his fellow captives to the Dey's palace. A scene of splendor and performance ensues:

> Here, after much military parade, the gates were thrown open, and we entered a spacious court yard, at the upper end of which the Dey was seated, upon an eminence, covered with the richest carpeting fringed with gold. A circular canopy of Persian silk was raised over his head, from which were suspended curtains of the richest embroidery, drawn into festoons by silk cords and tassels, enriched with pearls. Over the eminence, upon the right and left, were canopies, which almost vied in riches with the former, under which stood the Mufti, his numerous Hadgi's and his principle officers, civil and military; and on each side about seven hundred foot guards were drawn up in the form of a half moon."[78]

Underhill's verbal embellishment matches the lavishness of the Dey's appearance and costume, textually mimicking the impression made by the "fierce appearance of his guards, the splendour of his attendants, the grandeur of his court, and the magnificence of his attire."[79] Notably, Tyler's imagined scene displays more than just luxurious material; alongside the spectatorial bodies of American captives, it textually conjures up the presence of many hundreds of disciplined Algerian bodies, all cowed by the Dey's theatre of oriental despotism. Within the fictional world of the novel, the Dey creates the exhibition as an impressive demonstration of cultural and political power. Americans stand as ragged captives before the irresistible manifestation of power. It is a conventional scene, but one in which opulent display skewers the grandiose ethnocentric and nationalistic early American self-image. In effect, the scene both enables and deflates Rowson's final tableau of abjected Dey and triumphant Americans.

Tyler's narrative works in a vacuum of reliable firsthand information about Algerian court display—indeed, the scene requires neither accurate information nor mimetic realism. It does require, however, the context of the theatre. It constructs what amounts to a textual version of early American orientalist theatre. The novel captures the dynamic of forced exhibition that plays such as *Slaves in Algiers* and *Blue Beard* exploit. Like Rowson's play, the scene displays exotically arrayed Algerians alongside the forced display of captive Americans. Underhill narrates a scene of performance before patrons, and the eyes of captor and captive seem locked in a mutual gaze for the pleasure of the reader, the audience before whom the entire scene appears. The Philadelphia theatre could hardly have replicated such an extravagant exhibition, and Tyler's novel fantasy of orientalist grandeur far exceeds the reality of onstage Algerian acts. Nevertheless, the vision of captive and Algerian bodies on display seems strikingly conditioned by early American theatre practices.

Such displays trace early American assumptions of theatrical Algerian display. Anglo-Atlantic theatre operated alongside a wide variety of texts, scenes, and rumors of the Middle East to shape the scene's orientalist assumptions. Tyler quite obviously realizes this; in *The Algerine Captive*, Underhill disappointedly contrasts his captivity experience to a plot offered up by Aphra Behn or George Colman, two sources of the period's popular theatre.[80] Real Algerian captives are low, he laments, and captivity does not live up to romanticized scenes of common slaves amorously attracted to captive nobility. Despite that jaded response to eighteenth-century theatre, Tyler's novel remains ironically and deeply indebted to its forms. The overriding irony of Tyler's scene is its genesis in American theatre. Not the Algerian Dey, but an American author conjures the scene into existence. The paying spectators in early American playhouses become the patrons—if not literally producers—of such elaborate displays. The scene implicates audience members (novel reader as well as theatrical spectator), who replace the Dey as the agents that call the scene into existence. Although various other people (managers, actors, writers, prompters, musicians, laborers) helped shape it, the economic drawing power of the audience calls the exotic spectacle into being. They fantastically produce a seemingly effortless spectacle of oriental bodies just as Tyler's novel does and just as it imagines the Dey doing. Algerian display thus enacts an exotic fantasy of audience participation and patronage, revealing theatre as commanded by the audience, pit, boxes, and gallery. The ritual of playing Algerian produces not only Algerian characters; it conjures American audiences who appear simultaneously in the triple roles of oppressed captives, triumphant nationalists, and extravagant eastern despots commanding characteristic display of orientalist performance.

Early American Algerian performances troubled the relationships between acting and spectatorship in other ways. In 1805, when U.S. naval officers brought "Turkish" prisoners from the Tripolitan War to New York City, they displayed their captives at performances of *Blue Beard*, the "appropriate drama" for the occasion.[81] In scenes of captive bodies and spectatorship on command rivaling Tyler's novel for complexity, the prisoners attended the theatre repeatedly during the month of March 1805. The exotic visitors received top billing, listed by name in newspaper announcements that displaced the play with the show of their spectatorship. Those scenes reveal an early American theatre culture habituated to Algerian spectacles, but that also encountered them in complex and self-conscious forms. Playing at exotic North African and Middle Eastern identities, while always a political act staging identity and its boundaries, also produced inversive and unstable performances. They ranged from political satire and serious commentary to exotic spectacle and extravagant fun. Playing Algerian suggests the same sorts of performative habits that would produce other comic renditions of ethnic identities; stage Yankees, Native Americans, Irish, Jewish, and eventually blackface minstrelsy acts also embodied cagey and coded performances of layered identities. In these kinds of performances, suspicion and fear of difference coexist with the payoffs and pleasures of identifying across boundaries. Such acts perhaps even reveal emergent sympathies, empathies, and mutualities.

ROWSON'S COMIC CAPTIVE UNDERCLASSES

Rowson's play seems fraught with the desires and fears of alterity. It should come as no surprise that *Slaves in Algiers*, in the very act of resisting captivity and conversion, stages alternatives to the upstanding, virtuous citizens of the early republic. As the fears of turning Turk accompany the pleasures of playing Algerian, the categories of "American" and "Algerian" emerge as markedly unstable receptacles. The threat and charisma of alterity that destabilizes American national and ethnic definition also mobilizes the underclasses. Well before Marx's alarm over the downwardly mobile lumpenproletariat, Rowson cobbles together a mob of transnational enslaved characters to respond to the threat of Algerian captivity. Unlike Marx's later subclasses, however, Rowson's characters become an immediately useful and disposable imaginative tool, an antidote to the endemic categorical instability of playing Algerian.

The slave revolt, which gains momentum in the third act of *Slaves in Algiers*, is a literal subplot—a "low plot" in both senses of the phrase, an uprising and an often ludicrous parallel line of action. The subplot's star seems to have been Sebastian, the son of a Spanish barber and a laundress, who proves in speech and song that he is, indeed, not one of the "fine spoken folks."[82] The Spanish laboring-class character points to the motley character of the play's underclasses, the transnational ingredients producing Rowson's national identity. Although the patriotic ending overshadows these facts, its common characters—the plot's agents—do not simply represent a prefabricated American public-in-waiting. Instead, they present a more diverse collection of characters, including non-protestant southern Europeans, than those conventionally identified as constituting American identity.[83]

The comic Iberian character Sebastian in particular seems an unlikely inhabitant of scenes of Algerian captivity. He appears to be a dramatic surrogate drawn from earlier scenes of North African captivity. A Sebastian had appeared before in the context of dramatic Barbary captivity—Sebastian I, the sixteenth-century King of Portugal, who was defeated and probably killed in the 1578 Moroccan battle of Alcazarquivir. He later achieved legendary status as a rumored North African captive. Sebastian had also infiltrated Anglo-Atlantic theatre culture in John Dryden's late-seventeenth-century tragedy *Don Sebastian, King of Portugal*, which imagines Sebastian as the royal captive of Muley Moloch and Mustapha. Isaac Bickerstaff converted Dryden's play into a popular comic opera for the London Haymarket Little Theatre in 1769, and Frederick Reynolds's 1812 *The Renegade*, a revision of *Don Sebastian*, retained Dryden's North Africans and added an English captive named Olivia, perhaps borrowing a character from *Slaves in Algiers*.[84] This possible genealogy, although sketchy, suggests that characters can undergo multiple simultaneous acts of dramatic recycling and revision. Rowson's play seems to inherit the old Barbary captive tale; unmindful of national distinctions *Slaves in Algiers* converts the hero from Portuguese to Spanish, and downgrades the former king to clown. Rowson's Sebastian may thus have been part of a gradual transformation of elite captive into low buffoon. *Slaves in Algiers* relegates Sebastian from royal elite to mobile, laboring, underclass captive. It is a sign of the wished-for new dispensation, the desired reordering of the old order to accommodate America's rise in the Atlantic world system.

In the third act, as the slaves gather in a grotto on the coast, Sebastian demonstrates his indecorous comedy in a drinking song. Providing one of the play's pleasurable moments of roguish entertainment, the air extols the virtues of alcohol as a consolation for the inattention of women. Singing that nothing "helps the invention, or cheers the courage, like a drop from the jorum," Sebastian recounts his ill luck with the women. Finding them "frumpish and cold," the lad turns to drink, forming a jilted alliance with the "mirth-giving, care-killing jorum."[85] Echoing *The Beggar's Opera*, Sebastian evokes a fantasy of underclass alcoholic festivities and male conviviality that promises relief from the complex and intimidating world of sober heterosocial relationships. The scene, with its attractive sociability and catchy tunes, suggests that the play's common sorts are more than simply demeaned objects of the gaze. Sebastian perhaps displays an alluring fantasy of lowness. At the same time, he might simply evoke the intoxicated sociability that surrounded and encroached on the early American stage, whose customers enjoyed social drinking in and around theatres.[86]

Ultimately, Sebastian's jorum inspires courage and enables the revolt in *Slaves in Algiers*. Failing at the domestic conjugal rituals of courtship, Sebastian allies himself with booze and finds personal validation in other arenas. He soon finds himself conscripted into the military, "called into battle" against Turkish enemies by mysteriously invisible powers.[87] Fulfilling his low-comic function in the play, he "skulk'd away into the rear" when faced with the overwhelming "rattle" of cannon and bombs.[88] However, the underclasses have secret reserves—useful resources, it turns out, for Rowson's plot. Admonishing himself to gather his courage, he turns again to the bottle. In a display of alcohol-induced heroism, Sebastian and his fellow conscripts refuse to "give it up tamely and yield ourselves slaves," to the Turkish "rabscallion" and "knaves."[89] Sebastian sings the play's rousing call to arms:

> No, hang 'em, we'll bang 'em, and rout 'em, and scout 'em.
> If we but pursue,
> They must buckle too:
> Ah! Then without wonder,
> I heard the loud thunder,
> Of cannon and musquetry too.
> But bid them defiance, being firm in alliance,
> With the courage-inspiring jorum.[90]

Touting the virtues of intoxicated courage in lively rhythms and jaunty rhymes, Sebastian's song identifies his cohort as conscripts in xenophobic crusades. Theirs seems an ironic take on the holy wars of western civilization. Despite its focus on fighting North African opponents, the song hardly keys any sense of ethnic or religious solidarity. Nationalist or patriotic rhetoric also seems conspicuously absent. Sebastian's song takes class as its dominant imaginary; the less fortunate appear conscripted and thrown into battle. In the process, however, they negotiate their own forms of courage and collective action. Sebastian's drinking tune centers the underclass performances in *Slaves in Algiers*, singing their ethos in a bit of charismatic and class-conscious performance that perhaps becomes an inadvertent self-critique of the play's patriotic closure. Motivating the low with a blend of conscription, cowardice, and inebriation, the song counters elite inaction and abstract patriotism with alcoholically (and ironically) self-actualized underclasses.

Rowson's song, with its mannered and unusual slang terminology for alcoholic beverages, appears to have been an original, with no obvious offstage referents or predecessors. Audiences, however, had probably encountered significantly similar uses of the term "jorum" before. Jorum repeatedly appears, for example, in ballad operas by John O'Keefe. The Irish playwright uses folk music and underclass humor in popular comedies such as *The Poor Soldier* (1783) and its sequel *Love in a Camp* (1786), works that remained popular in Atlantic playhouses well into the nineteenth century.[91] O'Keefe's *The Son-in-Law*, which features a low-comic drinking song in Act 2 that celebrates jorum, possibly provides a direct source for Rowson's use of the term. The play had been performed in Philadelphia and published in 1794 as "corrected and revised by Mr. Rowson."[92] Likewise, *The Agreeable Surprise*, which appeared in Boston in 1794, uses the term in its own comic song.[93] Such performances probably introduced many audience members to the term jorum as a bit of Atlantic slang.

Even more suggestively, audiences enjoyed singing Sebastian's songs about the jorum themselves as much as they applauded them on stage. The number appeared in songsters such as the 1798 *Columbian Songster and Freemason's Pocket Companion*, which reprints Sebastian's number as one of its "*Sentimental, Convivial, Humourous, Satirical, Pastoral, Hunting, Sea and Masonic*" songs.[94] Audiences also encountered the term jorum in other low songs. A 1797 songster called *The Nightingale* includes a song entitled "The Farmer," which celebrates the humble classes' inebriate fondness for jorum.⁹⁵ More dangerously, readers of Thomas Mount's 1791 criminal confessions could also read and imagine the performance of a "flash song" that celebrated jorum. The presence of such musical numbers in songsters suggests admirers of theatre's comic underclasses could take pleasure in actively reenacting and singing such performances. Audiences could produce musical performances of sociability much as Alexander Graydon imagines when he writes of himself and the minister musically reproducing *The Beggar's Opera*. Early American leisure, as Graydon's account suggests, could return periodically to such acts of roguish sociability, and the song's low comedy scripts such practices.⁹⁶ Associated with criminal argot and secret subcultures, Sebastian's drinking song becomes not only an underclass or masculine convivial act, but also a performance of the stagey criminal cohort. Audiences could have found all three elements appealing in various ratios; singing about jorum provides access to bawdy comedy, male drinking culture, and otherwise obscure criminal rituals.

Despite their entertaining ways, the rebellious transnational low of *Slaves in Algiers* remains a suspicious cohort, as the American captive Henry remarks. "This fellow," Henry remarks of the Spanish slave Sebastian, "will do some mischief, with his nonsensical prate."⁹⁷ Frederic's defense that he "has an honest heart, hid under an appearance of ignorance" does little to alleviate the play's suspicions about the abilities and motivations of the mob.⁹⁸ In the next scene, the rebels surround Ben Hassan's house, freeing Rebecca and her son Augustus, but in the process confirming their (comically) unruly potential. Hassan, forced into his wife's clothing to escape the retribution of the slaves, is comically "rescued" and taken back to the slaves' grotto, where he narrowly avoids Sebastian's oblivious advances. Even so, *Slaves in Algiers* presents the rebellious low as proactively resistant, contrasting starkly with the inconsequential fretting of the other characters. The slave revolt operates decisively when Olivia and her family find themselves in an impasse of imperiled virtue. Olivia's feminized American virtue forces her to abide by her promise to marry the Dey, which ironically would negate her virtue. Olivia plans to avert this unthinkable outcome by committing suicide, and only the rebellious mob's entrance solves Olivia's dilemma, converting tragedy to comedy.

In the final scene, as Olivia contemplates her entrance in to the Dey's harem, Mustapha enters "hastily," reporting that the "slaves throughout Algiers have mutinied—they bear down all before them" threatening to "raze the city."⁹⁹ The rebelling slaves, signaled by amorphous offstage din, gradually intrude upon the scene. The violence happens offstage, and just in time to enable the play's decorous and patriotic ending. Led by the play's comic characters, the mob represents the transnational underclasses conscripted into the service of the theatrical national imaginary. They relieve the double binds created by the decorous characters' admirable but impractical virtues. The play needs its rebellious types, and accordingly redeploys them temporarily from the business of comic relief to the serious labor of emancipating American captives. In the process, Rowson briefly recuperates the mob from the insurrectionist tendencies demonstrated in events such as the recent Whiskey Rebellion, which had proven offstage that the young nation's domestic constituencies could perform their own struggle for freedom. Rowson's motley and unpredictable mob of theatrical extras stages America's disenfranchised internal cohorts, bringing the imagined marginal figures of early American publics front and center stage. Despite their critical role, the play allows the mob to take center stage only at the end, and only for a brief moment. In doing so, *Slaves in Algiers* suggests the vigor with which the early American national imaginary both relied upon and suppressed such characters.

ALGERIAN CAPTIVES RETURNED

Rowson's triumphant revolutionary mob seems a wishful and magical stage construction. The stage plot might have reminded Rowson's audiences of an episode in John Foss's account of Algerian slavery:

> In the month of October 1793, fourteen slaves of different nations, made an attempt to run away, with a boat, but were overtaken and brought back to Algiers. When they were landed, the Dey ordered the steersman, and the bowman, to be beheaded, and the rest to receive 500 bastinadoes each, and to have a chain of 50lb weight fastened to their leg for life, and a block of about 70lb to the end of that, which they were then obliged to carry upon their shoulder when they walked about their work.[100]

The onstage revolt against Algerian captivity was not unthinkable offstage, but it clearly performed an optimistic fantasy. Rowson's cohort contrasts starkly with Foss's narrated experiences, in which revolt only results in defeat, death, and further abjection of the survivors. Rowson's onstage cohort succeeds perhaps through the participation and guidance of all-American heroes such as Henry, who models a standard of decorum that an enslaved sailor such as Foss seems unlikely to achieve. The contrast between the two character types is instructive, and inflects offstage scenes of American captives. Algerian captivity evoked fears that American captives in Algerian hands would prove "low," and hence unworthy of redemption. Joel Barlow, poet, politician, and for a time American consul to Algiers, helped secure the release and return of a group of captives to Philadelphia in 1797. In a public letter, Barlow expresses relief that the American prisoners had not proven themselves a low and unworthy collection of captives:

> When we reflect on the extravagant sums of money that this redemption will cost the United States, it affords at least some consolation to know, that it is not expended on worthless and disorderly persons, as is the case with some other nations, who, like us, are driven to humiliation to the Barbary pirates.[101]

Barlow seems invested in imagining a reality consonant with the social expectations coded into *Slaves in Algiers*. American captives thankfully turn out more like Henry than like Sebastian. As in Rowson's play, Americans redeemed from slavery prove themselves worthy of the national imaginary. Nevertheless, captivity seems to have had lasting consequences. Although they had acquitted themselves well in captivity, Barlow continues, some of the captives returned unfit for work. The captives experience a distressing and lasting fall in class position—amidst all the threats of conversion to exotic Islamic identity, captivity produces the damaging conversion of class descent. Noting the futility of appealing to the national government for any other support, Barlow encourages merchants and fellow citizens to give "encouragement to their professional industry" and the "respect which is due to the sufferings of honest men."[102]

As Barlow's comments suggest, Algerian slavery makes the problematic of class falling evident in the abjected bodies of captive Americans. The captivity narrative of John Foss, one of the captives whom Barlow's efforts had ransomed, appeals to audiences by describing the experience of penury and hardship in a captivity narrative. His account recollects the spectacle of Christian slaves in "wretched habits, dejected countenances, and chains on their legs, every part of them bespeaking unutterable distress."[103] It seems a consistent strategy; the earlier letters he had written to his mother also describe the material impoverishment of captivity. Despite an allowance from the government, he writes in one letter, the clothing provided during the course of a year "won't last six weeks to do our work." "Our case," he continues, "is far worse than tongue can express, or pen describe, or heart conceive."[104] The troubling possibility of material want and wretched destitution also encourages spectators to demonstrate fellow feeling and public citizenship. Grateful for Barlow's redemption effort, Foss imagines the U.S. government as "an example of humanity to all the governments of the world," inspiring admiration among the "merciless barbarians."[105] The praise seems intent on inspiring further sympathy.

Against such spectacles of abjection, Americans produced counter-performances of dignity and nobility affirming that captivity only temporarily reduces captives to a helpless and pitiable condition. Foss's account counters its own abject pose, exclaiming, "Though we were slaves, we were gentlemen."[106] Ironically, Foss imagines the degrading experience as producing the dignity and status of "gentlemen." The return of America's Algerian captives in 1797 reiterated Foss's claims in thoroughgoing performances of class identity. Newspaper accounts told of the U.S. government supplying the slaves each with "a suit of wearing apparel" and a moderate allowance for visits to foreign ports on their way home.[107] Respectably costumed, Algerian captives performed further spectacles of redeemed captivity. Newspaper accounts tell of captives returning to America in February 1797 and making a public appearance at the Indian Queen Tavern in Philadelphia. The incident suggests the kind of mobbing excitement the Algerian captives' return could generate:

> Our late captives of the Algerines arrived in this city yesterday afternoon, under an escort of several hundreds of their sympathetic fellow-citizens of both sexes, who had gone to meet them on the road to town; upon their reaching the INDIAN QUEEN Tavern the crowd was so considerable as to render their passage difficult, and on their entering the house an ardent acclamation expressed the satisfaction of the people at their happy extrication and safe return.[108]

Three years after *Slaves in Algiers*, Algerian captives once again become the subject of spectacular display in Philadelphia. American captives, redeemed and respectably clothed, arrive home to demonstrate the recovery and reconstitution of American citizenry. Offstage, they achieve the ideological goals of Rowson's play. At the same time, public performance produces unscripted improvisations on the themes of

Rowson's play. Captivity conjures again the public spectacles of slavery and underclass mobs. The crowd, pressing in upon the returned captives, raises shouts in a boisterous scene of affirmative public ritual. The stagey spectacle, energized by the jostling, elbowing American public, offers a popular alternative to Rowson's wishful performance of decorous American subjects and their patriotic triumph. American slavery in Algeria, which Rowson counters with an unruly mob of underclass characters, becomes a performance of and for an equally boisterous American public.

A number of the redeemed captives did not appear at the Indian Queen Tavern. The newspaper account reveals that three of the redeemed captives had died before arriving in Marseilles, perhaps from mistreatment or overworking during their captivity. Redemption does not have a universally happy conclusion. Moreover, a captain and fourteen seamen had re-signed to "go up the Mediterranean on a trading voyage."[109] Perhaps goaded by the desire to continue traveling and adventuring, or unable to pass up steady employment, some of the former captives remained at sea. Perhaps their traumatic experience produced or strengthened a sense of outsider status, and they could not imagine a comfortable return to American soil. Whatever their reasons, a few of the American captives voyaged on, remaining in danger of a repeat Barbary captivity. The government's redemption of American captives, as those fifteen cast-out sailors demonstrate, did not always bring American citizens home to family reunions on American soil. Many Americans remained in foreign parts; some of the fundamental dilemmas of Rowson's play—the conversion of insiders into outsiders, and the redemption and reunion of American citizens abroad—remain unsolved problems for the young republic.

4. Treason and Popular Patriotism in *The Glory of Columbia*

If *Slaves in Algiers* performs the troublesome low in North Africa, other scenes depict them closer to home. Dunlap's 1803 patriotic spectacle *The Glory of Columbia; Her Yeomanry!*, which premiered at the Park Theatre on July 4, 1803, is one such play. Dunlap cobbled together his patriotic spectacular from *André*, a five-act tragedy that had failed to attract or hold Park Theatre patrons during its short and unprofitable run in 1798. *André* dramatizes the death of Major John André, one of Benedict Arnold's coconspirators, for treason and spying during the American Revolution. Centering on problems of punishment rather than treason and loyalty, even the play's patriotic figures seem troubling. The unnamed "General" (representing Washington, or at least embodying his authority) appears as a stern, even heartless father figure, and the young and hapless Bland finds himself caught between personal and patriotic loyalties. *André* produces a performance of vexed early national allegiances and identity, a troubled and troubling drama. Appearing at New York City's Park Theatre a total of three times, the tragedy made an increasingly bad showing night by night.[1] *The Glory of Columbia*, in contrast, earned $1,287 in its opening night, almost as much as *André* brought in through its entire short run.[2] Dunlap's patriotic spectacle, unlike his tragedy, remained popular for decades. Transforming the Revolution's high tragedy of mourning and memorial into low comedy, Dunlap's rewrite responds to the shifting demands of popular audiences.

Many estimates see Dunlap's revision, a process bridging the "Republican Revolution" of 1800, as a measure of either the democratization or the "dumbing down" of American theatre. Cued by Dunlap's own derision, such assessments perceive *The Glory of Columbia* as aesthetically deficient. His revision indeed reveals a theatrical and social paradigm shift, but I argue that it does not represent an artistic failure, or even, despite its broadened appeal, a radically democratic form. Instead, it makes a foray into new modes of producing and profiting from the popular. The play incorporates offstage performances into its act, reenacting and revising vernacular performances. Dunlap abandoned unprofitable loyalties to the niche market of Anglo-Atlantic cultural practices of tragic drama, revising the play's relationship to vernacular performance. The resulting infusion of popular forms produces a conflicted spectacle of order and disorder, of mock treason and equally mocking punishment, which nervously negotiates the potential unruliness of low comedy.

Like Rowson's *Slaves in Algiers*, Dunlap's spectacle makes patriotic performance popular by using plebeian characters. *The Glory of Columbia* turns to domestic settings, rather than exotic Algerian locales, recruiting and disciplining America's internal underclasses for the work of building the emerging theatrical mythologies of the early republic. Dunlap's revision alters *André*'s tragic plot, foregrounding the unruly potential of characters that seem by traditional standards unfit for tragedy. Their unruliness requires regulation, and Dunlap's comedy simultaneously displays and controls the rowdy low through the ordering structures of parades, processions, and onstage musters. In comic echoes of the Atlantic processions that brought condemned offenders to the Atlantic world's gallows, the play even imagines the punishment of rogues and traitors. Of course, the performance is a comedy, like Rowson's earlier play, and its explicit goal is not so much control of the underclasses as it is producing entertainment through repeated scenes of comic transgression and regulation. Nevertheless, Dunlap's revision reveals low comedy's cultural labors in service of the national imaginary.

Dunlap's two versions of André's plot, premiering in 1798 and 1803, bracket the rise of the Jeffersonian Democratic-Republicans neatly. Both *André* and its revision experiment with ways of staging the relationship of class to national mythologies. The plays also bracket a transitional period in New York City theatre. The John Street Theatre, home of the city's post-revolutionary theatre and host of Hallam's Old American Company, had become "outdated and old-fashioned" by 1794.[3] The 1798 opening of the Park Theatre, managed by Dunlap, represents a dual attempt to define new theatrical spaces and populate them with emergent publics—both on the boards and in the seats. The Park presented *André* on March 30, 1798, early in its first season with Dunlap at the helm.[4] The tragedy produced controversy, and Dunlap immediately rewrote one scene after Bland apparently disrespects the Continental military insignia, which also resembled the Federalist's black cockade. More damagingly, Thomas Abthorpe Cooper's halfhearted performance as the supporting character Bland exacerbated the play's poor reception.[5] Dunlap's initial adjustments failed, and the play departed the stage after only three performances.

Ironically, that commercial failure redeems the play for theatre historians and dramatic critics, providing implicit proof of its dramatic complexity. Compared to its plebeian rewrite, *André* seems "the more serious, more ambitious, more dangerous play."[6] The tragedy does indeed demonstrate conventional forms of literary complexity; it certainly seems more conflicted about depicting national loyalties in a rebelling nation. As Jeffrey Richards has shown, Dunlap's subtle citations of Otway's *Venice Preserved* allow his play to explore its relationships to English tradition, national identity, and Restoration conventions of sentimental expression. Likewise, Lucy Rinehart argues that the play also presents one of Dunlap's "manly exercises," renegotiating early American patriarchal authority.[7] The upshot of all this literary complexity, however, was theatrical failure; American audiences seemed resistant to certain political ideologies or certain kinds of drama, especially when acting quality fell below expectations.

By 1803, however, Dunlap had charted another course. *The Glory of Columbia* offered audiences a play that not only appealed to their patriotic sentiments, but also incorporated informal paratheatricals from other contexts. Dunlap cut many of *André*'s dramatic subtleties and added new scenes, characters, and subplots wholesale. The revision elides the tragedy's internal conflicts, replacing decorous sentiments with a brawling, noisy patriotic display. Patriotic songs replace thoughtful soliloquies, scenic spectacle pushes aside dialogue, and new characters drawn from the lower sorts replace Revolutionary political and military authorities. Dunlap replaced tragedy's formal unities and social decorum with variegated spectacle. The first productions of *The Glory of Columbia* even featured the acrobatic feats of a "Signor Manfredi" who "ascends, comes from the back of the stage, and walks across the whole house, over the heads of the gallery—a feat never attempted by any but himself before."[8] *The Glory of Columbia*, according to its many detractors presents a mixed bag of motley forms designed simply to titillate the senses. Dunlap himself had disliked precisely this kind of patriotic spectacle, deriding the radical Irish immigrant John Daly Burk's 1797 spectacle *Bunker-Hill, or the Death of General Warren*.[9]

Looking back, Charles Durang later concluded that the revision had "no merit as a drama," deriding it as "a sort of hybrid affair—fulsome in dialogue and pantomime, full of Yankee notions and patriotic clap-trap."[10] It seems a ruthless assessment. However, Durang's description, taken literally and ignoring the disdain, usefully articulates significant formal changes in the play. Durang's choice of descriptors is telling—he senses in the play's "fulsomeness" an offending profusion of popular performance forms. Hybrid, fulsome, "full of Yankee notions and patriotic clap-trap," *The Glory of Columbia* diffuses tragedy's traditional thematic and structural unities into the inchoate experience of comedic spectacle. Simply put, Dunlap's rewrite locates stage performance in the same motley and unruly realm as popular street displays. Over time, Dunlap's spectacle did not just incorporate those acts—it actually became one of them, entering the popular canon of patriotic festivities along with parades, toasts, fireworks, and patriotic concerts. In the end, it proved a successful adaptation to popular theatrical demands, cross-pollinating productively with the prerogatives of popular theatre culture. The very admixture that Durang and Dunlap disdain appealed to audiences to make the play a popular standby for the next quarter century.

The revision of *André* into *The Glory of Columbia* points to the increasing (or at least increasingly visible) multiplicity and heterogeneity of early American theatre after 1800. Decorous genres such as tragedy and elite political fêtes meet and intertwine with the more ephemeral and rowdy forms of popular performance. The idealized, stable script, the unified performance, the fixed meaning—if such things ever exist—conspicuously absent themselves from the stage history of André's death and his yeoman captors. The revision's hybrid variety troubles any assumptions that scripts can define or contain the performance's full meaning. *The Glory of Columbia*'s motley collection of other performance forms requires us to read the text between the lines (sometimes quite literally), extrapolating from a relatively thin body of written evidence to explain the performances that occurred on stage.

The Glory of Columbia also requires us to turn our attention to an expanding sphere of offstage paratheatrical performances familiar to Dunlap's audiences, some that the play reproduces directly and others it invokes in a performative conjuring of absent display. Dunlap's play recruits and restages street theatre for its patriotic performances, showing the complex and ongoing development of the early national public sphere. As a coproduction of elite drama and early American street theatricality, the play helped to theatricalize the political public sphere. It reveals the forms of politics and the playhouse interpenetrating in ways that persist today. Parades, songs, and toasts had entered the theatrical realm, and popular entertainment increasingly became a primary goal of what had once seemed elite occasions for ideological consensus building. Theatrical entertainments stage community and consensus as well, but in the early nineteenth century, such patriotic acts had become "the amusement of holyday fools," as Dunlap put it.[11] Although cultural elites with overtly political agendas might still produce and patronize such occasions, the performances also became occasions for the display of rowdy heterogeneity, both on and off the stage.

A MID-AIR SPECTACLE TO GAPING CLOWNS

André's death, the center of both the tragedy and the comic revision, stages underclass disorder and discipline. In Act 3 of *The Glory of Columbia*, André contemplates "the *manner*" of his death, fretting over the possibility of dying like "the base ruffian, or the midnight thief / Ta'en in the act of stealing from the poor."[12] Dunlap's choice to script André's punishment, although resulting in a relatively simple plot (he dies), also provides more complex moral predicaments than a play based on Arnold's straightforward and villainous betrayal. The tragedy's dilemma centers on the form of André's death, the question of whether the Americans would hang him ignominiously as a spy or shoot him honorably, as befitting a soldier. André's lines imagine his treason sliding from a politically motivated act to mere outlawry:

> To be turn'd off the felon's—the murderer's cart,
> A mid-air spectacle to gaping clowns:
> To run a short and envied course of glory
> And end it on a gibbet.[13]

Those lines, first uttered in *André* but retained in *The Glory of Columbia*, reveal (perhaps obviously) that spectacular punishment performs social status. Publicly punished as a felon, André would become no better than the rabble, joining their ranks even as he became a spectacle before them. Notably, neither play directly shows such a spectacle. The tragedy's final scene, however, does show André in a procession heading to his death, "cheerfully conversing" with his captors and demonstrating a "manly calmness" befitting his heroic status.[14] The "self-regulating gallows performance," as Rinehart argues, frames the play's internal debates about the status of theatre; it also reenacts the rogue convention of "dying game" that *The Beggar's Opera* had popularized on Atlantic stages.[15] Reenacting such conventional performances offers André a "short and envied course of glory," but at the potential cost of lowering himself to the level of the rogue. Ultimately, he fears, his death could become frivolous diversion, a "mid-air spectacle to gaping clowns."

The lines express deep disdain for and even suspicion of the rural low. In *The Glory of Columbia*, they seem deeply ironic, perhaps even insulting, to audiences who had just applauded Signor Manfredi's own mid-air spectacle. The label "clown," as Maya Mathur observes, had associated the comedic, the rustic, and the rebellious since the early modern period.[16] Clowns commonly appeared on both rural and urban early American stages, entertaining in playhouses as well as at the relatively itinerant circuses that competed with the Park Theatre. In the years leading up to Dunlap's revision of André, for example, John Durang (whose son Charles roundly condemned *The Glory of Columbia* as "clap-trap") had played the clown to popular acclaim. Durang, moreover, had literally been a rustic clown—between periods performing on the higher-profile stages of New York and Philadelphia, Durang's career led him into the rural and often German-speaking Pennsylvania hinterlands, where he combined feats of acrobatic skills and physical comedy in mid-air spectacles to spectators some might call "gaping clowns."[17] André's evocative phrase, which Dunlap retains in the comedy, thus codes deep cultural biases against rural culture and provincial theatre in the figure of the hanged and displayed rogue's body. Durang's phrase of "Yankee notions" perhaps rearticulates such semiconsciously held prejudicial associations.

Ironically, *The Glory of Columbia* took the very form that André bemoaned in 1798, rendering his dilemma a spectacle against his own objections. The play suffered the same fate as its erstwhile hero, both, as Dunlap later wrote, "occasionally murdered for the amusement of holyday fools" in the following decades.[18] Interconnected notions of class and style shape early American performances of disorder and order, treason and punishment. André's fears and Dunlap's derision highlight the politics of form, which evidently matters more than content. As André's worried speech suggests, spectacles of discipline and punishment pose significant problems of class and cultural prestige in the play. Only the manner of André's demise remains in doubt, freighted with increasing meaning. Fretting over the manner of his death in a form guaranteed to demean it, the tragedy's protagonist seems ironized and no longer precisely tragic. André's lamentable descent into the realm of the low signals a descent in genre as well, staging the failure of tragedy in early American theatre.

ANDRÉ, ARNOLD, AND PUNITIVE STREET PROCESSIONS

Both versions of Dunlap's play stage treasonous plotting, but they also commemorate public responses to the treasonous plot, recalling popular punitive spectacles that responded to the treasonous conspiracy of André and Arnold. The news of the plot arrived in Philadelphia a week after André's capture in 1780, provoking striking scenes of riotous parading and popular punishment. Philadelphians punished Arnold repeatedly in effigy, adapting the "Pope Night" processions that commemorated the Guy Fawkes gunpowder plot. The parades, like Dunlap's play, were memorial reenactments, although much closer to the original event, and of a different character. They also show the interaction of vernacular ritual and theatrical events shaped by elite patronage.

Two distinct parades occurred on the evenings of September 28 and 30, according to contemporary accounts. Residents first paraded a papier-mâché figure of Arnold to the gallows, ritually exercising symbolic vengeance on the traitor. One observer, Samuel Rowland Fisher, feared the mob might direct its energies against the Quakers, describing the event as informal and boiling over with potential violence. The next morning, however, he learned that it was a politically motivated act;

the "Mob had an Effigy of Arnold hanging on a Gallows," he writes, "the Body of which was made of paper hollow & illuminated & an inscription in large letters thereon, which they conveyed thro' many parts of the City."[19] The procession shows the signs of customary performances, the disorderly assertions of plebeian and pre-industrial traditions that Dale Cockrell describes in *Demons of Disorder*.[20] In 1780, such acts seem alive and well, although, as Peter Benes describes, their material manifestations became nostalgic mementos by the 1820s, and sanitized relics by the 1860s.[21] In 1780, however, they remained vital and expressive collective acts.

A day later, a second, more elaborate event was authorized and orchestrated by Philadelphia elites; this procession "appear'd not as a frolick of the lowest sort of people," Fisher wrote, but instead as an act authorized by "some of the present Rulers here," including artist and politician Charles Willson Peale.[22] Peale's patronage (and perhaps artistic handiwork) reconstructed the first performance, retroactively transforming the first into a demotic rehearsal for a more elaborate ritual display. The second parade thus seems a deliberately amnesiac reenactment predicated on revising and erasing the first display of unruly low energies. The ritual's restaging produces a backward construction of folk or vernacular culture as unsophisticated, ephemeral, raw, and manipulable. No images depict the first event—perhaps predictably, given its origins away from the formal mediations of print culture. The second spectacular display, however (with a day's notice and the assistance of Peale's cultural clout), produced at least two distinct graphic renderings. Indeed, Fisher recounts the second procession in his diary expressly because of its elite patronage, while the first one exists only in rumor and distant clamor. Two different Philadelphia almanacs, one in English, the other in German, featured folding prints that could be removed and perhaps displayed. The English version appeared with an elaborate caption and in broadside form as well. The scene's circulation in different forms reveals the wider audiences and traveling forms of localized performance. The restaging also shows social authority's ability to reshape and reproduce collective performance. At the same time, the transmission also suggests that informal forms of collective expression had already infiltrated other social spaces and practices. The original procession, although perhaps chaotic and less organized, provides serviceable and even compelling resources for Peale's authorized public demonstration.

The second parade presented onlookers with a much more elaborate scene. It featured Arnold with his wounded leg on a cushion and "two faces & his head continually moving," as Fisher recounts.[23] Arnold rides before the devil, his associate in evil. Various images and descriptions printed in Philadelphia, including a beautifully drawn depiction published in a German-language almanac, reveal the cart's function as a moving stage (figure 4.1).[24] The effigy substitutes for Arnold's embodied stage presence; Arnold and the devil, riding in the cart, hold or wear black masks, recalling the customary blackface masking of vernacular ritual. Featuring the labels "Spy" and "Traitor" above two hanged figures, the transparency accuses and displays the punishment of André and "I. Smith" (perhaps the Joshua Smith who assisted André before his capture).[25]

Newspapers also circulated other explicitly theatrical descriptions of the event. The cart, according to newspaper descriptions, supports a literal stage.[26] Arnold, costumed in his regimentals, bears the props needed to stage his treason—the gold

Figure 4.1 Philadelphia Effigy Parade of Benedict Arnold, September 30, 1780; *Americanischer Haus- und Wirthschafts-Calender auf das 1781ste Jahr Christi* (Philadelphia, 1780); Historical Society of Pennsylvania.

he supposedly desired, the traitor's noose he ought to receive, and a letter from "Belzebub" (a stage manager of sorts, one supposes) explaining the consequences of his actions. The description imagines the traitor's two-faced mask as "emblematical of his treasonous conduct"; theatre's masking provides a ready metaphor for treason's duplicity. A large "lanthorn of transparent paper" glosses Arnold's crimes, using the same sorts of techniques that late-eighteenth-century illegitimate theatre used to elucidate pantomime plots. The description also tells of the "Rogue's March" (apparently a common part of such processions) musically measuring the procession's progress.[27] The performance converts traitors into rogues, policing community norms. Ironically, the scene uses stage techniques to demonstrate displeasure with Arnold's theatrical duplicity.

As Peale and his cast of street actors transform the procession from impromptu demonstration to scripted official ritual, regimented mass participation overshadows the effigy's embodiment of collective outrage. The participation of local elites makes the procession above all a demonstration of corporate solidarity. Fisher's diary describes an impressively coordinated pyrotechnic display on the second night; about "20 of those called Militia & three of those call'd City Light Horse" escorted the procession, and "the Militia had each a Candle in the end of his Musket & perhaps about 100 Lads each having a Candle in his hand."[28] Elaborate published descriptions of the procession reiterate the procession's form:

> Several Gentlemen mounted on horse-back,
> A line of Continental officers.
> Sundry Gentlemen in a line.
> A guard of the City Infantry.[29]

The verbal order of the description scripts and regulates the ritual display. More importantly, it suggests the stakes of such public acts as they generate excessive description in order to define and stabilize the street performance.

Violence and disorder shadow the discipline secured by guards and reiterated in the verbal description. Behind the parade followed "a numerous concourse of people"; after "expressing their abhorrence of the Treason and the Traitor," they "committed him to the flames, and left both the effigy and the original to sink into ashes and oblivion."[30] The description simultaneously reveals the regimented theatricality of power and barely controlled popular performance. The coordinated groups of people, as observers feared, could easily turn into an unruly mob; as Fisher wrote in his diary, "the Boys were seen to be prepared with Brick Bats," menacingly advancing along Pine Street and readying for mischief.[31] Fisher, alert to anti-Quaker violence, may have been a bit sensitive, but his account expresses real links between theatrical ritual and chaotic violence. The image in the *Americanischer Haus- Und Wirthschafts-Calender* also suggests the disorderly potential of public procession, depicting the parade as an unruly and kinetic event. Although visually clean, the scene's composition disrupts the evenly regimented lines of figures appearing in other images; it seems entirely unlike another depiction of the scene, in which orderly lines of figures stand in disciplined visual stasis. Armed militiamen pace the procession, but the mob of onlookers dominate the scene. The few spectators facing the viewer gesture actively, disrupting the clean lines of the parade. Most intriguingly, lively figures transform the ritual of punishment into a scene of play. In scenes that resemble the broadside depiction of Levi Ames's hanging, children play and spectators interact amidst the processional discipline.

Although escorted by the regulating militia, the mob seizes its chance to act its own part in the theatricals. To some extent, the entire procession represents a popular claiming of authority—the description imagines the people punishing the effigy "for want of a body," taking disciplinary and punitive authority into their own hands after Arnold's escape.[32] Early Americans repeatedly took street theatre into their own hands in such ways. At the same time, the description implies that popular ritual, an empty form without the body, only offers a palliative for real authority, real action. It is a mistake, however, to assume that vernacular performances simply substitute for action. Rather than performing punitive displays of state authority, they enact the collective solidarity of those outside the centers of power. Such acts inevitably blend unruly riot with disciplined official performance, and various parties struggle to produce and manage those displays.

Such popular rituals act across class lines. Alexander Graydon, for example, recounts a ritual punishment that resonates with the Arnold's 1780 effigy procession. He tells of Kearsley, a loyalist Philadelphia doctor who

> was deemed a proper subject for the fashionable punishment of tarring, feathering, and carting. He was seized at his own door by a party of the militia, and in the attempt to resist them, received a wound in his hand from a bayonet. Being overpowered, he was placed in a cart provided for the purpose, and amidst a multitude of boys and idlers, paraded through the streets to the tune of the rogues march.[33]

Kearsley stars in a contentious ritual of performance and counter-performance, foregrounding rebellious energies that Graydon finds shocking.[34] The mob enforces its will through informally scripted performance. Carted to his punishment, the

"Rogue's March" casts the scheming loyalist as a sort of Macheath. At the same time, the ritual applies Macheath's critique of the powerful—making the high low, it reverses Gay's ironic observation that "lower sort of People have their Vices in a degree as well as the Rich."[35] The early American rich, it seems, can have their vices as well as the low—but they had better play their part properly.

Ritual can become real conflict. A "man of high spirit," Kearsley refuses to submit meekly or acknowledge the crowd for its "forbearance and civility," as Graydon sees a more sensible victim later do.[36] Instead, enraged, the loyalist calls for a bowl of punch and downs it in one gulp. Expressing his noncompliance with community norms, Kearsley reenacts gallows bravura, reclaiming Macheath's performance as his own. That seems to have taken the performance too far—observers found Kearsley's conduct so "extremely outrageous" that it merited exile from the community.[37] Graydon's description, although somewhat disdainful of the "boys and idlers," shows that such rituals do not simply produce chaotic violence. Although it incidentally wounds Kearsley, the mob neither tars nor feathers the doctor. After all, he embodies such a mild version of Macheath's roguery that his outrageous reenactment verges on self-inflicted satire. Instead, the structured, self-conscious, even subversive play produces and enforces new community norms. The mob seems intent on rehearsing roles and averting violence: you could act the rogue, they seem to say, but in turn, we could play the part of mob in earnest. Kearsley presumes to play the rogue, ending up in exile for his refusal to submit.

STAGE YEOMEN: LOW CHARACTERS AND LOW TREASON

Such scenes reveal the competing impulses of order and disorder that Dunlap uses in transforming his problem tragedy *André* into a comic spectacle. As heir to such street theatre, *The Glory of Columbia* twice displaces Philadelphia's 1780 street performance. First trading in its villainous Arnold for the tragic hero André, it then produces the procession's yeoman audience as its titular protagonists. *The Glory of Columbia* stages two full acts of newly invented low-comic stage business before any vestiges of *André*'s tragic plot appear. Comic variety upstages tragic drama. Those two acts introduce plebeian characters into the tragedy, inserting the mob into what had been an elite drama. As S. E. Wilmer asserts, the play advertises the yeoman as the "laudable" and admirable "individual hero," but the script reveals instead a rather troubled and troubling cohort.[38] Yeomen do not become heroic by inclusion in the play; instead, drama becomes roguishly low. The performance of Arnold and André's treason, with its mob in the background, contained the potential for such displacements all along. The cultural memory of the theatrical 1780 parade permeates *The Glory of Columbia*, actor replacing effigy and glorious yeomen substituting for the mob. Dunlap's play remakes ritual indignation as playhouse patriotism. Like the earlier acts centered on Arnold's treason, it too parades a tense standoff between discipline and play.

Glorious but also potentially mobbing yeomen embody the play's ambivalence about the low and their popular forms of public expression. The decision to foreground the play's yeomen resurrects the supposed sturdy rural character of André's captors. A pamphlet of musical excerpts from Dunlap's rewrite, for example, trumpets the "three glorious Columbian yeomen, whose incorruptible honesty preserved Westpoint and the American army."[39] Such claims capitalize on the political ascent of the Democratic-Republicanism, with their Jeffersonian pastoral tendencies to celebrate the humble farmer. At the same time, the play appeared in New York City economic contexts of artisanal craft labor. Its two decades of popularity in urban venues suggests that *The Glory of Columbia* remained popular even as transforming labor practices made offstage yeoman types increasingly distant memories.[40] By the 1820s, the play's low characters embodied residual and rural ways of life for audience members negotiating new forms of community in New York City's emergent industrial capitalist order. The yeoman character, while not a complete outsider, seems an oddly dissonant embodiment of popular patriotism, a wishful or nostalgic figure recalling residual forms of working-class life. Such figures had appeared before on New York's stages—Royall Tyler's Jonathan, for example, had emerged in 1787 from the rural hinterlands to offer his shrewdly foolish critique of urbanized theatre culture. *The Contrast*, however, also failed to drum up spectacular popular appeal.

Although yeomen served popular audiences more satisfactorily than André's tragic elitism, they also represent troubling and problematic characters. Perhaps because of the ambivalent relationship such characters would have had to urban New York City audience members, the text carefully avoids directly attributing any dishonesty or roguishness to its low characters. It protests their honesty perhaps a little too vigorously, revealing the play's own doubts about common types. Such misgivings played out in public repeatedly in the years after the Revolution. Controversy over the yeoman captors had persisted since the 1780 capture of André in the New York neutral ground, an area in which, as Robert Cray puts it, "simple questions about allegiances dissolved into a fog of ambiguity."[41] Audiences would have been vaguely familiar with the historical yeomen David Williams, John Paulding, and Isaac Van Wart, who themselves had enacted theatrically ambivalent scenes of disguise and misreading in 1780. André, for example, reportedly mistook Paulding's coat for the uniform of a Loyalist Cowboy. The original scene must have appeared more an act of highway robbery than of patriotism or even of combat—an armed party in irregular uniform stops André (on horseback and disguised in genteel civilian clothes) to strip and plunder the genteel traveler.

Controversy dogged the yeomen; when John Paulding, one of the captors, petitioned Congress in 1817 for a more generous pension, the scene became public again. Connecticut politician and Revolutionary officer Benjamin Tallmadge, opposed to the pension, brought up the old rumor that the three captors were not patriotic heroes, but were instead mercenary rogues, or even loyalist "Cowboys," who roved New York's neutral ground victimizing locals indiscriminately. Even Egbert Benson's 1817 *Vindication of the Captors of Major André*, too, although defending the yeomen, further publicized their problematic public personae.[42] The reality, as Cray writes, was probably less clear than either the captors' defenders or detractors would have it—"vengeance, patriotism, and profit went hand in hand" in the scene of revolutionary roguery.[43] André's captors were certainly not the only rogues to emerge from the ranks of the rebelling military—just some of the more prominent. Thomas Mount confesses in his 1791 *Confessions* to a life of petty crime that had begun in the ranks of the Revolutionary service.[44] As Louis Masur notes, ex-military personnel constituted a steady stream of criminal offenders who ended up on the gallows.[45] This occurred for various reasons; the sheer numbers of former soldiers after the Revolution could have increased their numbers on the gallows. Trained in and habituated to violence, soldiers might have also acted criminally more frequently. More complexly, former military personnel may have experienced trouble reintegrating into civilian life, reclaiming jobs and readjusting to the rituals of peacetime life. Finally, as outsiders to the community, former soldiers (along with foreigners, racial and ethnic outsiders, and transient people) were perhaps at some increased risk for capital punishment. With many of these cases appearing in broadsides and confessional accounts, Americans would have known spectacular examples of soldiers convicted and executed as criminals during the Revolution and the early national period.

Dunlap's play, staging the ambivalent role of the common soldier, attempts to rehabilitate such rogues into glorious yeomen. In the opening scenes, plebeian characters appear admirable but dismissible. The patriotic commoner Williams would prefer to serve his country in combat, rather than as Arnold's servant.[46] The dilemma contrasts with Arnold's "internal war" between the urges of loyalty and treason, and the play easily solves his problem, parsing national loyalties along class lines. A truly patriotic American, as Williams shows, may be lower class, common, or simple, but he can yet exercise agency, becoming martial and manly. It works out nicely, too, that the master whom Williams repudiates happens to be Benedict Arnold, the play's center of treasonous gravity. In the end, *The Glory of Columbia* attempts to render the underclass as neither heroic nor dangerous, but as simply comic. Arnold tellingly ventriloquizes Dunlap's approach when he calls the yeoman Williams an "honest clown."[47]

Nevertheless, the halfhearted attempt at historical rehabilitation flirts with contrary prospects, and the play's staging of the lower sorts subtly echoes Tallmadge's suspicions. They seem ever liable to turn into a mob, or worse, a band of outlaws. Alongside Williams, the play incorporates two other characters of significance to stage underclasses: O'Bogg, a stage Irishman, and Sally Williams, the yeoman's sister, who becomes a cross-dressed female adventurer. Neither character is properly a "glorious yeoman," and both offer troubling categorical transgressions. They enact a series of "low treasons," transgressive but comic counterpoints to André's tragic high treason. *The Glory of Columbia* reinstates the integrity of gender and class in its affirmative patriotic ending, but between the first act and the last, Sally and O'Bogg hint at the disconcerting underpinnings of the play's humor.

Played at the Park Theatre by Miss Hogg, Sally recalls André's theatricality treason in a scene of cross-dressed mock espionage. Comically rehearsing the yeomen's capture of André, the scene jostles and reinstates the boundaries of closed masculine military culture. It begins innocently enough. In her first appearance onstage, she brings a basket of fruit to her brother, who has neglected to write letters home; as he explains, "I'm no dab at a pen: my fist be made for a gun or a pitch fork."[48] Sally's hand seems made to deliver baskets of fruit. At the same time, however, she rebels against her domestic, feminized role, wistfully exclaiming, "I have a dreadful mind to see the officers, and soldiers, and guns, and fortifications; and yet what has a woman to do with them?"⁴⁹ Her brother protectively insists that she stay far from the dangers of a camp full of soldiers, making it clear that the only role a woman has in camp is the domestic business of supporting the troops.

Sally's cross-dress adventure begins after that scene, spurred by her desire to watch and even to participate in the masculine militia experience. Admitting that women "have a natural propensity to be meddling with what don't belong to us," Sally nevertheless resolves to carry out her desires.⁵⁰ She saucily foreshadows André's treason as she infiltrates the American encampment, and her gender transgression becomes a literal travesty of espionage. Entering camp in cross-dress in the second act, she tries the efficacy of her disguise. Her laughter gives her away, and the soldiers, including her own brother, detect her act of mock-espionage. They determine to teach her a lesson by pretending not to recognize her and threatening punishment. Williams (seeing through her disguise) playfully identifies her as "as great a rogue who ever lived."⁵¹ He calls her a spy and threatens summary execution, drawing his sword; "shall we take him prisoner and hang him, or cut him to pieces on the spot?" he asks, comically rehearsing and condensing the play's later capture and execution of André.⁵² On the surface, it seems a simple bit of comedic stage business, but the scene also anxiously evokes the more serious possibilities of treason and punishment. Sally's attempt to infiltrate the closed masculine culture of the military camp also places her in the equally masculine (and far more dangerous for her health) cohort of spies, rogues, and outlaws. As cross-dress, Sally's mock-espionage performs low comedy over and against the lavishly dressed gender performance of the rogue. Her act is a multiple travesty, a mockery of not only the honest yeomanry, but also of the lavishly clothed, disdained and admired—and ultimately hanged— Atlantic outlaw.

Sally, like Rowson's Fetnah, technically plays a disguise role, rather than a true breeches part, like the comic disguise of Rosalind, for example, in Shakespeare's *As You Like It*. In whatever form, as Marjorie Garber has argued, the appearance of cross-dress on stage "signals a category crisis."⁵³ In Dunlap's play, the category crisis begins as a challenge to gender roles and ends as low treason that it can only solve through the speedy reinstatement of accepted gender roles. Of course, the scene rehearses the comic *failure* of theatrical cross-dress. As Elizabeth Reitz Mullenix has observed, such acts ultimately resist the theatrical subversion of gender difference in transvestism.⁵⁴ The fact that the audience and her brother quickly see through her act shows the detection and category-enforcement that counter the conceptual crisis. If the revolution sets in motion women trying to play men—and explicitly trying to infiltrate closed masculine associations like the military—then it becomes the duty of patriotic yeoman Americans to restore order. Offstage, Deborah Sampson had made cross-dressing in the service of patriotic performance famous in the years immediately preceding *The Glory of Columbia*'s premiere.⁵⁵ Her story provocatively suggests the connections between class location and gender crossing during and after the Revolutionary period. Born in poverty, Sampson had disguised herself as a man and enlisted in the Continental Army in 1782, presumably in part for the enlistment bonus. Sampson participated in a well-established tradition of militant gender crossing in the Atlantic world that Dianne Dugaw has described, taking advantage of distinctly early American opportunities to first refashion her identity and then publicize it in a sequence of interleaved disguise and display.[56] She served ably, and despite the fact that passing as a man was illegal, her officers protected her identity until her discharge. More intriguingly, her cross-dressing became publicly theatrical at the end of the century, when she petitioned for a pension and back pay in 1797. The Republican editor Herman Mann published her account as *The Female Review* and Sampson embarked on a speaking tour, delivering orations at theatres in 1802 and 1803.[57]

Sampson's act would become less publicly accepted as the nineteenth century witnessed increasingly limited opportunities for the performance of gender, but the theatre remained a space of gender play well after her public performances. Audiences could see numerous plays, some thematically centered on cross-dressing (like Mordecai M. Noah's 1819 *She Would Be A Soldier*) and others straight plays in travesty. The male rogue had cross-dressed in popular travesty versions of *The Beggar's Opera* at the end of the eighteenth century, suggesting that audiences perceived plebeian characters as particularly well suited (no pun intended) for such acts. A broad culture of low comedy infused travesty and cross-dress both on- and offstage. The differences between Deborah Sampson and Sally Williams—between comic travesty and military masquerade—are instructive. Where Sampson served honorably, petitioned for a pension, and told her story initially to applause, Sally has no realistic hope of participating in masculine military culture. At best, she can try to become a spectator to the male camaraderie, and at worst, she becomes a punch line, the butt of a cruel joke of threatened execution. Whereas Sampson, at least for a time, could parlay her cross-dressing into patriotic service and gainful employment, Sally provides only comic relief and a negative example.

Sally's mock treason occurs alongside another comedic treason enacted by O'Bogg, a stock Irish character played by the Park Theatre's comic actor Joseph Tyler. O'Bogg's comic singing and ne'er-do-well temperament mask a more serious and suspicious performance of underclass identity. An Irishman who deserts the British army, he appears, according to the stage directions, "with arms and accoutrements as a British soldier"—he wears the costume and holds the props, but that is all.[58] As the play shows, he plays a role that he can easily shed, joining the Americans with almost literally no questions asked. O'Bogg thus displays the shifty and suspect, even theatrical, allegiances of the Atlantic low. He also enacts sexual transgression and undisciplined rowdiness. The yeomen find him away from his unit, eating and drinking while on patrol. He is on the run from his multiple wives, and sings of infiltrating a convent and debauching its nuns. His defection, too, reveals his own presumed allegiances as contingent and shifty, even theatrical. His singing, his brogue, and his comic lowness mark O'Bogg as an outsider and potentially transgressive, even as they exempt him from serious punishment.

American theatre inherits its stage Irish from English theatre, where, as Jeffrey Richards observes, the Irishman had been "the comic foreigner of choice for a century."[59] Influential actors and playwrights (some of them actually of Irish origins) such as Charles Macklin, John O'Keefe, and William Macready had made Irish types increasingly popular in plays such as *Love a la Mode* (1759), *The True Born Irishman* (1762), *The Poor Soldier* (1786), and *The Irishman in London* (1793). In 1803, the later shifts in American theatre forms and attitudes toward Irish immigrants had yet to reshape American stage-Irish conventions. Nevertheless, O'Bogg shows the stage redeployment of the type to help define American, rather than English, identity.[60] Irish types appear early in American plays, including the American-authored but unstaged 1767 *The Disappointment*.[61] The stage Irishman, while transferring old-world conflicts and stereotypes onto American stages, also participated in the work of performing American ethnic and national identity. In literary fiction, for example, Hugh Henry Brackenridge's 1798 *Modern Chivalry* most famously deploys the stock Irishman Teague O'Regan as a low buffoon reenacting the absurd excesses of democracy. Dunlap's O'Bogg performs a slightly different kind of labor, relocating to America his performance of resistance to English cultural colonialism. He distills stage-Irish moves into concentrated Anglo-Atlantic lowness, wheeling about to add his anti-English energies to the American yeomen's rebellion.[62]

Dunlap had adapted Irish characters before for American nationalistic ends. His 1789 *Darby's Return*, for example, popularly reinvents O'Keefe's Irish-born comic lead from *The Poor Soldier*. The play presented the stage-Irish in celebratory performances of American nationalism, most famously before George Washington himself.[63] Intriguingly, Dunlap's most interesting move in *The Glory of Columbia* is not reinvention, but recycling—he imports O'Bogg's songs wholesale from a 1790 English pantomime, *The Picture of Paris*, which pokes conservative fun at the French Revolution. Showing traces of its genealogy, O'Bogg's military masculinity emerges in tones distinctly different from the American yeomen. His songs recount misadventures wooing French nuns and celebrate his ability to best any number of other national types in amorous adventures. In *The Picture of Paris*, Irish braggadocio and lower-class unruliness articulate the French Revolution's radical violence. O'Bogg carries those revolutionary overtones forward in his songs and stories, retroactively infusing the American Revolution with radical French republicanism. Associating Irish types with the French Revolution (via their Catholicism, perhaps) was not unique. As Rinehart notes, Dunlap had earlier cut lines from *Darby's Return* in which Darby observes and reports on the violence of the French Revolution.[64] O'Bogg also embodies one of the rhetorical strategies of politically exclusive Federalists, who frequently accused their Democratic-Republican opponents of the triple crime of lowness, immigrant status, and French radicalism in the early 1800s.

As a member of the Atlantic criminal underclasses, O'Bogg seems a threat on stage, but he had perhaps become less dangerous in 1803. The rise of Jeffersonian republicanism and the repeal of the Alien and Sedition Acts had failed to produce the radical effects some feared, and O'Bogg can assume merely comic tones. His rogue acts appear the stuff of humor rather than of treason. That humor can nevertheless radically invoke low roguery, as Dunlap's fourth act shows with a morbid and stereotyping joke:

> *O'Bogg.* Arrah then put on your spectacles. Talking of spectacles puts me in mind of my mammy.
> *Will.* She wore spectacles, I guess.
> *O'Bogg.* No, she couldn't; she had no nose.
> *Van V.* How did spectacles remind you of her then?

O'Bogg. Be azy and I'll tell you. When my brother Teddy was hanging up in a hempen necklace, "oh, what a spectacle!" says my mammy. "Don't blubber and howl so, mammy," says I, "see they're just stringing brother Phelim, and then you'll have a pair of spectacles, and all of your own making."[65]

O'Bogg aggravates the gleefully heartless pun when he reveals that he narrowly avoids the same fate by being a mere stereotype. "When my two twin brothers broke into the church," he relates, "I was so drunk I couldn't go with them."[66]

The hangings become twice-performed spectacles of punishment as the stage Irishman transforms the scaffold's theatre of state-authorized terror into a new, but equally political, theatre of Irish buffoonery. The morbid and telling bit of gallows humor shows the extent to which stereotypical Irish clowning also marks disorder and resistance. First conscripted by British imperialism and then discarding his forced allegiance, O'Bogg moves blithely through the criminal and transnational underbelly of the Anglo-Atlantic world. O'Bogg's "spectacles" of hanging also recall the Philadelphia parade of Benedict Arnold's effigy, a mock performance of punishment that continues to shadow performances of the treason a quarter of a century later. The gallows persisted as a transatlantic site of popular memory, the stage on which authority performs its power over the transgressive lower classes.[67] *The Glory of Columbia* imagines the scaffold as an absent but contested site of performance, which O'Bogg casually reclaims for low humor. For all its calloused stereotyping, Dunlap's play captures O'Bogg cagily acting multiple roles in the shadow of the gallows pole. As (nearly) one of the Dublin hanged, he avoids the scaffold, and after conscription remains ready to desert at a moment's notice. Joining the revolutionary colonists, he clearly picks his sides cannily. As a comic relief character, he avoids execution for treason (the punishment for desertion that he logically deserves) through appealing performance—much the way Macheath did. As one of the play's newly defined Americans, he figuratively marches with the Philadelphia crowd alongside Arnold's effigy in 1780—both in the line of regulatory troops and among the wielders of brickbats. Ultimately, he represents the audience as well, one of the gaping clowns that André despised and feared.

Together with Sally, O'Bogg becomes one of the most transgressively stagey comic characters in *The Glory of Columbia*. Those two outsiders, attempting to break into America's patriotic yeomanry, reveal a scripted unease about what it means to be and act American. American yeomen capture traitors, but they also play at such roguishness to the point that they might become such characters. Such ambivalence about the vulgar sorts was not entirely a product of the theatre. Offstage, the war made rogues as well as heroes out of its soldiers.[68] Dunlap's rewrite remains vividly concerned with such problems as André ponders his fate as a mid-air spectacle to gaping clowns. Hanging is for the low, and André's execution as a "base ruffian, or the midnight thief," remains a poetic injustice, even in the comic revision. *The Glory of Columbia* wards off such spectacles by performing comic versions of the punishments that the 1780 effigy procession and Dunlap's 1798 *André* doled out to treasonous rogues. Sally and O'Bogg's twinned acts of mock treason displace the historical betrayals of Arnold and André. Ultimately, those are playful and inconsequential scenes, revealing the broader cultural work of Dunlap's revision as it swaps high tragedy for low comedy and turns decorous commemoration to motley spectacle.

MUSTERS, MUSIC, AND MOBS

The Glory of Columbia, however, is not all morbid joking and narrowly averted treasons. The play's patriotic spectacle draws on the broadening of popular political culture that scholars such as Simon P. Newman have detailed.[69] As David Waldstreicher, Susan Davis, and others have also shown, performance permeated the political public sphere—patriotic and political expression relied on displays, parades, spectacles, exhibitions, and the stage itself.[70] The common or low cohorts, acting within and against the patronage and authority of cultural elites, claimed their places on the early American stage as they had begun to do in political public arenas. *The Glory of Columbia* helps stage those new modes of public theatricality, paving the way for the elaborate public theatricals that Rosemarie Bank describes in the 1820s and beyond.[71] Dunlap's revision of *André* thus participates in early American theatre's gradual renegotiations of elite and popular patronage, staging membership in the public political sphere alongside outlaw acts. *The Glory of Columbia* ultimately works through the underclass presence in the political sphere.

The lower-class types radically alter the form of *The Glory of Columbia*, introducing not only mock treason, but also the carnivalesque hybridity that commentators such as Durang noticed. While for the most part the play's yeomen seem low-income versions of the elites, they also act out a comically transgressive, perhaps carnivalesque theatricality. They disguise, play-act, cross-dress, sing, and dance as the play's elites never do. Most importantly, the play transforms acts of street patriotism and political participation into stagings of order and disorder. Although the play avoids enacting real unruliness, its ordering of popular forms introduces the shadow of disorderly performance and mob action. It seems quite commonsense to note that processions of disorder (charivari, carnival, burlesques, and riots) satirically invoke their orderly opposite numbers as disruptions of, for example, official justice, processions of power, decorous drama, and peaceful gatherings. *The Glory of Columbia*, however, suggests the converse; performances of order and hierarchy, especially amidst the mobbing and rioting of early American culture, evoke their own disorderly shadows. The act of ordering follows and responds to disorder—perhaps even requires the threat of unruliness simply to make sense. Dunlap's revised focus on the lower echelons of American society seems to conjure ordering structures, warding off the ever-present disorderly acts of mobs and rogues.

The parade functions in *The Glory of Columbia* as a key tool for controlling low disorder. Dunlap's yeoman play restages the patriotic parades and musters occurring outside the theatre, turning them into spectacles of theatrical variety as well as ordering structures. Predictably, its martial plot requires the frequent on- and off-stage movement of troops and dignitaries, appropriating the popular forms of early American parades and militia musters. In the second act, for example, characters enter, relieving the sentinels "in due form" and participate in an official review of the troops.[72] In effect, the performance features a series of onstage parades, orderly processions enacting forms expressive of the hierarchies and the structures of military culture. The scenes enact collective action, transforming the individual tragedy of *André* into ritualized (and apparently disciplined) group movement. Enacting the poetics of patriotic display, Dunlap's parades order (in the dual sense of commanding and structuring) yeoman actions. They muster the troops, constituting performers and even its audience for patriotic acts. A chorus in the fourth scene, with calls of "to your arms, boys!" and "to your ranks, boys!" literally bodies forth the troops in a stage version of the militia training days familiar to most Americans in the late eighteenth century.[73]

Performance produces orders, but also the idea of order itself. Music produces the play's processions (as the "Rogue's March" had done for 1780 Philadelphia effigy-burning procession), moving characters on and off stage. The last scene of the second act, for example, features no dialogue, only a chorus accompanying soldiers entering in formation; in lieu of words, music dictates the order of the military parade. Announcing the approach of its "heroes," the song's verses gloss the sequence of the procession. First comes the "hollow sounding drum," then the "stately horse," and finally the musketeers and cannoneers along with the standards and trumpets.[74] The succession of the verses dictates the order of the procession, its effect structured by the multiple formal demands of military reenactment, ritual parade sequence, music, and rhyme. Such scenes seem military reenactments turned parades, military violence twice removed from war by means of theatrical ritual. The ordering of that scene extends in a figurative sense from the stage yeoman outward, into the audience. Published texts of the play's songs allow (even encourage) the play's audience to sing along with the glorious yeomanry, playing the parts in a bit of vicarious theatricality. The songs, published in a standalone pamphlet "intended for the celebration of the Fourth of July," could also provide a template for offstage patriotic festivities.

Dunlap knew well the ways in which musical and processional performances could create order and disorder in public performance. On July 27, 1798, for example, partisan groups of New York youths paraded in public, singing patriotic songs in a public ritual of group affiliation. Dunlap himself noticed and remarked upon the event, which occurred not long before he would restructure *André* using the forms of low street theatre.[75] After parading down Broadway, the patriotically inflamed youths returned to the Battery, where, as Paul Gilje puts it, a "singing contest began."[76] It was a highly public and publicized ritual; a report in the *New York Spectator* reveals the stakes of such acts in terms of class identity and mob tendencies. Initially, the songs simply signaled group membership, "Hail Columbia" identifying the Federalists, and the French revolutionary import "La Carmagnole" expressed Republican affiliation. Although the songs overtly expressed party loyalties, in the elitist and apparently Federalist-sympathetic account, a class conflict develops between "a company of five young gentlemen" and a cohort of "the lowest class of mechanics."[77] Predictably, perhaps, such performances inspired accusations of foreign attachment—fighting words, in view of each side's patriotic performance. Amidst exclamations of "Damn all Jacobins" and "Damn the British faction," ludic scenes became agonistic as a "cowardly villain" from among the mechanics attacked one of the Federalists. Performance continued to articulate the violence, however;

according to the account, the "democrats sang the carmagnole" during the mêlée.[78] Dale Cockrell comments that to "act before common audiences was generally to promise violence, but not necessarily to deliver."[79] Dunlap, however, witnessed a more aggressive underclass alternative, in which public performers seem entirely willing to deliver on their promises of violence.

New Yorkers witnessed numerous such acts firsthand. As David Waldstreicher has shown, ritual performances such as songs and toasts shaped plebeian political participation in the early American public sphere, providing fodder for Walt Whitman's later celebration of "chants democratic." The public amplitude and frequency of such performances (and the reports of their observers) increased in the early 1800s. The partisan singers reveal changing modes of public political expression, but importantly, these modes of political expression were not limited to the emergent Jeffersonians most often identified with the unruly low at the time and afterward. The Anglophile (and apparently Federalist) participants respond in kind, both musically and violently. They enact rowdy, public politics alongside the Francophile low, staging acts virtually identical in form. Each individual performer has his own relationship to the collective act; if the performers differed in class makeup in the ways that the account claims, then the two groups participated on very different terms. Nevertheless, competitive singing offers a mediated space in which ritual order shapes expressions of identities and loyalties. Even if the account simply reflects the paranoid fears and biased perceptions of those less rowdy, the scene suggests the powerful imagined relationship between performance and violence. The watch breaks up the proceedings, but not before the combatants transmute the ordering, regulatory power of musical procession into mob disorder. Order and disorder, words and deeds, expression and violence can emerge simultaneously from the same theatrical forms.

PLEBEIAN POLITICAL PARTICIPATION AND MOCK ENFRANCHISEMENT

Dunlap's play only hints at such unruly actions, and staged mob disorder remains an absent shadow in *The Glory of Columbia*. At the same time, such actions constitute a richly textured contextual field whose equation of performance and disorder inevitably influenced the play. In its parading order, *The Glory of Columbia* restages the militia musters, parades, and triumphal processions of the early nineteenth century, many of which featured a complex mix of discipline and disarray. The performances also extended into print culture; newspapers, for example, reproduced parade and drill sequences, rehearsing the authority of local militia officers to command order. At the same time, such events hosted low-intensity tumult and insubordination, especially among the lower sorts. Dunlap's disciplined processions share their form with a wide variety of street processionals, many of which could become much more disorderly in the crowded, kinetic, and chaotic early American streets.

Well before Dunlap's revision of *André*, the underclasses publicly enacted patriotism in their own diverse ways, especially in musters and militia parades. In his 1811 memoirs, Alexander Graydon reveals his perception of yeoman military service during the revolutionary period. Encountering "very few gentlemen and men of the world," in arms, he describes the plebeian military presence in disdainful tones. Anything, he writes,

> above the condition of a clown, in the regiments we came in contact with, was truly a rarity. Was it, that the cause was only popular among the yeomanry? Was it, that men of fortune and condition there, as in other parts of the continent, though evidently most interested in a contest, whose object was to rescue American property from the grasp of British avidity, were willing to devolve the fighting business on the poorer and humbler classes?[80]

Graydon's lament identifies the rustic and comedic (and ultimately demeaned) low as patriotic performers. He also seems to presume that their presence indicates voluntary and coequal participation in the patriotic opportunities of military service.

Among such lower sorts, order, uniformity, and good behavior seemed frequently in short supply, especially when local laws forced the hardship of musters upon working-class residents. The lack of uniforms and weaponry (yeoman soldiers sometimes carried agricultural tools or broomsticks rather than regulation firearms) created burlesques of orderly militias. When local resentment against exploitative elite practices existed, musters and election days could become deliberate satires of military processions. The "clowns" who had staked their claim to enfranchisement with military service also frequently contested the imposition of order on the body politic. Such plebeian spectacles stage competing urges. They enacted inclusion; participation in the ritual bestowed membership, although not only on the terms set by authorities—the roles ranged from extravagantly costumed officers to ragged, disgruntled conscripts toting garden tools. Participants could and often did perform individual variations on the script. With such a diverse array of participants on the street, even disciplined performances invoked their burlesque and riotous doubles.

As Shane White pointed out, the related rituals of General Training Day, Negro Election Day, and Pinkster enacted burlesque and complexly satirical displays of mock underclass enfranchisement.[81] While these events seem predominantly acts of racial inversion, they code oppositions of class, authority, and power conveniently in black-white oppositions.[82] The socially and politically disenfranchised—the northern slaves, servants, and underclasses of the late eighteenth century—staged mocking exercises in power, elections, coronations, and military training. Each with their own distinct styles and performative provenance, these festivals satirized authorized forms of political participation as they rehearsed the low's own disenfranchisement. Importantly, such acts provided spaces for underclass play, undisciplined socialization, and even resistant self-expression. In 1803, for example, reports of Albany's Pinkster festival, a traditional Dutch spring rite, reached New York City, and readers of the New York *Daily Advertiser* could vicariously watch a spectacle of Albany's black servile population choosing one among their number to be a ritual king for the day.[83] The acts were hardly static retentions of African culture reenacted in the New World. Instead, the ceremony had absorbed a remarkable variety of artifacts and cultural influences—the "chief character," as White notes, in a "ceremony on a Dutch holiday in America was an African-born black wearing a British brigadier's jacket of scarlet, a tricornered cocked hat, and yellow buckskins."[84] The performance appeared before outside observers and seems to have encouraged fraternization between "blacks and a certain class of whites."[85] Pinkster (as White describes and as contemporary observers may have understood) presented compelling scenes of authority and resistance, its alternating, exchanged mimicry veering cagily combining self-abasement and subversive mocking. Such performances, albeit stabilized by commentary, reached wider audiences through the print culture of Albany and New York City newspapers. Within a few years, Pinkster seems to have begun featuring a variety of commercial theatrical acts, hosting equestrian feats by Ricketts's circus riders. As a discrete phenomenon of "folk" culture first commercialized and later proscribed, Pinkster appears to have had a relatively short life. The theatrical forms it participated in, however, continued to travel onward.

SPECTACULAR PATRIOTIC FESTIVITIES

Dunlap's onstage processions seem generally orderly; nevertheless, like the 1780 Philadelphia effigy parade, militia muster, or turn-of-the-century Pinkster, they stage intrusions of the low into regulatory scripts. Such incursions, although regulated and performed from positions of cultural authority, can still destabilize. Although Dunlap's play works mightily—and successfully, for the most part—to recuperate the mob of yeoman for respectable patriotic performance, the disorderly conduct of the lower classes continues to shadow such performances. People could and did convert parades into burlesques, decorous patriotic displays into street riots—even while expressing patriotic sentiments. Ultimately, as Durang observed, *The Glory of Columbia* itself came to display theatrical disorderliness, mixing the stage conventions of musical numbers and dance with the street forms of parades militia musters. It simultaneously invokes their disorderly and burlesque shadows, along with mock-transgressive ethnic stereotypes and cross-dressing performance. The play's mongrelized variety of vernacular forms baffles the pedigreed conventions of elite drama.

The Glory of Columbia's long theatrical afterlife only augmented its social and generic unruliness. Patriotic spectacle came to celebrate heterogeneous lowness, and even transgression. In the early decades of the nineteenth century, the play took a central place in the early American calendar of popular festivities. Dunlap's revision remade patriotic performance into entertaining displays of spectacle and variety, and such festivities turned away from national tragedy and toward vernacular theatre. The cover of the play's songbook proclaimed it as "intended for the celebration of the Fourth of July at the New-York Theatre." At the same time, *The Glory of Columbia* served equally well for a variety of holidays, appearing in numerous towns and in assorted celebratory guises. On Evacuation Day, November 25, 1803, the New York *Morning Chronicle* announced the production of *The Glory of Columbia* as a "commemoration of that Happy Event, which restored our exiled citizens to their fire sides, and gave assurance of those blessings we now enjoy." It would continue on that occasion, on and off, into the 1820s.[86] In New York City, the play celebrated the American acquisition of Louisiana the same year as its premiere. In commemoration of the purchase, *The Glory of Columbia* appeared with a new afterpiece by Charleston playwright James Workman, *Liberty in Louisiana*, which featured a "Representation of the cession of Louisiana to the United States" in Act 2.[87] At the beginning of the War of 1812, *The Glory of Columbia* appeared belligerently subtitled "What We Have Done We Can Do," turning Revolutionary memorial theatre into a topical performance of nationalist sentiments.[88] Dunlap's spectacle dominated patriotic occasions, playing at Washington's "Birth Night" in Baltimore, honoring Lafayette's visit in Newport, Rhode Island, in 1824, and with a new subtitle commemorating Cornwallis's surrender at Yorktown in 1829.[89] Unlike its failed tragic progenitor, *The Glory of Columbia* appears a remarkably versatile offering, coming to articulate patriotic performance for a wide array of celebrations.

Settling into America's festive calendar, *The Glory of Columbia* marks the spread of patriotism in comedy, song, and spectacle. As patriotic celebrations became more popular, and acts of civic culture moved out of elite dining rooms and into plebeian public spaces, they became increasingly heterogeneous in form. One of the striking features of *The Glory of Columbia* in these theatrical resurrections is the presence of transparencies, tableaux, and songs in the play. The first performances prominently advertised "music by Mr. Pelissier," supplemented by the visual spectacles of a "British ship at Anchor," the "American Encampment," and the military confrontation at Yorktown. The 1803 advertisements (which reappeared almost verbatim in others ads years later) elaborately describe the mechanics of the spectacle. The play represented advancing troops using "artificial figures in perspective"; in a later scene, Dunlap's spectacle enacted a fixed-bayonet charge using "boys completely equipped, and of a size to correspond in perspective with the Scenery and Machinery."[90] Figures, miniature soldiers, scenery, and machinery—*The Glory of Columbia* clearly relies on novelty and spectacle more than narrative thread. As if to underscore the point, audiences witnessed a finale in which a large-scale "Transparency ascends, and AN EAGLE is seen suspending a crown of Laurel over the head of WASHINGTON, with this Motto, Immortality to Washington."[91] In a thoroughgoing reversal of *André*'s tragic ambivalence about patriarchal authority, Dunlap's revision produces a mechanical, musical display of the Revolution centered on Washington's symbolic apotheosis.[92]

The transformation of Dunlap's patriotic drama into spectacle heavily reliant on visual, musical, and pantomimical stage techniques marks a broader shift—national drama had become a variegated and vagrant national spectacle. In the play's transparencies, military mockups, and musical numbers, in afterpieces and entr'actes, the theatre remade patriotic performance into variety entertainment. Multiple offerings had long been standard fare in eighteenth- and nineteenth-century theatre, with serious mainpieces traditionally preceding comic afterpieces, and songs or novelty acts in between, and the pattern applied to patriotic spectacles as well. Increasingly, however, "serious" and literary drama appeared alongside a variety of shorter popular acts in the nineteenth century. In 1803, for example, *The Glory of Columbia* preceded a comic opera (*Lock and Key*), a conventionally heterogeneous and vernacular theatrical form.[93]

For the 1804 celebration of Independence Day, newspaper notices assured the audience that "every preparation in the manager's power has been making to celebrate the *anniversary of our nation's independence* in a manner suited to the importance of the occasion." Strikingly, the "principal novelty bought forward" alongside *The Glory of Columbia* seems distinctly unsuited to the occasion. The "new grand pantomime" of *Black Beard*, which had just arrived in America from the popular theatres of London, has no apparent patriotic value. What it did have was sensational adventures of piratical usurpation, mutiny, and endangered femininity, all pantomimed in music and gesture. Clearly, such celebrations demanded not patriotic decorum, but vernacular variety. *The Glory of Columbia* itself featured a hornpipe, a "pas suel" and an "allemande de trois."[94] Alongside the rogue spectacles of Blackbeard's piratical cohort, Dunlap's patriotic play had quickly become the occasion for a heterogeneous mix of song, dance, scene, and spectacle. The Fourth of July, and Dunlap's patriotic yeomen, would continue to appear in motley guise and alongside variety entertainments throughout the early nineteenth century. The revision of *André*'s tragedy seems complete. Harlequinades, dancing exhibitions, visual displays, pantomime, and musical numbers had become the stuff of patriotic performance.

Such popular performances do not merely show the retreat from complex drama into mindless entertainment. Instead, they indicate the simultaneous and interdependent politicization of entertainment and spectacularization of politics. In hindsight, *André* and its popular revision seem to stage a sort of long-term debate over the constituencies for membership in the national polity. The two plays stake their success on the questions of who stars in patriotic plays, whose forms shape such acts, and ultimately who orders such scenes. Patriotic plays, like parades and punishments, enact the interplay of high and low, order and disorder, treason and punishment. The long-range shift also affected how audiences related to performances; where *André* required its spectators to admire its elite heroes from a distance, the spectacular forms of *The Glory of Columbia*, perhaps, allowed its audiences to identify with the common patriot. The shift also, however, represented a danger to residual tragic characters such as André (and even Dunlap himself), who seem to resent the prospect of dangling before a crowd of "gaping clowns." The ordering and disordering impulses of the play thus become more than mere effects on fictional characters; they become structuring and disciplinary techniques, albeit with built-in ways of imagining and managing their own failure.

Elite characters recede behind the antics of low types, and generic unity apparently gives way to anti-genre in the popular and unregulated mix of forms. At the same time, performance can transact among even apparently opposed positions. Dunlap's revision process itself condenses and reenacts the intertwining of vernacular acts and traditional drama in early American performance spaces. "Elite" and "yeoman" cultural forms were not simply contestatory or oppositional, and *The Glory of Columbia* accordingly displays regulated orderliness alongside disorderly impulses. Most importantly, perhaps, the yeoman celebration reveals early American theatre's relationship to the already-theatrical practices of plebeian ritual and unruly festival. *The Glory of Columbia* mimics those offstage rituals of community back to its playhouse audience, ultimately producing new audiences. In the end, *The Glory of Columbia* took on a life of its own, reshaping popular patriotism as mock treasons and mid-air spectacles.

5. Pantomime and Blackface Banditry in *Three-Finger'd Jack* ✧

American theatre flirted with the forms of racial and ethnic difference in its underclass theatricals. When the stage's overtures erupted in full-scale racialized performance, blackness seems to have licensed radical scenes of low rebellion. New York City's Park Theatre staged a dramatic performance of black banditry and slave revolt on May 27, 1801. First produced at London's Haymarket Theatre on July 2, 1800, John Fawcett's pantomime *Obi; or Three-Finger'd Jack* was a mobile and circumatlantic performance. It stages the story of Jack, a black bandit in Jamaica who terrorizes whites and slaves alike with his "obi," or obeah, a vernacular magical practice of the black Atlantic. Its action proceeded through music, dance, and gesture, rendering its various characters expressive in song, but almost completely silencing its black rogue. Perhaps trying to distinguish its black rebelliousness from the popular forms of holiday pantomime that dominated the stage at the end of the eighteenth century, the play billed itself as a "seriopantomime." Although the form mutes him, Jack's banditry remained the center of the plot. Jack's offstage presence tensions the first act as the English captain Orford arrives in Jamaica and begins to woo Rosa, the planter's daughter. Emerging from his cave in the Jamaican hills, Jack captures first Orford and then Rosa, who cross-dresses in sailor's garb in order to rescue Orford. Her breeches act defeats Jack and secures the drama's successful conclusion, but Rosa returns to her properly passive and matrimonial place at the play's end. The play poses an important revision of John Gay's 1729 *Polly*, the sequel to *The Beggar's Opera* that appeared on the London stage briefly in the 1770s and 1780s.

Jack's threat continues until two willing slaves, Quashee and Sam, volunteer to track him down and undergo a ritual baptism to counteract Jack's obeah magic. With Jack's violent and spectacular demise, Rosa and Orford can marry, and the island's happy residents (both slave and planter) end the play by celebrating the restoration of colonial order.[1] Within a year, the play appeared in New York City, Boston, and Philadelphia, whereupon it became, in Errol Hill's words, a "staple on both sides of the Atlantic."[2] The story left the stage, entering American folk knowledge soon after its initial debut in theatres. Its story was versatile and trans-generic—for example, an epistolary novel by William Earle entitled *Obi; or, The History of Threefingered Jack* appeared in England in 1800 and in Massachusetts in 1804, leading a fifty-year-long onslaught of prose versions.[3] Earle's novel gave Jack a surname, a lineage of African royalty, and the moral high ground, aligning his revolt with abolitionist rhetoric. Even those audiences with limited access to professional theatre or formal literature could read of Jack's rebellion. John Nathan Hutchins's 1802 *Almanack*, printed in New York, for example, reprints the first edition of the theatrical wordbook, disseminating Jack's scenes among almanac readers.[4] Within a very short time after the debut of Fawcett's pantomime, multiplying versions circulated through Atlantic literary and theatrical culture. It retook the stage intermittently throughout the nineteenth century, notably becoming a vehicle for black performers when William Brown's African American theatre in New York daringly staged the play in the early 1820s. The famous black tragedian Ira Aldridge also included the play in his repertoire after 1830.

Three-Finger'd Jack vaulted slave rebellion into spectacular visibility alongside early American culture's constant reminders of dangerous circumatlantic blackness. First, Jack stages his pantomime rebellion in the also-muted presence of African Americans, whose labors extended from the stage wings outward, along circumatlantic currents to the very Caribbean plantations that *Three-Finger'd Jack* imagines. Second, Jack appeared amidst constant fears of black violence lurking, as it were, just over the imaginary horizon. Observers saw such violence most spectacularly in the example of Saint Domingue, but they could also link slave revolt to other sites of coerced black labor in North America and the Anglo-Caribbean colonies. *Three-Finger'd Jack* performs a complex act; it seems on one hand a ritual exorcism of black Caribbean unrest, a rehearsal of the ways in which dominant or governing culture could turn outlaw and outcast characters against their own kind. The pantomime theatricalizes the history of slavery and slave revolt, telling a decidedly conservative story by transforming revolt into banditry and threats into entertainment. On the other hand, the pantomime shows a deep and formal ambivalence about its black outlaw. *Three-Finger'd Jack* presents a pantomime rogue, a theatrical response to the Black Atlantic's "exotic" and mysterious cultural peripheries. He appears mysterious and magical as well as conventional and performative, and above all, compelling, inspiring repeated reenactments in the Atlantic world. *Three-Finger'd Jack* continues the rogue racial masquerade of *The Beggar's Opera*, realizing the potential of the rogue to become a blackened actor. Fawcett's pantomime thus imaginatively extends John Gay's *Polly*, in which Macheath masks in blackface to pursue a career of Caribbean piracy. When George Colman, Jr. staged *Polly* in London in 1777 and 1782, the play embodied a newly self-conscious Atlantic blackface. *Three-Finger'd Jack* adapted this tradition, but naturalized the black masking. The bandit appears not as a blackface pirate, but as a black rogue. Inheriting Anglo-Atlantic theatre as well as vernacular forms, *Three-Finger'd Jack* presents a scavenged mixed bag of performance. Jack's rogue blackness appears alongside the black vernacular Caribbean figure of Jonkonnu, which embodies a surge of collective and insurgent popular performance even as the play condemns its black outlaw to defeat.

STAGE MARRONAGE AND THEATRICAL SLAVE REVOLT

In London, *Three-Finger'd Jack* "sustained a long vogue at the Haymarket theatre," seeing seventy-nine performances in its first four seasons.[5] With the play's premier in the London theatre world, English audiences set the tone for responses. The *Morning Post*, with the usual zeal of theatrical announcements, announced a "new grand

Pantomimical Drama in two acts" for the play's premiere. The *Morning Herald*, divining the play's promise, trumpets Jack as the "Black Hecate, of Jamaica"; the character seemed "likely, from the force of its magic, to throw [its] spells over all the town." The "Pantomimical Drama," with music by Samuel Arnold, indeed found immediate success; a review in *The Dramatic Censor* observed that *Three-Finger'd Jack* "appears to have gained a fast hold on the favor and partiality of the town" upon its debut. Its success appears due to a combination of timely subject matter, star performances, and novelty. Identifying the play as a "Pantomimical Novelty," the *Whitehall Evening Post* reviewer wrote that the pantomime "resembles several representations of a similar kind, yet it is well put together, and excites a strong interest throughout." Maria Theresa DeCamp, the same review notes, "displayed great powers of gesture in Rosa," and Charles Kemble "invested Jack with terrific importance."[6]

Three-Finger'd Jack quickly gained novel transatlantic popularity, seeing American performances within the year. The play, promoted in advance by its English celebrity, was a boon to New York City's Park Theatre. The theatre, newly opened and at the end of its fourth season under William Dunlap's management, had succeeded the John Street theatre, which had become "outdated and old-fashioned in its outlook," by the early 1790s, according to Heather Nathans.[7] If the Park still had the advantage of novelty in 1801, it also found itself forced to compete with a widening array of popular entertainment forms. Equestrian and acrobatic acts at Ricketts's Circus, visiting performances at the Greenwich Street theatre, and the "New Vauxhall" pleasure garden all began to cut into the Park Theatre's market share in the preceding decade.[8] The Park responded to such competitors by attempting to position itself as the home of elite drama, which seems to have meant importing well- (or at least widely-) reviewed and financially successful offerings from London. The seriopantomime, aesthetically reworking and "elevating" a popular Atlantic genre, fit the bill for its American producers.[9]

Three-Finger'd Jack appears a somewhat incongruous offering for one of America's prestige theatres. Black banditry and Jamaican slave performance seem provocative, especially in light of the refugees from the Haitian Revolution and the rumors of slave revolts in North America, some as near as Gabriel Prosser's rebellion in tidewater Virginia.[10] To some extent, however, attention to form seems to have drowned out the threat—American theatres reconstructed the "novelty" of black banditry into an aesthetically and culturally sanctioned event. The buzz over *Three-Finger'd Jack* began before its New York City premiere. The *Weekly Museum*, for example, extracted London reviews for the play, trumpeting the "magnificent spectacle" while it was still in rehearsal. The common claim of the play's British success (it was "as now performing with unprecedented success and applause in London, and Edinburgh"[11]) seems to drown out any misgivings about the play's depiction of black unrest. With two well-known dramatic figures in the play's lead (John Hodgkinson playing Jack and Lewis Hallam, Jr. as Captain Orford), the pantomime and its "negro dance" (by Laurence and Martin) appears to have only delighted audiences and reviewers.[12] The play entered the repertoire in the following years as a season-ending benefit performance, calculated to please audiences and collect large receipts. The Philadelphia theatre followed suit at the end of the year, first staging *Three-Finger'd Jack* on December 26, 1801—the heart of the traditional

Anglo-Atlantic pantomime season—and giving elaborate descriptions of the scenes in order to entice audiences. By April 2, 1802, the Federal Street Theatre in Boston produced its own version of the "grand pantomimical drama."[13]

Precisely what *Three-Finger'd Jack* meant to its transatlantic audiences—if and how it might have evoked real-life slave revolts—remains a vexed and open question, even with reasonably good documentation of its reception. It seems probable that the play provoked fears of slave unrest in early American audiences, but it is not a foregone conclusion. It inspired no public debates about slave revolt, for example, and reviews make little or no comment on its subject matter. Despite the frequent claim that Fawcett based the play on historical events, most commentaries evade the issue of its mimetic qualities. In one of the more detailed responses, a reviewer for the *Port Folio* explains:

> We should not notice the representation of Obi, were we not actuated by a wish to pay the tribute of our applause to the gentleman who personated Three-Fingered Jack. We own ourselves partial to a story, well told in action; and we wish that this amateur may favour the public with a display of his talents, on many future occasions of the same kind. His action was spirited, appropriate, and energetic, in the highest degree; and his desperate conflict with his pursuers, was sustained with such truth and nature, that it almost excited sensations of horror in the spectator.[14]

The review suggests a fundamental repression governing interpretation of the play. Theatrical form competed for attention with the play's appalling content, and Anglo-American pantomime seems to foreclose attention to the play's mimetic staging of slave revolt. The reviewer's use of the word "horrors" is significant, repeating a scare word widely used (as Leonora Sansay did in her novel *Secret History; or, the Horrors of St. Domingo*) to describe circumatlantic slave revolts.[15] The affective response to black violence, however, disappears behind an aesthetic and formal response to the play. On the surface, the review implies that Jack's onstage rebellion simply did not register on the same level as fears of contemporary Caribbean slave revolt. In historical hindsight, the commentary has the ring of avoidance, if not deliberate repression. Those "almost excited" feelings of horror emerge with the play's mimetic truthfulness, suggesting that audiences encountered the play conditioned by a fugitive awareness of slave revolt. The pantomime's overt artifice, its self-conscious theatricality, and the conventions of stage spectatorship, perhaps, kept reviewers from explicitly responding to its rebellious content.

Three-Finger'd Jack restages the Caribbean haunting of Atlantic culture at the end of the eighteenth century. Government documents and newspaper reports show that for nearly two years, from 1779 into the beginning of 1781, an actual Maroon bandit dubbed Three-Finger'd Jack engaged in a guerrilla war in the hills of Jamaica. Only sketchy details of his rebellion survive in colonialist lore corroborated by a few government proclamations and newspaper announcements. Jack, however, seems no ordinary runaway; accounts describe him as physically imposing, terrifying even without his mysterious obi charms. Thus aided, Jack and a "desperate gang of Negro Slaves," as one official proclamation identified the outlaws, waylaid travelers in the Jamaican backcountry.[16] As he robbed travelers, resisted the law, and attempted to instigate an uprising, his actions caught the imagination of popular culture and inflected performances of underclass and slave rebellion for decades.

Jack's banditry spurred circumatlantic cultural production. His story first reached English audiences as a short but intriguing anecdote in Dr. Benjamin Moseley's 1799 *Treatise on Sugar*, a study of natural history, tropical medicine, and colonial plantation management. The story that Moseley transmitted forward from the hills of Jamaica was undoubtedly biased and incomplete, partial in both senses of the word as it worked from the monocular point of view of a terrified but ultimately triumphant Jamaican plantocracy. Fawcett's script as well as the wordbook (*Songs, Duets and Choruses in the Pantomimical Drama of Obi, or, Three-Finger'd Jack*, sold to theatre-going audiences) openly quoted Moseley's narrative. Reviews such as Charles Dutton's lengthy commentary on the play focus explicitly on Moseley's account and the play's mining of historical detail. Moseley's authoritative recital of historical "facts" attempted to stabilize Jack's story within ethnographical and scientific discourses, but also popularized the bandit.

Moseley's history of Jack ends with an exercise in theatrical punishment and discipline, adapting Anglo-Atlantic punitive ritual to the black Caribbean body. In Moseley's account, a party of free blacks known as Maroons led by Sam and Quashee took up the legislature's offer of a three-hundred-pound reward for Jack's capture or killing. After tracking the outlaw down, his pursuers engaged him in hand-to-hand combat. The advantage finally turned to the Maroons, and the three combatants tumbled down a steep bluff, where Sam and Quashee shot and bludgeoned Jack to death. Jack died on January 27, 1781, but his story continues. Apparently determined to display the fruits of black rebellion, Moseley's account ends with a literary tableau of slaves and Maroons alike celebrating over the fallen outlaw's severed head and hand. A "vast concourse of negroes," Moseley recounts, carry the trophies in a pail of rum to Morant Bay, "no longer afraid of Jack's Obi, blowing their shells and horns, and firing guns in their rude method," they paraded on to Kingston and Spanish Town.[17]

The identity of Jack's killers as Maroons shapes Moseley's telling in key ways. The Maroons, as Moseley understood, were populations of free blacks descended, at least in theory, from motley breakaway communities of escaped slaves, indigenous people, and fugitive Europeans, including pirates. They had a long history of precarious relations with both the white slaveholders and the enslaved population. Orlando Patterson, speaking of the Jamaican Maroons, observes that the first "eighty-five years of the English occupation of the island (1655–1740) were marked by one long series of revolts."[18] Two agreements, the Leeward and Windward Treaties of 1738 and 1739, ended the open rebellion and calmed relations between Maroons and whites. These treaties, though they would never ally the rebellious communities wholeheartedly with authorities, placed Maroons in an ambivalent relationship to other blacks. While the treaties promised peace, for example, they also stipulated that they "use their best Endeavors to take, kill, suppress or destroy, either by themselves or jointly, [...] all Rebels wheresoever they be throughout this Island, unless they submit to the same Terms of Accommodation" of that treaty.[19] Similarly, the ninth and tenth terms of the Leeward treaty specified that the Maroons must return all runaway slaves, including the ones they had set free during their struggle. Noting the difficulty in imagining black rebels who would assist in capturing runaway slaves, Campbell suggests that the treaty may have been fraudulently presented, but she finally concludes that the "promptness with which [the Maroons] assisted in dealing with runaways" indicates their willing cooperation.[20] Indeed, for whatever reason, the leeward Maroons "had shown more than ordinary zeal in hunting down" runaway slaves; R. C. Dallas's 1803 *History of the Maroons* corroborates the bleak picture of Maroon complicity, and related that authorities sometimes had to offer monetary encouragements to bring slaves in alive.[21] Maroons exemplified racial and political liminality: they were of mixed black and indigenous heritage; they were free, but descended from slaves, and were ultimately politically coerced and removed from the island. They seem at times aggressively independent and at others eager to collaborate.

By all historical accounts, marronage posed real problems for Anglo-Caribbean authorities, cropping up throughout the Americas, including in the slaveholding territories of North America.[22] Marronage (sometimes simultaneously) operated as counter-colonial resistance and participated in plantocracy's economic and cultural hegemony. In 1780 and 1781, the original Jack performed a form of Jamaican marronage, hiding away in the hills and pursuing guerilla-type banditry. He continued the episodic, unstable, violent, and self-sustaining resistance to English authority. Jack's insurgency thus becomes a story of Maroon unrest that authorities could only solve through the troubling recruitment of other Maroons. Governor Dalling's proclamation for Jack's capture or execution thus seems an almost desperate plea for assistance in maintaining order, and the Maroons who take advantage of its offer perform yet another opportunistic and mercenary act that undercuts even as it enforces colonial authority. Moseley's account, however, seems most invested in narrating violent heroics that result in the triumph of government. Accordingly, it construes Jack's acts as outlawry rather than resistance, converting Maroon-on-Maroon violence into banditry and corresponding police action. If the scandal of Moseley's narrative is authority's reliance on the very subjects that resist it, its triumph is the story of governmental hegemony. Moseley's account celebrates power's ability to fragment and use subordinated populations willingly against themselves. The histories and official documents most likely fail to access significant portions—perhaps most—of the lived experiences of either Jack or his Maroon pursuers. Narratives constructed from such partial evidence almost inevitably misconstrue motivations and restructure history. Moseley's image of the celebratory cohort of slaves seems a ritual act of wishful thinking. It certainly may have happened, but almost as certainly does not mean precisely what Moseley seems to think it does.

Such magical thinking permeates Fawcett's *Three-Finger'd Jack* as it rewrites the conflict again. Using plantation slaves to dispatch Jack even more neatly, the play elides the bothersome presence of the island's Maroons, centering black revolt and its solution on the plantation. The play transforms a troubling cohort into submissive slaves, producing docile, loyal black characters to protect their master and his extended family. The play opens with the festive songs of the satisfied slave cohort: "Good Massa we find / Sing tingering, sing terry,— / When Buckra be kind, / Then Negro heart merry."[23] Despite the threat of sale and familial separation articulated in the opening act, slaves assume carefree, grateful, and ultimately docile attitudes in the play. The play's "kind massa" becomes a unifying character whose overarching paternal presence reconfigures the slaves, erasing the black-on-black violence of the historical narrative.

THREE-FINGER'D JACK'S MAGICAL PANTOMIME

Jack appears onstage not just as a blackened rogue, but rather as a pantomime embodiment of blackness. A maimed and magical escaped slave turned bandit, the black rogue represents a physically absent threat, an offstage sound threatening disruption of Jamaican plantation life's festive order. The play separates Jack (an exotic, marginalized presence in the play) from the docile slaves by his obeah magic, the sign of which even supersedes Jack's name as the play's title. Obeah's vernacular ritual, and its objectified charms containing "Grave dirt, Ashes, the Blood of a black Cat, and human Fat," enables Jack's rebellion.[24] Christian magic, meanwhile, counters obeah magic, differentiating insider and outsider slaves as the pantomime signals Jack's impending end by having the slave Quashee christened. Wielding his new religious affiliation as a counter-obeah, Quashee compares his new powers to those supplied by Jack's "obi-bag," promising in song to "tear the charm to rag."[25] In the midst of a "Negro Dance" in the first act, the sound of a gunshot brings his threat to center stage, and the manuscript libretto show Jack's impact: "Tuckey screams out at a distance—then enters and acquaints them his master has been attacked by 'Jack." At the sound of "Jack" the Negroes are greatly terrified."[26] The published description of the play indicates "Panick of the Slaves, at the name of Jack;—and the superior courage of the two negroes, Quashee and Sam."[27] The scene indicates the theatrical force of the black rebel—his sign alone provokes the others to mime terror, reacting in their own gestural, unspoken ways. In a quintessentially pantomimic stage moment, actions substitute for words. When the pantomime does use words ("Jack!"), they operate as flashing signs, in hisses, gestures, and poses, rather than in rational verbal communication. Jack's whispered name—apparently, one of the few words not sung in the play—operates less as verbal communication than as a sort of auditory icon of the stage antihero.

When he does appear onstage in the play, Jack does not speak, and only rarely sings. He revolts in pantomime, defined by the elements of music and dance. Tempo and tone, gesture and pose, rather than language, shape his act. The published score, for example, indicates in an otherwise wordless stretch of music that a "Noise by Jack is heard"; the score reacts with a frenetic, repetitive series of higher and faster notes, the musical signs of Jack's presence.[28] Aided by his occult rituals, he emerges from his cave (identified in the manuscript as an "Obi Woman's Cave," a feminized and darkened site of magic and ritual) to terrorize the island's inhabitants.[29] In Fawcett's play, Jack moves slickly on and offstage, down traps into cavern scenes and up ladders, escaping his white pursuers. His skilled use of "private entrances" (as in his magical entrance in the play's third scene) becomes a source of wonder for both his English pursuers and his own cohort of black robbers. Jack practices a magical, performative mobility that recalls the excarceral acts of other circumatlantic rogues. Jack himself follows the same pantomimic and emblematic logic, operating within

Figure 5.1 Hand-colored etching entitled "Mr. Smith as Obi, in Three-Fingered Jack, drawn and etched by J. Findlay," n.d. Billy Rose Theatre Division, The New York Public Library for the Performing Arts, Astor, Lennox, and Tilden Foundations.

the play as an embodiment of black roguery rather than as a developed character. Images based on his performance display Jack in a characteristically spectacular and heroic pose (figure 5.1).

Jack's three-fingered hand, heroically raised in revolt, effectively condenses the bodily signs of his lawlessness.[30] The hand, like his fetishistic charms, functions as an emblem of his presence, an image persistently conjured in attempts to stabilize the rogue's elusive, flickering hyper-mobility. By Moseley's account, Jack had lost his two fingers in an earlier encounter with one of the Maroons who later killed him. It thus stands as dramatic sign of his outlaw status and eventual defeat. It also presents a corporeal reminder of the history of the physically dismembered Atlantic criminal.

In 1719, the English penal system resolved the dilemma of hanging or branding on the thumb with the Transportation Act.[31] As a result, a certain number of offenders arrived in the West Indies with manual disfigurements forecasting Three-Finger'd Jack's later display. Like Crook-fingered Jack in *The Beggar's Opera* (whom Peachum ironically calls a "mighty clean-handed fellow"), Jack's disfigured hand becomes evidence of his equally disfigured relationship to capitalist industry.[32]

Drawing on the Atlantic resources of such deformed roguery, Three-Finger'd Jack displays his physical defect as a sign and a means of empowerment. Ultimately, his executioners redeploy those signs in the pail of rum containing the bandit's severed head and hand as a deterrent (and still pantomimic) sign. Jack's maimed hand, his fetishistic emblems of obeah magic, and his underground lair produce a pantomime shorthand for the spectacular revolts happening throughout the extended Caribbean at the end of the eighteenth century. Jack renders spectacular the signs of rebellious Atlantic blackness. Jack's slippery circulation also echoes the mode of popular theatre—Fawcett's pantomime was trendy and portable, traveling under the motive power of its hit songs and entertaining dances. Appearing first in London and later in New York City and Philadelphia, the prolific circulations of Jack's stage act, impelled by the Atlantic economy of entertainment and theatrical celebrity, echoes the bandit's dangerous mobility.

POLLY'S MORANO: REENACTING ATLANTIC GENEALOGIES OF MARRONAGE AND BLACKFACE

Blackface marronage, though rare, had appeared before on the Anglo-Atlantic stage. George Colman the Younger's Haymarket Little Theatre premiered John Gay's *Polly*, a sequel to *The Beggar's Opera*, on June 19, 1777. Colman's revision and resurrection of *Polly* was only a minor hit, but it brought a radical plot of underclass mobility and racial masking to the stage. Polly Peachum, the female star of *The Beggar's Opera*, ventures to the Caribbean in search of her fugitive outlaw husband Macheath, who has disguised himself as the black pirate captain Morano. Morano's gang of pirates, threatening to attack a West Indian island, creates the chaos that indirectly enables Polly (dressed in male attire) to continue searching for her husband. Perhaps predictably, Morano's gang captures Polly, who bribes her way out of captivity with the Indian prince Cawwawkee, still unaware of the pirate captain's identity. The heroine then joins the planter elites and their native allies in repelling the pirates. The play concludes hastily with Morano's (offstage) execution, performing the punishment that the charismatic Macheath had avoided at the end of *The Beggar's Opera*. The play finally reveals his underlying identity, but too late to make any difference to the plot. The play's finale abruptly turns away from the work of rogue punishment to call for festivities over Polly's intended marriage to the Indian prince Cawwawkee. The planters, teamed with tractable Native Americans, have repelled piratical attack and reinstated the old order in the nuptials of European and noble savage.

Morano's revolt extends the roguery and punishment of *The Beggar's Opera* into the circumatlantic world. The thieving urban gang has become a band of pirates, reenacting the mobility of the Atlantic criminal low. *Polly* imagines pirate culture as a self-contained society that redefines its motley origins in terms of a new goal, a war on society. The revised script of Colman's 1777 production suggests the forcefulness of the blackened pirate and his doomed crew. Macheath's self-imposed separation from the world enables a "heroic" assault on "the whole world," as his lieutenant Hacker proclaims.[33] The second act calls for a rousing song voicing the outcast ethos as Morano declares, "Let us on to battle, to victory!" "Let despair lead to battle! No courage so great: / They must conquer or die who have no retreat."[34] Although the play does not simply celebrate the soon-to-be defeated rogues, this moment of song compactly figures Morano's resistance, his struggle fueled by a despairing courage. It represents the play's desire to display the appeal of such a heterogeneous, outcast crew, thrown together by their desperate war on the world. Morano faces precisely the alternatives of conquering or dying at play's end. Unable to conquer, he dies, punished for his blackface piracy. The stirring display of "keen and awing" rebellion vanishes as quickly as it appears, just a flash of lightning on the Atlantic stages. For a brief moment, however, he embodies the bravery and audacity that audiences sought in outcast characters.

Morano's name, as Calhoun Winton has observed, is probably a corruption of "Marrano," the term that would designate New World Maroons.[35] In *Polly*, then, marronage provides a compelling theatrical trope for imagining circumatlantic underclasses. Morano, reenacting an earlier sense of marronage, finds himself cast away from both the metropolitan center of English society and the culturally privileged forms of whiteness. Like New World Maroons, his was an outcast rebellion of murky fugitive origins. Morano condensed into one theatrical character the confluence of concepts reserved for castaways (outcasts from circumatlantic voyages or metropolitan streets), rebels, and fugitive slaves in the Caribbean.

If Morano's revolt seems a uniquely Caribbean affair, existing English theatrical forms crucially enable it. Although a key feature of Morano's Caribbean piracy, black masks had also traditionally marked lawless English renegades both on and off the stage. The "Waltham and Windsor Blacks," a group of rural English poachers, blacked up in the early eighteenth century, as E. P. Thompson has observed.[36] Authorities proscribed not only the poaching but also the performing—the Black Act of 1723 that "made it a capital crime to 'black' one's face," as John O'Brien has noted, "stayed on the books well into the nineteenth century."[37] Perhaps most strikingly, the Act criminalized blackface offstage just as John Rich's theatre at Lincoln's Inn Fields began to popularize pantomime with its frequently blackfaced protagonist. As Dale Cockrell has argued, the mask of blackness was long present in Anglo-Atlantic folk performances, a sign of preindustrial alterity that gradually accrued conscious racial connotations.[38] Morano perhaps represents the onstage culmination of such processes. *Polly* (and later *Three-Finger'd Jack*) represents formal stage practices converging with those vernacular forms, emerging form the wings as customary embodiments of outsider identity and underclass disorder.

Perhaps as the eighteenth-century enactments of lawless blackness had always done, *Polly* presents a self-aware and deliberate masquerade of resistance to Atlantic capitalism. Morano, for example, hides his war on society behind the simultaneous publicity and secrecy of a blackface mask. *Polly*'s political energies, like Morano's war, appear only in disguised form, and masking has become a key theme and form in the ballad opera. *Polly*'s blackface piracy responds to emergent transracial cohorts in the Americas, but perhaps more importantly, it emerges as an overtly theatrical product of Anglo-Atlantic stages. Macheath/Morano (for audiences knew the role as self-consciously doubled) thus represents a momentous character. His blackface disguise stages an early—perhaps the first—instance of self-conscious racial mimicry on Atlantic stages. Earlier roles had acted out various versions of blackness; Shakespeare's 1603 *Othello* and Thomas Southerne's 1695 adaptation of *Oroonoko* had presented noble African figures, and Isaac Bickerstaff's 1768 *The Padlock* had presented the famous comic black character of Mungo. They all inevitably left their mark on *Polly*'s performance; Morano, however, presented something newly threatening—deliberately masked underclass rebellion. As a self-consciously theatrical act, Morano's blackness broke new ground in 1777. The burnt-cork makeup on Colman's stage represented a theatrical act rather than actual blackness. Morano, masked onstage as a black pirate, acted out blackface itself.

In the late eighteenth century, Morano appeared before audiences who knew Macheath's banditry as residual, a legendary cultural survival like the Waltham Blacks. Moreover, *Polly*'s blackface masking appeared in an Atlantic world increasingly attuned to racialized scenes and characters. Macheath/Morano *plays* pirate, sliding theatrically from one position to another, refusing to sit still or inhabit any one role exclusively. His overt theatricality, while perhaps signaling popular indecision about how to represent the Atlantic low, also ironically links theatrical celebrity to a performance of social death. "I disguis'd myself as a black," Macheath/Morano tells his consort Jenny Diver, becoming "dead to all the world."[39] In that phrase, Macheath locates the ironic popularity of such characters: they may achieve stage fame, thriving before their public, but ultimately at the cost of symbolic death. The highwayman, playing pirate, adds to his outlawry an ironically doubled theatricality: Morano plays noble leader and outcast rebel, white bandit and black pirate, both and neither at the same time.

Morano's blackface theatrically signs his rogue audacity, performing a dualized, mimicking position. Audiences can hardly forget that Macheath/Morano signifies both blackness and whiteness, both admired theatricality and outcast rebellion. *Polly* imagines theatricality as a tool of the resistant underclass, and blackface operates in a self-conscious mode that reveals the mechanisms of representation. Macheath/Morano acts out his blackened rebellion onstage, in the place where "speaking for" meets "speaking for oneself." As Morano, the rogue claims the privilege of creating and acting his own part. This act of self-representation causes his downfall as much as his rebellion, for Polly, the one character who would stay his execution, does not know his true identity in time to rescue him. For blacking up—for becoming theatrical—more than for thieving, whoring, or for his frontal assault on the plantocracy, the blackface pirate must die in the end.

In this self-conscious performance, *Polly* stages a version of what Peter Linebaugh and Marcus Rediker call the "hidden history of the revolutionary Atlantic." At the same time, the emphatic use of blackface as a means of underclass resistance in 1777 points to the early emergence of the "stereophonic, bilingual, or bifocal cultural forms" that Paul Gilroy identifies with the later formation of the Black Atlantic. *Polly* acts out the convergence of class resistance and emerging racial emancipation, imagining what David Armitage would later identify as the Red and Black Atlantics—the mobile worlds defined by the experiences of lower-class and darker-skinned populations.[40] *Polly*, limited though its productions were, represents eighteenth-century theatre developing new forms in an attempt to respond to the circumatlantic cultures of race and class. *Polly*'s fatal ending, however, does not tell the whole story. Such character types survived beyond the limited scope of Colman's productions, finding continued representation on the stage. The charismatic posing of the pirate captain has a much greater effect on the early-modern Atlantic world than Morano's binary pronouncement ("they must conquer or die") indicates; the pirate crew certainly did not conquer, but neither did they disappear from Atlantic theatres. The outcast hero's masking, while rendering him dead to the rest of the world, also has the paradoxical effect of making him ever more attractive. His theatrical rebellion exercises options other than victorious conquest or fatal defeat. Macheath/Morano represents the resourceful underclass thieving and scheming of *The Beggar's Opera* gone circumatlantic—escaped, connected, layered, mobile, and rebellious.

Appearing in the wake of Harlequin, Morano, the Waltham Blacks, and the Jamaican Maroons, *Three-Finger'd Jack* makes distinctly theatrical claims about the nature of blackness. Blackness, though rendered exotic, is ultimately (and literally) white underneath—the play imagines slave revolt as a version of traditional English insubordination. Newspaper accounts of the play, variously referring to Jack as a "valiant robber," a "ruffian," and a "freebooter," replay the long-standing association of blackness and blackface with criminality. Jack, like Macheath and Morano before him, heads a gang of lowlife rogues. As Macheath had been, Jack and his gang are performing characters, although they operate in the forms of musical pantomime rather than ballad opera.

Fawcett's pantomime naturalizes the radical perforation of the diegetic frame that had governed the self-conscious ending of *The Beggar's Opera*. Beggars and players no longer ironically intervene to explain the discipline of rogues. Instead, the play imposes magical acts upon the play's invented world; the songs and dances demanded by musical pantomime's form make Jack's lawlessness a matter of performance. Fawcett's pantomime, unlike Gay's ballad opera, does not self-consciously point out the fact. Conventionally played by a white actor in blackface, Jack performs a version of Morano's earlier racial layering, but one that ultimately transfers the act of racial mimicry from rogue to actor. That is *Three-Finger'd Jack*'s major departure from *Polly*. Jack, unlike Morano, does not appear as a blackface character within the play's diegesis. Instead, he embodies a de-ironized and "authentic" blackness, even as a white actor played the character. Morano's blackface outlawry had become, in the words of the New York reviewer, the horrifying "truth and nature" of black revolt. *Three-Finger'd Jack* thus invests rogue blackface with increasing urgency even as it moves the act of mimicry into the space of the theatre.

Three-Finger'd Jack thus performs Caribbean slave revolt in the familiar image of English outlawry. At the same time, the nonverbal form of the pantomime could also estrange the act from those familiar contexts—even the most familiar of forms could operate in unfamiliar registers. Pantomime, as *Three-Finger'd Jack* attests, could also render slave revolt mysterious and horrific—beyond the expressive power of words. Audiences labored to interpret actions (at the end of the eighteenth century, managers frequently resorted to banners and supplemental wordbooks to explain the action to audiences) available only in song and gesture. Scholars have generally taken the existence of at least eleven London editions in nine years of *Three-Finger'd Jack*'s wordbooks as evidence of its popularity; the documents also indicate the degree to which Jack's rebellion needed interpretive regulation—audiences required some degree of explanation, glossing, retelling.[41] The pantomime's insistence on Jack's magic (signaled in the displacement of Jack's name by "Obi" in the title) gives his revolt more than simply an exotic twist. It claims a share of the attention from the rebellious black body, performing Caribbean slave unrest as unexplainable, mysterious, and supernatural. In that sense, Jack's acts take the masking of the Morano or the Waltham Blacks to their logical conclusion. Those acts of disorder and disobedience seem shaped by an internalized, traditionally English—and, for its practitioners, eminently commonplace and understandable—"moral economy."[42] Racial masking, and Jack's insertion into the English theatrical genealogies, performs the utter difference—a socially registered class difference, perhaps—of such acts. *Three-Finger'd Jack*'s magical, mobile pantomimed revolt thus performs both the mundane and the marvelous, portraying blackface roguery as at once eminently expressible and beyond representation.

"OPTIC BLACK" PERFORMANCE

Audiences encountered Jack as a performing palimpsest, the compounded and layered sum of the discourses of roguery circling Fawcett's pantomime. Jack appears onstage as a product of cultural processes that W. T. Lhamon has called "optic black." "Optic white" was Ralph Ellison's trope in *The Invisible Man* for the production of whiteness; Ellison's narrator, employed by the Liberty Paint Company, adds almost undetectable amounts of black paint to the larger batch of white in order to produce a brilliant, "more pure" white dye—optic white. Activated by carefully dosed bits of theatrical whiteness, the overlays, as Lhamon writes, produce a "contrapuntal cultural style" of blackness that "embodies a persistent countermemory of historical opposition."[43] The concept usefully characterize Jack's compound performance, but also points out another central aspect of the Jamaican bandit's stage presence—never speaking, in pantomime, he is a primarily visual presence. Jack represents the complex display of blackness that reveals through masking—a form of the "spectacular opacity," perhaps, that Daphne Brooks describes in later contexts.[44]

Three-Finger'd Jack's optic black performance may have been entirely inadvertent, a side effect of attempts to make the play's characters comprehensible within Anglo-Atlantic traditions. At the same time, the play reveals the undercover allure of blackness persistently permeating other practices in Atlantic theatre culture. The wordbook and prospectus of the play sold at the theatre gestures toward the Anglo-Atlantic genealogies of Jack's complex, racially layered performance. The description, taking poetic license, compares Jack to Robinson Crusoe, rather than the nonwhite Friday.[45] The published score further elaborates upon *Three-Finger'd Jack*'s debt to stage versions of *Robinson Crusoe; or, Harlequin Friday,* indicating a "Subject from the Savage dance, in Robinson Crusoe," probably indicating a borrowing from a popular song in Richard Brinsley Sheridan's pantomime version of Defoe's novel that had premiered at Drury Lane in 1781, and which had seen republication in 1797. Shifting registers, the wordbook also praises Jack as having "ascended above Spartacus."[46] Jack becomes rhetorically equivalent to the rebellious Thracian rebel Spartacus, who, not coincidentally, enjoyed his own Atlantic fame beginning with the 1831 premiere of Robert Montgomery Bird's *The Gladiator*. Associating Jack with classical history, popular pantomimes, and novels of eighteenth-century colonial adventurism perhaps warded off the realities of Caribbean racial rebellion, repositioning Jack in a world of legendary fictions.

Optic blackness, emerging from interwoven discourses of black Atlantic rebellion and heroic rogue whiteness, seems to have created the Jamaican bandit as a complex, powerful, and even attractive embodied display. Audiences and producers alike saw the antihero in a complex and relatively positive light. The play's prospectus, using Moseley's overblown and melodramatic language to promote the play, transforms his dangerous rebellion into heroic "exploits" and his villainy into gallantry, a reenactment of Macheath's charismatic roguery. The literature waxes eloquent at times:

> These [Jack's charms], with a keen saber, and two guns...were all his Obi; with which, and his courage in defending into the plains, and plundering to supply his wants, and his skill in retreating into difficult fastnesses, among the Mountains, commanding the only access to them, where none dared to follow him, he terrified the Inhabitants, and sat the Civil Power and the neighbouring Militia of the Island at defiance, for nearly two years.[47]

Such statements suggest that *Three-Finger'd Jack* aimed at inspiring a pleasurably panicked response of terror and admiration for the black rebel's exploits, and the play's long production runs suggest that it succeeded. While it remains hard to prove that those responses actually drove the play's popularity, the pantomime certainly presented scenes that allowed, even presumed, such responses.

Imagining Jack as a racially layered character, such rhetoric places him in a pantheon of heroes drawn from the Anglo-Atlantic world's reserves of political and cultural mythologies. Such modes create schizophrenic or stereoscopic effects— association with western culture's mythologies does not simply erase the violent threat of blackness, especially as unrest throughout the Caribbean reached its crescendo in Saint Domingue. The problem relationships of whiteness to blackness constitute a focal point for the play. Its supporting materials also obsess over the artificial, unnatural, and even constructed nature of racial categories. The pantomime's prospectus, drawing heavily upon Moseley's account, provided a pseudo-scientific description of certain "Obi-Men," a cohort of "abandoned Exiles" who "survive a general Mutation of their Muscles, Ligaments, and Osteology; becoming also hideously white in their wooly Hair and Skin."[48] These deformed, bleached figures, banished to the hills, often sheltered "robbers and fugitive Negroes."[49] The obi-men represent a sort of unnatural theatrical whiteness—the deformed whiteface counterpart, perhaps, to Jack's blackface performance. The text revealingly attempts to produce a medical and sociological explanation for Jack's banditry, identifying it as a mysterious pathological effect of obeah magic, compounded by the sequestering of fugitive black outlaws in the Jamaican hills.

Reprinting Moseley's pseudoscientific claims, the prospectus supplements the play's imaginative performance with developing pseudoscientific discourses of race. Significantly, the interplay of performance and text did not only become clear in historical hindsight. The title page of *Songs, Duets, and Choruses* (which saw eleven editions in nine years in London, and was still being republished after a decade in the United States) claims to have been sold at the performance, and presumably audiences could have referred to it for interpretive guidance. The text displays twin strategies of associating Jack's rebellion with whiteness—on the one hand, the legendary or fictional whiteness of Spartacus and Crusoe, and on the other hand, the equally fictional whiteness of deformed obi-men. Unnatural whiteness, it would seem, authorizes a black rebellion that is unable to act or represent on its own. At the same time, such tropes suggest trans-racially (mis)understood rebellion and the difficulty of sorting out the threats of shifty circumatlantic underclasses. The textual obsession with tropes of artificially whitened blackness overcompensates for racialized discourses unequal to the task of comprehending black rebellion. The submerged suspicion of multiracial affiliation pervades *Three-Finger'd Jack*, an ironic counterpart to its white actors in burnt-cork makeup. Such strategies of representation complicate the assumption that the play stages fears of blackness. Ultimately, the racial layering of roguery underwrites *Three-Finger'd Jack*'s presentation of binary racial difference. The multiple and contending modes of representing Jack point to converging genealogies of Anglo-Atlantic outlawry and Black Atlantic resistance.

JONKONNU IN *THREE-FINGER'D JACK*

The play punctuates the threat of slave revolt with adaptations of black vernacular performances that draw heavily on Caribbean contexts. One of the play's moments of complexly overlaid performances occurs at the end of the first act; "Rejoicings of the Slaves" conjure up a character, unique on formal Anglo-Atlantic stages, which Fawcett labels "Jonkanoo." The singular character stands in for a constellation of Black Atlantic performance practices that had gathered under the conventional (though variously spelled) name "Jonkonnu."[50] *Three-Finger'd Jack* presents one of the first known instances of the black vernacular act of Jonkonnu appearing before Anglo-Atlantic audiences. Although he seems a rare occurrence on Anglo-Atlantic stages, the play and its contextual materials assume some familiarity with Jonkanoo and his cultural origins. Charles Dutton's review in the *Dramatic Censor* describes him as a "grotesque character, equipped with a ludicrous and enormously large false head, [which] presides at the negro balls in Jamaica, in the capacity of master of the ceremonies."[51] At least some audience members, and probably those with direct experience of Jamaican plantation culture, would have been familiar with the practices of slave festivities and the mild carnivalesque of black masters of ceremonies. To the extent that the character Jonkanoo signaled such circumatlantic practices, he marks the circumatlantic movement and return of practices originating on the colonial margins—the forms of the colonial periphery intertwined with, and even interpenetrating, those of the metropole. The pantomime rendition of the figure represents more that a simple theft of black Jonkonnu forms for white entertainment. Jonkanoo suggests a fascination with complex, subaltern, and unruly forms of black play.

It displays the formal interplay of white and black spectatorship and performance that shaped Jonkonnu's broader practices before Fawcett's pantomime conscripts the form. Jonkanoo (identified in song not just as master of ceremonies, but also as a "master of all") appears onstage as a theatrical substitute for the plantation owner, a Caribbean version of the carnivalesque that Bakhtin found in early modern European rituals.[52] The pantomime's brief summary specifies none of Jonkanoo's onstage actions—indeed, he seems to have functioned as an iconic stage body who contributed little to the plot. Formally, however, he embodied the pantomime's conflicted investments in black performance. Jonkanoo enacts a temporary and limited insubordination merging in and deflected by comedy. With the slaves' singing of the "funny big man" (the prospectus refers to him as ludicrous), Jonkanoo enacts a temporary, limited, and comical insubordination.[53] It seems to come to nothing in Fawcett's play. In fact, Jonkanoo helps celebrate the coming restoration of order. Ultimately, his only temporary inversion becomes the centerpiece of celebrations of authority's restoration.

However, the character radically embodies a vernacular Black Atlantic presence in the play. Jonkanoo reenacts the ritual inversions of Caribbean vernacular performances for English and North American playhouse spectators. The character's strategic location within the play is perhaps more important than any specific acts he performs. He makes his entrance as Jack's rebellion moves toward its inevitable failure, filling a growing gap in the play's performance of black roguery. He operates as a sort of formal substitute for the master, but also for Jack, sublimating theatrical rebellion in celebratory performance. Displaced by the squeezing pressure of colonial authority, individual banditry reemerges as collective dance, and Jonkanoo embodies revolt by proxy. He reenacts the logic that makes rogues performers, which saw, for example, Macheath sing and dance, Morano mask, and Jack's gang sing and carouse. Jonkanoo's lasting contribution is perhaps his radical sublimation of Black Atlantic roguery into the genealogies of Anglo-Atlantic theatre.

The pantomime did not invent the Jonkanoo character. Fawcett drew the Jamaican act from persistent vernacular Black Atlantic practices. As scholars such as Judith Bettelheim and Elizabeth Fenn have shown, Jonkonnu acts evolved from Caribbean and North American planter society performances, through Atlantic street performance, to the stage, and eventually found their way into manuscript and published memoirs.[54] The collective performance histories of Jonkonnu, then, reveal geographical mobility linked to generic mutations. As the act traveled, it jumped generic boundaries, repeatedly insinuating its characters and scenes first into institutionalized theatre and eventually into literary culture. Jonkonnu's movements also point to elaborate and complicated dynamics of attraction and exploitation, rebellion and collusion in eighteenth- and nineteenth-century representations of race. Jonkonnu may have survived partly because of its enactment of coded liberatory potential, even if it only enacted masked and temporary resistance to oppression. At the same time, accounts of slaveholders and members of the colonial plantocracy reveal that the act exercised a compelling attraction to outside observers, despite its riotously disruptive potential. Jonkonnu clearly had multiple audiences, with manifold contending understandings of the act.

Harriet Jacobs's retrospective account of antebellum American slave life provides one of the best-known descriptions of Jonkonnu, which she calls "Johnkannaus."[55] Her narrative, looking backward at practices that occurred well after Fawcett's pantomime, can give us a sense of some of the act's persistent vernacular elements. Jacobs remembers collective, group-oriented performances, elaborately staged and thoroughly prepared. "For a month previous," she asserts, participants compose songs for the occasion. The performance event itself provided a pretext for barely controlled revelries hinting at threatening reversals of slavery's power structures. Jacobs recalls the license granted on these occasions as "companies, of a hundred each, turn out early in the morning, and are allowed to go round till twelve o'clock, begging for contributions." The actors use the contributions of rum and money for "carousals" authorized but not always supervised by the masters.[56] The North American rituals Jacobs witnessed took place during the winter holidays, tolerated and even encouraged in the break between Christmas and the New Year.[57] Dancers (usually male), costumed in a variety of disguises, performed the act on the street or the grounds of a plantation. In descriptions, spectators frequently use the term Jonkonnu to refer to the leading character as well as the entire performance; other characters ("Jaw Bone," "Warrior," and "Pitchy-Patchy" Jonkonnus) perform variations on the theme.

Led by the primary Jonkonnu figure, a rhythmical procession of slaves interacted, often in mock-antagonism, with slaveholders and other white citizens, singing songs and playing musical instruments to solicit money and other small favors. The songs, especially if the audience does not meet Jonkonnu's demands, turn oppositional and unruly; Jacobs recounts one such verse:

Poor massa, so dey say;
Down in de heel, so dey say;
Got no money, so dey say;
Not one shillin, so dey say;
God A'mighty bress you, so dey say.[58]

The ironic "so dey say," deriding the tightfisted master as "down in de heel," as well as the use to which the slaves put the money (carousing), suggest Jonkonnu's potential as oppositional performance. Indeed, Roger Abrahams characterizes such "verbal play directed at a powerful figure" as "perhaps the keynote of African American expression."[59] Likewise, *Three-Finger'd Jack*'s onstage version of Jonkonnu codes disdain within obeisance; the offstage song recalls the opening chorus of the play ("Good Massa we find / Sing tingering, sing terry,— / When Buckra be kind, / Then Negro heart merry") with relentless irony.[60] Jonkonnu creates a carnivalesque performance space of licensed inversion, where once or twice a year, slaves could follow their own leader, dictating their entertainments and shaming masters into giving them gifts. The contributions would then fund the revelries (again, officially tolerated) of the slaves' holiday.

JONKONNU'S VERNACULAR TRANSATLANTIC HYBRIDITY

Although Jonkonnu probably reenacted gestures originating in African ritual, the archive also reveals Jonkonnu as an always-already hybridized act, a performance of the kind of translocal "ex-centric" cultures that James Clifford discusses.[61] Edward Long's 1774 account, like many others, identified Jonkonnu as a commemoration of John Conny, a "celebrated cabocero at *Tres Puntas*, in *Axim*, on the Guiney coast; who flourished about the year 1720. He bore great authority among the Negroes of that district." As Long reports, the "cabocero," or chief, defended Fort Brandenburgh against the Dutch after the Prussian colonial authorities withdrew.[62] If the searches for Jonkonnu's roots take us to African culture, they also lead back to an Africa that participated actively in the commercialized Anglo-Atlantic. Moreover, Long's account suggests (almost in spite of itself) that Jonkonnu does not simply play a passive, transported role in the Atlantic world. Instead, the performance might have been an actively maintained commemoration, a celebration and perhaps even a reenactment of self-empowerment.

Conny, with his potentially subversive "great authority," exercised significant power in the conflicts of a multiracial, radically heterogeneous Atlantic world. His resistance to colonial authorities precedes by three decades the Surinam Maroons who sent a Dutch colonial governor home in defeat. As a figure of authority and self-determination, he also symbolically forecasts, for example, the Haitian slave revolt secured by Toussaint L'Ouverture in 1804. If Caribbean Jonkonnu acts do indeed memorialize Conny, the performances might retain fibers of the figure's militant agency, commemorating resistance to colonialism and even slavery. The derisive songs in Jacobs's memoir might signal the performance's distant potential, inherited from a figure like Conny, for insubordination. If Jonkonnu emerges from early interracial interactions in colonized Africa, such transactions continue in later versions of the performance. Jonkonnu sometimes featured "Koo-Koo" or "Actor Boy" characters, for example, that self-consciously mimic white English popular theatre in their extravagant costuming.[63] Black Jamaican culture created itself in the appropriation of white English theatricals in other scenes as well. On New Year's Day, 1816, for example, Anglo-Jamaican planter Matthew G. Lewis (perhaps best known for having shaped gothic melodrama back in England) witnessed a procession of black Jamaicans featuring a "glittering tawdry figure, all feathers, and pitchfork, and painted pasteboard." The character, played by a black maidservant, "turned out to be no less a personage than Britannia herself, with a pasteboard shield covered with the arms of Great-Britain, a trident in her hand, and a helmet made of pale-blue silk and silver."[64] Later Jonkonnu scenes participate in the mimicking and borrowing transactions that appear in *Three-Finger'd Jack* and persist well after.

Whether encouraged, forced, or merely tolerated by the English, Caribbean performance featured the interplay of Black Atlantic and Anglo-European theatricality. Jonkonnu characters featuring houses, houseboats, or ships on their heads perform some of the most compelling of these forms. Isaac Mendes Belisario's striking depiction of one such character is one of the few images we have of this type (figure 5.2).[65] Although it pictures Jamaican popular culture after emancipation rather than during slavery, the act displays processes that had long been at work in black Caribbean performance. Belisario's rendering reveals Jonkonnu performance as thoroughly engaged with, even pilfering, English forms. Complex layering reveals cultural imbrication in nearly every detail of the costume. The whiteface mask and wig mimic the island's English residents, reversing Anglo-Atlantic theatre's blackface

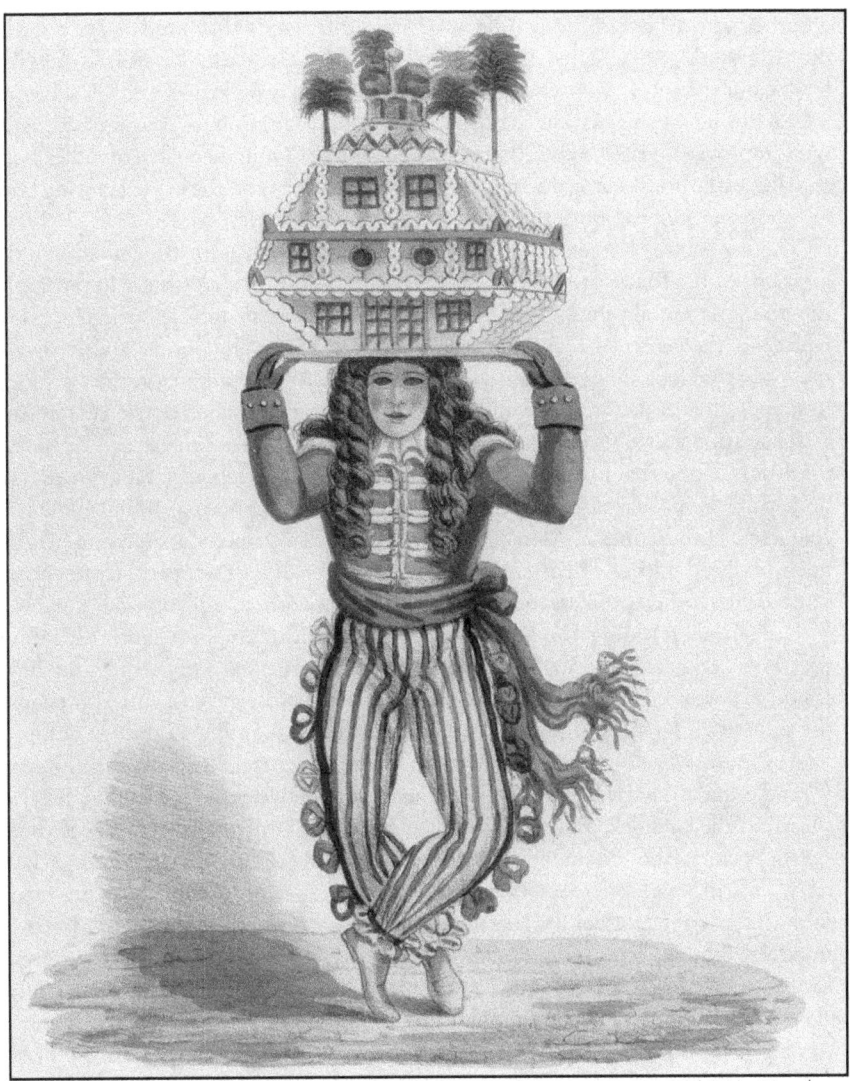

Figure 5.2 Isaac Mendes Belisario, "Jaw-bone, or House John Canoe" (1837); Lithograph in Belisario, Sketches of Character. Yale Center for British Art, Paul Mellon Collection.

with perhaps deliberate irony. Gloves protect soft hands that need not engage in dirtying work—they signal irony in the context of a black Jamaican's everyday labors, but they also evoke the event's history of holiday performance. The clothing appears to be a combination, perhaps scavenged, of an army coat and naval pants with improvised tassels and fringes. The performer simultaneously embodies both the elite authorities that had so recently enforced slavery and the other laboring bodies whose conscripted work sustained English colonial enterprises.

The Jonkonnu character thus actively plays with English forms, embodying the locations of the Black Atlantic's greatest social tensions. Perhaps most importantly, the headgear reveals clues about the performance's social significance. The house represents the center of plantation labor and discipline—Jonkonnu bodily signals the forms and locations of English slavery, even after they had begun the transition to supposedly free labor. If the message were not clear enough, a crown centered on the plantation house visualizes the royal power that authorized and sustained slavery until 1834—and that until 1840 maintained slaves in an "apprenticeship" with only limited freedoms. Two decades earlier, Matthew Lewis noticed a similar Jonkonnu character. He describes a "Merry-Andrew dressed in a striped doublet, and bearing upon his head a kind of pasteboard house-boat filled with puppets, representing, some sailors, others soldiers, others again slaves at work on a plantation."[66] Lewis's account, understanding the act less as an extension of Anglo-Atlantic clowning, describes a character bearing totems of the Great House and ship at the same time. In forms borrowed from English popular entertainments, the costume also peoples the scene with puppets, figures of the laborers (nautical, military, and agricultural) who keep the wheels of the Atlantic colonial and commercial systems turning.

Jonkonnu's headdress literally stages the tension between the nautical mobility of ships and the landlocked immobility of houses. Both forms, of course, represent the sites of Atlantic labor discipline. "Plantation slavery and the vastly expanded merchant navy," as Erin Skye Mackie observes, both "qualify as total institutions and as precursors to the industrial factory of the later eighteenth and nineteenth centuries."[67] Moreover, these disciplining institutions also have a long tradition of inspiring resistance—both pirates and Maroons, Mackie argues, "both constitute sustained and organized refusals of participation in the two central institutions of the colonial machine."[68] The costuming ironically embodies the institutions against which Morano and Three-Finger'd Jack fight. It also shows "stubborn codes of optic blackness," as Lhamon writes, converting the forms of Atlantic labor organization and discipline into play.[69] While the open rebellion of pirates and Maroons achieved a mythic status not granted to black performers, Jonkonnu's sideways critique of colonial oppression persisted. House on head, the Jonkonnu performer carried the colonial labor systems, replicating in play the labor depicted in his headgear. In the act, leisure slides into work—Jonkonnu traditionally occurred at holiday time, when masters magnanimously granted a week of labor-free time to slaves. Enacting play by carrying that weighty headgear collapses the distinction between work and recreation that slaveholders and employers used to mollify workers and justify hard labor. The act, occurring during "free" time, offers an ironically hard-working critique of Atlantic labor discipline.

Insurgent genealogies and subtextual performances may have lent Jonkonnu staying power, but white fascination with the act brought it to Anglo-Atlantic theatres.

Observers such as Matthew Lewis, a playwright who seems especially attuned to the power of drama, sensed Jonkonnu's allure in the first decades of the 1800s. He found the act just as irresistible as Jacobs did; Lewis observed in his diary that "John-Canoe is considered not merely as a person of material consequence, but one whose presence is absolutely indispensable."[70] For slaveholder as much as slave (although certainly for different reasons and in different ways), the act represents a compelling form of interracial interaction. Jonkonnu transforms the powerless, disposable bodies of the black Atlantic low (if only briefly) into unique, spectacular, powerful, and indispensable characters. The shifting reinventions of race and identity fuel the broad appeal of acts such as Jonkonnu, Morano, and Three-Finger'd Jack. Perhaps, Lewis responded to such alluring theatricality when he delayed his travel plans in order to witness one such performance in 1816. We might speculate that theatrical mutability and interracial traffic, rather than the simple spectacle of stark difference, fascinates Lewis when he suggestively writes, "there was no resisting John-Canoe."[71] Lewis's response, although penned a few years after *Three-Finger'd Jack* premiered, gives voice to the act of Anglo-Atlantic spectatorship. Importantly, Lewis's response to Jamaican vernacular acts could actually have responded to *Three-Finger'd Jack*— "John-Canoe" captivated him a decade and a half after Fawcett's play had entered the Atlantic repertoire.

THE PERSISTENCE OF PANTOMIME REBELLION

Three-Finger'd Jack's multiple debts to Jamaican and English acts suggests the vigorous and intimate nature of the traffic among Anglo-Atlantic theatricals and their multiple contexts. The quick traversal of Jack's story through modes and genres also suggests the vigorous routing of such acts through the circumatlantic world. *Three-Finger'd Jack*'s 1800 London popularity only hints at the dispersal that followed. The black bandit's performances continued circulating and redeploying Jack's rebellion in an antebellum culture that remained fascinated and fearful of black revolt. The play, as some historians have claimed, resurged alongside rebellions and threats of uprising.[72] It seems possible; slave revolts, like theatre, are historically tricky. They were neither everyday events, nor unthinkable, subtly permeating antebellum American culture in rumor and reputation as much as in reality. Talk of revolt, like onstage depictions of Three-Finger'd Jack's black banditry, can serve the purposes of abolitionists or slaveholders, generating subtle sympathies as well as suspicions.[73] The actual experiences of slave revolt inevitably affect *Three-Finger'd Jack*'s appearance on stage, but those experiences were themselves the product of representation and transmission, news and rumor, rather than direct experience for most audience members. The relationships between such contentious realities and the equally conflicted performances of *Three-Finger'd Jack* hardly seem predictable.

Although the play almost certainly restaged the recurring white fears of slave revolts, it inevitably turned to other purposes as well. The play, with its optic black blend of heroism, masking, and rebellion, even became a powerful vehicle for black performers. William Brown's New York City "African theatre" at Mercer Street performed the play in 1823, and later, William Henry Murray transformed the pantomime into a melodrama around 1830, according to Charles Rzepka, which the black actor Ira Aldridge performed in England.[74] Significantly, the melodrama added speaking lines, giving Jack an articulate and politically and socially forceful voice. As James O'Rourke has argued, Aldridge helped rewrite the part, performing it in England during the years immediately preceding the abolition of English slavery.[75] Aldridge's act capitalized upon the residual charisma of rogue acts, converting that energy into abolitionist sentiment. As a Black Atlantic actor in a role conventionally played by a white actor in blackface, Aldridge imparts new resonances to Jack's rebellion. The black actor reclaims a forceful performance of black and underclass culture. To Jack's newly invigorated abolitionist message, Aldridge also added his reputation as a rising Shakespearian actor of considerable talent. Aldridge (born in New York in 1807) attended the African Free School, performed in New York City, and may have even rehearsed the role of Jack. While we have no proof of Aldridge's involvement, it appears at a time when he could have been associated with the theatre, as Shane White speculates.[76]

Aldridge, however, played the role later to Atlantic audiences. In Europe, the New York actor became widely known as the "African Roscius," playing both traditionally darker-skinned roles and white characters. As the epithet suggests, his fame rested at least partly on the fact that he had emerged, like Jack himself, from the circumatlantic processes of transporting and disciplining African bodies and cultural forms. Aldridge made a career out of rogue blackness, performing the Caribbean bandit and other similar characters throughout his life. By 1860, Aldridge's "usual programme" prominently featured a range of black characters, including some suggestively rebellious or unruly types: according to his biographers, he starred in "*Othello, The Padlock, The Slave, The Black Doctor, Macbeth, Obi, Bertram, Titus Andronicus,* and *Robinson Crusoe, or The Bold Buccaneer*," a mix of outsider, outcast, and racially darkened characters.[77]

Jonkonnu's vernacular infiltration of English theatre suggests that black Caribbean theatricality contributed to the development of American blackface performance. Before blackface minstrelsy coalesced into racist forms, as W. T. Lhamon, Jr., argues in *Raising Cain*, slippery, insouciant forms of racial masking coded the mutuality and resistance of the low to the Atlantic world's class exploitation. Likewise, Eric Lott, finding more envy and tension in blackface performance, nevertheless recognizes racial mimicry's ability to consolidate underclass identities.[78] *Three-Finger'd Jack*'s Jonkonnu character suggests a covert, tense mix of admiration and disdain, amusement and fear; it theatricalizes race in ways that would shape T. D. Rice's equally slippery and inversive Jim Crow character. Fawcett's play performs an uptake and transmission of popular Caribbean rituals; in its bid for rogue authenticity—the realistic mimesis of outlaw acts that the play both desires and rejects—*Three-Finger'd Jack* cannot help but transmit forward elements of black circumatlantic culture.

Audience demand brought these performances to their publics. The charisma and appeal of Atlantic blackness and of vernacular performances pulls along the circulations of theatre culture. Each of these performances contains kernels of earlier performances, "surrogations" or "restored behaviors," as Joseph Roach has called them, of intercultural and interracial encounters. As Roach recognizes, the forms of remembering also seem "particularly subject to forgetting."[79] In the case of *Three-Finger'd Jack*, the overwhelming spectacle of blackface in revolt supplants the rogue acts that gave form to such scenes. Acts of surrogation rely upon outcast theatrical bodies to help explain, fabricate, and rationalize the realities of the Atlantic world. The interracial underclasses lurk within the performances of blackness that dominate Fawcett's *Three-Finger'd Jack*. As it traveled, Fawcett's pantomime staged the darkened lower classes and ultimately their relationships to other mobile and disdained groups of the Atlantic. Melding blackness and roguishness in the forms of musical pantomime, the play thus continues the fugitive work of cultural transmission along Atlantic routes. Its performances mark way stations for loric elements, imparting new spins to acts as they continue on their way.

"NOT FAR BEHIND THREE FINGER'D JACK"

Voiceless acts of unruly and rebellious blackness cast a long shadow in early American culture. Occasionally, they assume spectacular offstage forms, echoing Jack's popular and persistent stage appearance. The 1818 appearance of a black bandit named Tom in the Virginia countryside shows the ways in which on- and offstage performances intertwined and shaped each other. News of Tom's roguery traveled quickly and widely, appearing in newspapers before New York City's broad audiences and to readers in distant New England. The reports present intriguing scenes of black rogue theatricality:

> This ferocious bandit had long been the terror of the country between this place and the Great Bridge. Of almost Herculean stature and strength, and possessing great intrepidity and cunning, he contrived to elude the paths of those who were prepared to seek and capture him, while he frequently pounced unawares upon the unarmed traveller, and made him the sure and easy victim of brutal violence and plunder. Numberless are the instances recorded of his outrages and robberies, and scarcely a plantation within his range escaped the ravages of this marauder.[80]

The admiring rhetoric of the account and its focused attention on Tom's rogue body suggest that this was no ordinary bandit. Indeed, Tom's notoriety exudes a peculiar charismatic quality, commanding a respect reminiscent of other rogue scenes. The notice takes an imaginative turn, framing Tom's acts in explicitly theatrical terms. "If common report may be credited," the report continues, "Tom was not very far behind his great prototype, three Finger'd Jack, in the number and atrocity of his offences, or in his 'hair breadth escapes.'"[81] Echoing long-popular accounts of the Jamaican bandit, Tom shows epic criminality and excarceral performance. Tom stalks the Virginia countryside, inflated in the popular imagination to larger-than-life proportions.

The incident reveals the cultural persistence of such theatrical models and the shifting purposes such acts can serve when staged outside of their original theatrical contexts. Tom, like his pantomime forebear, ultimately dies at the hands of local authorities. After being hunted down and captured much like Jack was, Tom stood trial and was sentenced to hanging. His execution, surely as scripted as the Massachusetts hanging of Levi Ames decades before, depicts his ability to act within the conventions of rogue performance:

> While awaiting his doom, he evinced but little concern for the future. His feelings, however, were evidently softened by the influence of pious individuals who visited him in his confinement. When asked by the sheriff, a few days before his execution if he was prepared for death, he coolly replied, "Yes, better than I should ever be if I was permitted to live. I have no wish to live, because I know too well how it would be with me: I can't trust my temper. If you don't hang me now, I know you will have to do it some other time; and as I have made up my mind to suffer, I had rather go now than not." At the gallows he was quite cheerful, and called out to the officer in attendance to let him know when he was to be turned off! This intimation was accordingly given him, and he had just time to articulate, "God bless you all," when the fatal cord stopped his breath forever.[82]

Tom's execution performance takes the pantomime's urge to display and discipline black banditry to its logical conclusion. Compelled by the logic of disciplinary display, the drama no longer seems restrained by the propriety of displaying such a scene before playhouse audiences. Accordingly, the Norfolk hanging enacts a scene that Fawcett's pantomime sublimates in the celebrations of docile stage slaves.

Although the newspaper reports do not describe his audience, as a public and widely reported event, Tom's execution could have attracted a motley mix of persons. Some would have been intent on seeing black criminality punished, and others may have sympathized or even identified with the bandit. The more powerful in the audience may even have forced others to observe the execution as a deterrent. The Norfolk execution performance appeared before its audience unfiltered by the forms of entertainment; the act no longer appeared inside the imaginative zone of the playhouse, observers overlooking the pit or meeting fellow box-members for a social evening. Although the spectacle and the crowd may have lent an air of recreation to the proceedings, open-air spectators, unlike playhouse audiences, could no longer interpret the black banditry exclusively in the forms reserved for entertainment and make-believe.

The 1818 Norfolk hanging was drama, but also a deadly serious demonstration of the power of legal authority to define and punish a transgressor. In other ways, however, the event became a dramatically heightened alternative to the plot the authorities probably intended. The informal script of Tom's 1818 execution allowed—or even called on—Tom to speak, and the archival evidence transmits his voice forward—indistinctly and indirectly, but he speaks. The scene suggests a performance of charismatic, even compelling black unruliness, but also a performing black bandit composed and in control of the scene. Tom died game, performing a role that audience members would have recognized, if only subconsciously, as the Atlantic rogue Macheath. Whether or not deliberately or knowingly, Tom plays the part of the pantomime's rogue antihero. He invokes the cultural memory of Jack's circumatlantic outlawry, casually demonstrating the inability of power to terrorize him as he had terrorized the countryside. Tom in a sense outperforms the pantomime, becoming the performing figure at the center of his own execution. Acting in ways that recall both Three-Finger'd Jack and Fawcett's Jonkanoo, Tom demonstrates that punishment does not dampen his spirits.

It is impossible to settle definitively the question of Tom's conscious agency in reprising the role of Three-Finger'd Jack, only in part because of the multiple biases of the account. We can hardly know with certainty whether he actually acted as the account narrates. As the execution performances of the eighteenth century show, the various empowered parties who help produce such scenes can have motives for seemingly undercutting the totalizing grasp of power. Even if Tom did act as the report says, we can hardly know precisely why he would have behaved that way. As Levi Ames and Thomas Mount showed, executions had developed relatively stable conventions in the eighteenth century, and numerous examples suggest that the condemned had, in sense, a choice of roles to play. As Tom's gallows performance suggests, the conversion of a common thief into a charismatic, daring stage bandit might open up a space for some sort of resistant self-expression—even if it occurred only in the moment before Tom was to be "swung off."

The report also fits unusual and threatening events into a Fawcett's narrative and performative model that comprehends, criminalizes, and ultimately contains Tom's actions. Terrorizing the local countryside becomes a version of Jack's banditry, which (not too long after Prosser's rebellion, and only four years before Denmark Vesey's rebellion) probably seemed vastly preferable to the terrifying possibility of widespread slave revolt in the Virginia countryside. Overlaying the pantomime's template onto real life makes Tom's actions comprehensible as banditry (and hence governed by traditional conventions), averting the radical chaos of slave revolt. There are certainly other possible reasons for Tom's apparent performance, and reasons that audiences might seem invested in perceiving and retelling such a performance. Performing revolt as theatrical banditry could have multiple payoffs for both actors and audiences, and various parties could manipulate scripts of pantomimed black rebellion, even if the rogue still must die in the end.

Whatever Tom's internal motivations, he chooses to act—and/or his viewers choose to see him—according to time-honored scripts of condemned cheerfulness and the more recent acts of black Caribbean banditry. His terse chroniclers, like Matthew Lewis, seem to have found the performance in some way compelling, worth valuable column-inches of faraway newspapers. The performance of Tom's hanging remains invaluable precisely for its exquisite inscrutability, proof of the complex cultural work of theatre in early American culture. The events reveal the persistent centrality of the theatrical imagination, and the complex functioning of such models within systems of unruliness and discipline.

Tom, alongside his "prototype" *Three-Finger'd Jack*, helped shape the ways in which American audiences viewed spectacular black rebelliousness. In *Three-Finger'd Jack*, pantomime rebellion compelled audiences with its ability to simultaneously evoke and avert the revolutionary extended Caribbean. *Three-Finger'd Jack*, the stage version of Tom's Norfolk hanging, reveals the stage fascinated but flinching before the threat of rebellious Caribbean blackness. Fawcett's pantomime contains and redirects the black violence that it cannot ignore, transforming the revolutionary Caribbean into local acts of black outlawry. Although on their face constructed through the white pilfering of black cultural forms, such acts traffic in the complex counter- and co-performances that continually renovate the blackened or blackface theatrical rogue as a cultural resource. The intertwining circumatlantic genealogies of performance and spectatorship stock the play with ambivalent energies that invigorate vernacular ritual, onstage revivals, and even the bandit career of an escaped Virginia slave.

6. Class, Patronage, and Urban Scenes in *Tom and Jerry*

In the face of the threatening popularity of figures such as Three-Finger'd Jack, the stage also imagines mechanisms of controlling and producing the urban and racialized low. Such acts always seem to carry competing charges. Theatre subjects its undersiders under surveillance, but that proximity places outcasts in America's own neighborhoods. The stage imagines ways of controlling with the low, but that often requires intimate interactions with disdained types. Sometimes it even means imitating the underclasses. Ultimately, the stage seems ready to resurrect and surrender to their charismatic appeal. In the 1820s, American audiences encountered such dynamics by vicariously walking London streets and viewing the contrasting scenes of high and low life in W. T. Moncrieff's *Tom and Jerry, or Life in London*. Moncrieff's play scripts the theatrical underclasses through processes of mobility, observation, and participation.

Tom and Jerry follows the adventures of three sporting rakes—"Corinthian Tom," his country cousin Jerry Hawthorne, and Bob Logic, their sidekick with a talent for navigating the fast life. Jerry, who leaves his country estate to tour the city, progresses from novice to urban initiate in various "rambles" and "sprees"—presumably the same trajectory that spectators follow by watching the play. Armed with their superior understanding, the sophisticated urbanites drag Jerry in and out of scrapes. All the while, unbeknownst to the bachelors, three of their young female friends (Sue, Kate, and Jane) follow in disguise, monitoring the rakes' misadventures. Urban voyeurism, it seems, is not exclusively a masculine practice—but the play clearly uses feminine voyeurism as a foil, and the female characters supplement, contrast with, and ultimately foreground the play's rituals of masculine belonging. Although it does not feature the same sort of spectacular rogues as *The Beggar's Opera* or *Three-Finger'd Jack*, Moncrieff's play brings the high perilously and fashionably close to London's cagey, criminal, and theatrical underclasses.

New York City's Park Theatre, managed by Stephen A. Price, first imported *Tom and Jerry* to America on March 3, 1823; it subsequently became a staple in American playhouses.[1] By the time American theatres staged *Tom and Jerry*, popular depictions of the duo's urban rambles had virtually inundated English audiences. Pierce Egan's popular 1820 novel *Life in London*, with illustrations by George and Robert Cruikshank, preceded numerous London stage versions of the ramblers.[2] W. T. Moncrieff's *Tom and Jerry, or, Life in London*, which premiered at London's Adelphi Theatre on November 26, 1821, became the most popular of those plays and the version that American playhouses restaged.[3] The play popularized the already-existing elite subculture of slumming, in which Regency "bucks"—young, single men of leisure—moved fashionably amongst the "lower" orders of London society.[4] These images and scenes produced tangible representations of underclass culture while defining London as itself "a hedonistic, panoramic stage of pleasure."[5] The Adelphi, like the Haymarket some forty years before, hosted a varied selection of illegitimate theatricals, pantomimic spectacle, and conservative high drama. Its popularity and its position in the center of London's thriving popular theatre business brought Moncrieff's characters and scenes to an enormous number of people—*Tom and Jerry* appeared at the Adelphi Theatre ninety-four times in its first season.[6]

The play appeared seventy-five more times in the following season, dominating the stage along with its sequels, pantomimic adaptations, and imitations for years. American theatres never matched such numbers, but they did capitalize in their own ways on the play's legendary London popularity. The play represents a transatlantic performance event, constituting itself in the routes of transnational cultural traffic. Its history, as this chapter shows, refuses to abide within a linear sequence in which London premiere precedes American imitation. London stagings of *Tom and Jerry* are thus not strictly originary, but instead local elaborations of a transatlantic theatre culture in which acts, actors, and audiences circulated continually. London culture did produce thicker and more detailed archives, which, while drawing attention to the metropole, can also help fill archival gaps in American performance practices.

American playhouses have left less evidence of Tom and Jerry's rambles than did English theatres, but we can recover a suggestive and relatively reliable picture of *Tom and Jerry* in America by reading English acting editions against the less detailed American publications; in addition, American playbills, broadsides, and newspaper announcements sometimes note retentions and revisions. An 1824 printing of *Life in London*, published in Philadelphia and claiming to represent the play "as performed at the Boston, New York, and Philadelphia theatres," gives a sense of how the "burletta of fun, frolic, and flash" appeared in the New World. Behind the main actors, actors popular in comic and clowning roles appear in *Tom and Jerry*'s underclass parts, and behind them, a supporting lineup of anonymous "Jockies, Beggars, Sportsmen, Watchmen, Millers, &c." and "Boxers, Debtors, &c. &c." flesh out the scenes of urban life. Such characters form the heart of the performance, actively shaping the play's scenes and offering audiences their glimpses of low life. They converge in a "Grand Carnival," a finale with everyone masked and performing "Tumbling, rope-vaulting, and all the eccentricities of a masquerade."[7]

Its structure relies upon scenic progression, rather than plot, and ironic contrasts of high and low life. As George Odell concluded, the "gist of the matter is found in rapidly shifting scenery, great diversity of city types of character, and a large amount of consequent spectacle, dance, and song."[8] More specifically, those quickly changing scenes presented spectacles of lowness and the linked theatrical processes that produced and shaped them—voyeurism, patronage, and ultimately reenactment. Charles Durang later barraged the play with insults, calling it a "chaos of dramatic inconsistencies and improbable incidents clothed with vulgarity."[9] His disdain for the popular shows the class commitments of aesthetic prejudices; it also reveals the moralizing anxiety that the play engendered on both sides of the Atlantic. Durang's unease was not entirely groundless; *Tom and Jerry* spawned entire subcultures of underclass imitation and reshaped local theatre institutions. In American cities, as much as in London, the play consolidated cohorts who admired Tom and Jerry's chameleonic mobility and access to hidden scenes. The play shaped and articulated the affiliations of "genteel rowdies," giving the slumming young culture a cultural kernel around which to base their own offstage social performances. A form of the emergent Atlantic youth culture that would later form the core audience for blackface minstrelsy becomes visible in the wake of *Tom and Jerry*, claiming and celebrating access to low acts as its own mode of social performance.[10]

OBSERVING, PATRONIZING, AND REENACTING LOW PERFORMANCE

Tom and Jerry scripts several scenes that notably link mobility, observation, and the production of underclass performance. The first of these occurs in a scene conventionally identified as a "Beggar's Hall in St. Giles."[11] The scene features a banquet with a number of notorious low characters, including Billy Waters, a black London performing beggar of transatlantic fame. Like *The Beggar's Opera*, Moncrieff's *Tom and Jerry* imagines performance as a particularly underclass behavior; such performance has come to be perhaps more duplicitous than charismatic, however. Audiences witnessed, for example, the "cadgers" of St. Giles, who proudly perform, "gammoning a maim" (faking a disability), for example, for charity.[12] Billy Waters (in what seems a forerunner of the twentieth-century stereotype of the welfare abuser) ostentatiously displays his poverty while demanding luxuries such as sausages with his turkey. The play consistently equates lower-class status with theatrical tendencies, and both with the urge to deceive or cheat. The acts show deep suspicions of characters with unconventional modes of profiting from their public self-representation. In a sense, such scenes reenact the conventional early modern suspicions about theatricality that had labeled actors as beggars and vagabonds. More complexly, the play imagines theatricality as a strategy of the underclass, the same skill set that had notably enabled the escapes and struggles of Macheath, Morano, or Three-Finger'd Jack. Disempowered characters use performance for survival, evasion, and even resistance.

Tom and Jerry imagines deceptive performance as "authentic" underclass behavior, producing a robust conundrum of real shamming. Ironically, the play claimed authenticity in part by pilfering performances from the London streets. In the 1820s, English performances of *Life in London* featured the character Billy Waters, a black beggar who performed as a busker outside London's Adelphi Theatre at the end of the eighteenth century. Waters's factual history remains obscure—mythmaking and theatrical publicity began to define him during his own lifetime. A newspaper obituary appearing in the early 1820s, for example, reputes him to have been a sailor who fought in the American Revolution. He then supposedly made his way to the fashionable Strand in the 1780s, entertaining London audiences on their way to and from the theatre. After that, his only claim to fame seems to be through the theatres outside which he played his fiddle, and that would eventually reenact his character.[13]

Precisely because the Atlantic culture industries' mythologies overlay his flesh and blood presence so thoroughly, his circumatlantic performances transition rapidly into popular memories and reenactments of those memories. After Waters died in the mid-1820s, Moncrieff's play continued to stage his act, making a sort of virtual Billy Waters intermittently visible in iconic, theatrical traces. Waters and *Tom and Jerry* itself stage performances of memory, the surrogations of earlier circumatlantic acts. The acts of memory circle provocatively around a retroactively constructed (and absent) center of original and authentic performance. George Daniel's preface to the Cumberland edition of *Tom and Jerry*, for example, delights in claiming that audiences thought an actor playing Billy Waters was the real thing.[14] Waters might have played himself in some of the early stage productions; similar characters would in other contexts. However, the real significance of the comment lies in its bid for authenticity—its acts supposedly mimic street theatre so well that even sophisticated London audiences, it boasts, could not tell the difference. The claim resembles the assertions of authenticity that bolstered later urban underclass acts and eventually become part of the mythology of American blackface minstrelsy. Ultimately, *Tom and Jerry* made such authenticity its stock-in-trade. Low performance styles—the "authentic mimicry" of the stage—offered compelling acts to audiences looking for access to hidden and charismatic sectors of society.

The character of Billy Waters displays some durability in the United States, reappearing as an offhand verbal reference in Benjamin Baker's topical urban play *A Glance at New York in 1848*.[15] Waters remained a familiar theatrical type two decades later, and *Tom and Jerry*'s American versions self-consciously reenact the underclass theatricality once patronized by audiences going to and from the London stage. Audiences of the American productions seem less likely to have confused Waters for a real street performer, given his London origins. Instead, the character became an iconic character, a stage embodiment of transplanted and circulating Atlantic underclasses. If Billy Waters reveals anything authentic about urban underclass theatricality, his presence shows the self-conscious exchanges among patrons and performers, the prosperous and beggars, occurring both inside and outside the theatre. After his transatlantic movement, Billy Waters no longer embodies (if he ever did) mimetic realism or staged authenticity. Instead, his presence in New York indexes the tendency to see various kinds of acts as already theatrical, as scripted, shaped, and ready to make the jump onto the formal stage. It also bespeaks the larger interconnections of Atlantic theatre and the cachet that low performance exercises within that system.

The play offers another striking scene of underclass performance set in the low dive "All-Max," an underclass version of London's exclusive and fashionable social club "Almack's." The scene, a popular cavorting blackface musical number by African Sal and Dusty Bob, rehearses the blackface conventions that would later vault to Atlantic popularity from these same stages. African Sal and Dusty Bob appear prominently in most announcements of American performances of *Tom and Jerry*. The Philadelphia acting edition indicates a "minuet and comic dance" by the blackface characters Dusty Bob and African Sal, and an 1823 Boston newspaper likewise lists a "Pas de Deux by African Sal and Dusty Bob" as the sixteenth scene in the play's sequence of spectacles.[16] In American theatres, male comic actors played African Sal and Dusty Bob; the characters represent forerunners of minstrelsy's cross-dressed and dancing "wench" characters. English editions of *Tom and*

Jerry provide the most detailed description of this scene. Unfortunately, little evidence remains of specifically American stagings of this scene, but at least the early American productions probably worked from detailed English texts. The lack of description might simply indicate audience habituation to such acts. Moreover, the gestural nature of the scene does not translate well into verbal descriptions.

The scene's core elements are worth examining, even if later American performances did improvise upon the English script. It imagines blackface dance as constructed and patronized, conjured into being by upper-class observers, who call for a "double shuffle" even as Tom labels the occupants of All-Max "unsophisticated sons and daughters of nature." The English editions of the play reveal a complexly meaningful dance:

> Sal, by way of a variation, and in the fullness of her spirits, keeps twirling about; at the same time going round the Stage—Bob runs after her, with his hat in his hand, crying, "Sarah! vy, Sarah, 'ant you well!" &c.—the black Child seeing this, and thinking something is the matter with its mother, also squalls violently: stretching its arms towards her; at length, Sal, becoming tired of her vagaries, sets to Bob, who exclaims, "Oh! it's all right!" and the dance concludes.[17]

Black performance appears spontaneous and improvisatory, a product of emotional excess and always liable to turn unruly. Sal's random movements gyrate out of control, and the scene shows threatening as well as entertaining black performance. The unsophisticated daughter of nature becomes unnatural in her excess, frightening her own child. Dusty Bob's concerned cry and the child's violent "squalls" alert audiences to the threat of black theatricality. In addition, some stagings of the play augmented the dance's "unnatural" quality by casting a male actor in the part of the maternal Sal. Gender crossing compounded racial crossing, as it would later in blackface minstrelsy. Once again, underclass performance occupies a conflicted location in Atlantic theatre culture. The dance itself steps out the tension between "natural" and socially constructed performances. The instinctive overflow of emotion contrasts with the dance's frightening inexplicability; cross-dress complicates maternal qualities, and burnt cork both produces and belies racial performance as authentic.

Sal and Bob's scene stages its own construction, enacting a sort of metatheatrical joke. Logic, addressing the blackface musician as "Snow-Ball," urges him to play a danceable tune. As he does so, he gives the already-inebriated musician "gin and snuff, and begrimes his face."[18] Bob re-blacks the fiddler, a white actor already in blackface. Logic recreates blackface on stage, self-consciously pointing to the conventions of blackface as a constructed performance rather than a natural act. Alcohol, snuff, and the urgings of the upper-class observers conjure the performance into existence. Belying Tom's delight in its natural naïveté, the scene represents a complexly staged theatricality, hidden behind smokescreens of authenticity. Dusty Bob and Sal perform at the urging of the white, upper-class gatecrashers who dominate and manipulate the scene until it produces entertainment. Far from revealing low performance as the unsophisticated and natural event, Moncrieff's script actually acts out the performance's construction. The scene highlights the tensions in patronage and patronization—the patron-observers seem openly both disdainful and fascinated. The scene exploits this awareness, celebrating, rather than covering up, the conditions of its own production. In the process, *Tom and Jerry* acts out the conflicted transactions of mobility, observation, patronage, and production of performance in broadly comedic strokes.

Such acts pervade popular representations of urban underclasses. In these scenes, theatricality jumps the boundaries of class and ceases to be the production simply of plebeian entertainers and upper-class patrons. Read against the grain and perhaps despite itself, *Tom and Jerry* deflates the imagined artlessness of underclass theatricality. The more privileged observers continually appropriate "low" acts, creating them as natural, innocent, and worthy of celebration—but at the same time, theatrical, staged, artificial, and somewhat disturbing. Theatrical productions foreground these tensions, rather than erase them, and that contributed to their popularity.

TOM AND JERRY'S ANGLO-ATLANTIC GENEALOGIES

Tom and Jerry's 1823 arrival in New York City makes visible the play's transatlantic processes of becoming. Comedian Joe Cowell, who played the part of Jerry at the Park Theatre in 1823, had been performing in London when Moncrieff was writing the play. According to Cowell's memoirs, he arrived in New York City in 1821, soon after Moncrieff had written the character of Jerry Hawthorne "exclusively" for Cowell's style.[19] The immigrant comedian perhaps even brought a copy of the London script to America, since he later writes that the Park Theatre management had a copy long before they performed the play. Manager and actor Edmund Simpson, apparently "under the apprehension that American audiences would never tolerate the vulgar slang nonsense," had held the script for some time before producing it at Cowell's "earnest solicitation."[20] Ironically, the London leading actor who shaped an English-authored play depicting London's hidden scenes first enacts the role in New York City. It would clearly be a mistake to dismiss the play's American runs as mere imitations of the London productions. Authors and scripts rapidly travel transatlantic routes, and arguably, the seemingly imitative Park Theatre production featuring Cowell as Jerry represents one aspect of the playwright's intentions better than do the London productions. Atlantic theatre's mobile actors, scripts, and scenes vex attempts to locate originality in production and privilege dramatic "ur-texts."

Despite the Park's initial managerial apathy and lackluster promotion, competing American performances of *Tom and Jerry* soon appeared. The play debuted in Philadelphia on April 25, 1823, where prominent actors, including Henry Wallack, Joseph Jefferson, and Francis Courtney Wemyss (another leading actor recently arrived from the British circuits), appeared in the lead roles. When the Chesnut Street Theatre's managers cut short its performances of *Tom and Jerry* and took their troupe to Baltimore and Washington, as Wemyss relates, Cowell's "circus corps" profited from the demand by showing the play on Walnut Street for the remainder of the 1823 season.[21] In Boston, too, *Tom and Jerry* saw production in 1823 and 1824, in both the theatre and the circus, where it followed displays of horsemanship and rope-walking skills.[22] As a performance of London scenes, the play remained popular for the next few years; Philadelphians, for example, saw the play in 1824 with an abbreviated sequel, *The Death of Life in London*, as an afterpiece. In 1826, the Boston theatres again presented *Tom and Jerry*, proving it could last at least a few seasons.[23]

In American theatres, *Tom and Jerry* probably seemed at first an act of international tourism—an exotic picture of "Life in London." Cowell himself, traveling from London to American theatrical province, probably seemed the seasoned representative of London urban ways to New York audiences. The play probably also benefited from the recent American tour of Charles Mathews, whose comic routines capitalized on travel accounts and unusual characters and scenes. Such appeal, however, seems to have had a short shelf life, however. In Philadelphia, for example, *Tom and Jerry* earned well above the theatre's nightly average in its first run, but returns quickly diminished in subsequent seasons.[24] That lackluster success, especially compared to the multiple imitators and persistent popularity of *Tom and Jerry* in London, suggests a fundamental difference in the way American audiences encountered the London scenes.

American theatre responded to the play's flagging novelty value, however, by tapping the urban lore of American cities. Glimpses of the distant metropole's underworld gave way to representations of a newly localized urban culture, and *Tom and Jerry* helped constitute a specifically American urban public sphere built upon complex rituals of underclass performance and elite patronage. Later derivatives of the play clearly show the fascination with local scenes. Odell describes, for example, a *Life in New York, or, Firemen on Duty* on January 24, 1827. Jonas B. Phillips, too, penned a version entitled *Life in New York*, which appeared in that city in 1834, and later incorporated blackface scenes written for T. D. Rice. *Life in Philadelphia* and *Tom and Jerry in America* appeared in Philadelphia in the 1830s and 1840s. Likewise, Boston's Adelphi Theatre adapted the play with local themes in 1847, subtitling it *Life in Boston*. Other derivative versions, such as John Brougham's *Life in New York; or Tom and Jerry on a Visit*, continued to respond to popular demand for such scenes.[25] Benjamin Baker's immensely popular *A Glance at New York in 1848*, with its "Bowery B'hoy" working-class hero Mose, also inherits a fascination with urban underclass acts.[26]

Tom and Jerry traffics in privileged access to trendy, localized scenes of urban low life. Although it advertises the contrasts of class ("ups and downs"), the play primarily trades in energetic, rowdy, and barely controllable acts of underclass characters in low settings. Moncrieff's play sets its popular centerpieces in dives, places where beggars drank, prostitutes plied their trade, and mixed-race couples danced together indiscriminately. Clearly, the play delights in watching its protagonists crash seedy coffeehouses, gamble illegally, and cavort with the characters inhabiting such places. Audiences seem to have prized this knowledge of how the "other half" lives for the insider status and privilege it confers. As a preface to the popular Cumberland edition of Moncrieff's script observes,

> we ought to express our grateful acknowledgments to Mr. Moncrieff for having introduced us to characters and scenes, of which we might never have known the existence but for his helping hand. Taking into account the vast popularity of Tom and Jerry, who shall now say, "One half of the world don't know how t'other lives?"[27]

The play offers audiences the same kind of unflappable urban competence that Corinthian Tom and Bob Logic can boast, promising initiation into urban mysteries along with the country squire Jerry. Fluency in underclass ways is clearly difficult to acquire and hence prized. At the same time, the play's scenes have the quality of an open secret. Certainly, after a few weeks of successful London performances, the play had become a known quantity, recycling previously viewed scenes. *Tom and Jerry* seems thus not precisely staked on depicting low secrets. Instead, it repetitively rehearses acts of surveillance and social differentiation in guided tours of familiar places. Couching those scenes as entertainment, *Tom and Jerry* produces knowledge of the society's undersides for consumption by others whose very watching defines them as "not low."

Observers understood *Tom and Jerry* as the resurrection of earlier scenes of the unruly underclasses. English commentators such as the minor aristocrat and detailed diarist Harriet Arbuthnot, for example, describe the play as "a sort of very low *Beggar's Opera*."[28] An introduction to T. Richardson's 1828 acting edition of the play similarly connects the play to earlier depictions of London's rogues:

> *Tom and Jerry* must undoubtedly be regarded as the *Beggar's Opera* of the present century; its scenes certainly do not possess any of the brilliant wit and pungent satire which sparkle so plentifully throughout the pages of Gay; but, on the other hand, they are more generally true to nature, and have none of the disgusting depravity and undisguised profligacy, that so greatly alloy the gratification we receive in the company of Macheath and his associates.[29]

The substitution of *Tom and Jerry* for the still popular but aging *Beggar's Opera* is telling, revealing new values of representation and new class relationships. The function of the low seems to have changed—in the 1820s, they simply represent themselves in "true to nature" acts, instead of offering comic inversions of class division with "brilliant wit and pungent satire." If the *Opera*'s older style of criminal underclass stands on scaffold stages to perform their escapes with "undisguised profligacy," the new rogues remain safely in their lairs as privileged flâneurs guided audiences into London's underworld. Under the supervision of elite characters, *Tom and Jerry* presents, at least to some reviewers, a more controlled version of the Atlantic low.

The commentators rightly notice the persistence of *The Beggar's Opera* in transatlantic performances of *Tom and Jerry*, which subtly recycled songs and memorable phrases from the century-old *Opera*. Proclaiming in song that "all men are beggars 'tis very plain to see, / Only some they are of low and some of high degree," Moncrieff's characters openly evoke Macheath's inversive social critiques. The play's beggars, for example, allude to famous lyrics from *The Beggar's Opera*: "With our doxies round great as a Turk, / We taste all life can give, / Why none but a slave would work, / When he at his ease might live."[30] The phrases evoke Macheath's memorable final scene, in which the song "Thus I stand like the Turk, with his Doxies around" punctuated Macheath's escape from hanging. In Gay's ballad opera, the song celebrates Macheath's excarceration, but also marks the irony of his imprisonment in multiple marriages. In Moncrieff's play, the song allows sporting rakes to celebrate without irony their imagined freedom from women. The play's elites also appropriate vernacular musical numbers. In the play's second scene, Tom, Jerry, Sue, and Jane celebrate their privileged mobility with "Over the Hills and Far Away," a ballad that had helped popularize Macheath's rogue performances.[31]

Despite one English acting edition's claim that "*Tom and Jerry* and *The Beggar's Opera* are Dramas of nearly the same *genre*," they are decidedly not.[32] Elite voyeurs replace singing thieves and rakes pilfer the charisma of Macheath and his gang. Moncrieff's gentlemen claim the right to act like the charismatic low; celebrating their access to underclass antics, they invert the premise and the social critique of *The Beggar's Opera*.[33] Formally, Moncrieff's panoramic urban ramble (with songs and dances inserted for variety and spectacle) represents a clear departure from the compound ironies of Gay's tightly structured ballad opera. In contrast to Gay's ballad opera, which rendered underclass performance in song and dance, Moncrieff's play and its American offshoots imagined the low as spectacle. Adelphi playbills, for example, advertised *Tom and Jerry* as an "entirely new Classic, Comic, Operatic, Didactic, Moralistic, Aristophanic, Localic, Analytic, Terpsichoric, Panoramic, Camera Obscura-ic Extravaganza."[34] The description exposes the play's satirically signaled debt to forms of visual spectacle such as the panoramic display, which Irish showman Robert Barker popularized in Edinburgh and London at the end of the eighteenth century.[35] *Tom and Jerry* structures its performance as a series of scenes that evoke the simulated movement of panoramic displays; its playbills even mimic this structure, visually depicting sequences of the play's scenes.

Most strikingly, the play reorients the relationship between underclass performance and its audiences. Whereas spectators of *The Beggar's Opera* had gone to the theatre to enjoy a stylized and satirical inversion of London underworld culture and elite politics, *Tom and Jerry* offered something rather different: the foregrounded spectacle of elite slumming and carefully chaperoned access to the underworld. Where a Beggar and a Player mediate access to Gay's London underworld, Moncrieff imagines the rakish Logic as a "complete walking map of the metropolis," a guide to the slum areas of St. Giles. Elite protagonists, knowingly performing the flash and cant languages of the criminal underworld, provide a sort of magically privileged access.[36] As the panoramic form suggests, self-conscious mobility and observation lies at the heart of *Tom and Jerry*'s stagings of the urban underclasses. The ramblers' observations, as Dana Arnold has shown, enact a version of the emerging urban surveillance and discipline popularized by Jeremy Bentham's proposed panopticon, the all-seeing disciplinary mechanism.[37] The scaffold display so central to Gay's *Beggar's Opera* has receded in Moncrieff's play, and elite surveillance and discipline replace scenes of entertaining punishment. The move reflects broad trends in English culture, which had begun to vacate the stage of crime and punishment in London, moving public hangings from Tyburn in 1783.[38] It also suggests that a broad shift had occurred in the ways in which audiences watched the low on stage. The stage still operates as a symbolic scaffold, a site where condemned criminal actors perform defining final acts of resistance or submission to discipline. Even so, the intensity and focus (not to mention the stakes) of such performance has changed. Instead of witnessing onstage hangings and prison celebrations, audiences viewed comic songs and dances in smoky back rooms. Social control and entertainment intertwine, producing each other as familiar systems, and the acts trace the exercises of power in various cross-class interactions.

Those comparisons between *The Beggar's Opera* and *Tom and Jerry*, at once perceptive and deeply misguided, are not simply casual synaptic associations of vaguely remembered scenes. They mark the complex workings of theatrical transmissions and cultural survivals. Thematically related forms spawn successors, and *Tom and Jerry* indeed becomes the new century's *Beggar's Opera*. In the transmission from Macheath's gang to elite flâneurs, audiences witness again the theatricality of the underclasses even as the upper-class voyeurs push their way to the front of the scene. Old songs remain popular, evoking residual structures of feeling. Macheath's inversive critique of social hierarchies persists as a compelling performance, although the privilege of its new performers sharply undercuts its satire. Performance inherits, but it also restructures its theatrical pasts in its own image; posing Gay's ballad opera as a forerunner of *Tom and Jerry* reimagines the earlier play as an entertaining display of low characters for elite voyeurs. As the *Beggar's Opera* for the new century, Moncrieff's play restructures Gay's ironic and inversive cohort, transforming relationships among actors, audiences, and underworld scenes.

TOM AND JERRY AND AMERICAN URBAN SCENES

Tom and Jerry capitalized on urban demographic changes, as well as emerging offstage performances of urban class relationships. The first two decades of the nineteenth century saw drastic demographic shifts in American cities. Along with denser populations and shifting neighborhoods came changing perceptions of the various classes that inhabited those cities. The Five Points, for example, although it would become the focal point of reputed crime and squalor in mid-century New York City, had only recently materialized as an underclass space in the 1820s. As Tyler Anbinder observes, "overlapping waves" of diverse immigrant populations contributed to the formation of neighborhoods as increasingly ethnically and class-defined areas.[39] In addition, events such as the financial panic of 1819 and a yellow fever epidemic in 1822 helped shape the perceptions, if not the realities, of class-based identities and relationships in *Tom and Jerry*'s first decade.[40] As Paul Erickson has argued, antebellum urban growth created a greater "qualitative alteration in urban experience" than Americans would see even at the end of the nineteenth century.[41]

New York City at times seemed a vast urban stage that ritually revealed scenes of poverty. In the antebellum period, for example, the beginning of every May saw tenants move in droves as leases expired in a traditional moving ritual. Although not all movers belonged to the lower sorts, many must have moved out of necessity or in search of lower rents. The process, Elizabeth Blackmar writes, "dramatized the power of the price to determine the social organization of space."[42] The ritual, widely commented upon in newspapers, brought the material reality of less privileged life into the open air, piled high on carts and sidewalks. Low scenes defined the city streets. In what became to his own yearly ritual of underclass observation, Harper's editor George William Curtis commented upon the "grand unveiling of a thousand household economies."[43] Playwright Royall Tyler had noticed the dramatic potential of the scene, staging it in a 1787 play entitled *May Day in Town, or New York in an Uproar*. Although the script has been lost, lyrics survive, and they suggest that the play depicted urban African Americans and cross-class relationships in what seems a distant forerunner to Moncrieff's urban voyeurism.[44] One of *Tom and Jerry*'s basic premises, the scenic revelation of the underground low, seems built into the city's ritual public life well before the 1820s.

The urban underclasses, of course, did not appear only once a year, and fascination with such scenes appears throughout the Anglo-Atlantic world well before the nineteenth century. Vernacular texts, broadside ballads, newspaper accounts, visual arts (like William Hogarth's street scenes), and plays (such as *The Beggar's Opera*) all trafficked in the entertainment value of urban underclass scenes. In the eighteenth century, London supplied what Americans could not or would not see in their own cities. By the beginning of the nineteenth century, however, Americans seem to have grown increasingly conscious and invested in the entertainment value of their own urban low scenes. We might trace American popular culture's fascination with stagey urban underclass characters in the persistently popular series of "city cries" texts. City cries, an older genre, regained popularity in London and then became a transatlantic genre at the end of the eighteenth century. English editions of *The Cries of London*, for example, featuring elaborate woodcuts appeared in American bookseller's lists and newspaper advertisements in the late 1770s. Such texts produced increasingly elaborate scenes for audiences distant from the London sites of the acts in which they traded. American business lagged, but by the first decade of the 1800s, American publishers were printing the *Cries of London*, and by 1810, *The Cries of New-York* and *Cries of Philadelphia* appeared in bookstores and advertisements in those cities.[45] American audiences, while still fascinated with London's urban environment, had begun to represent their own cities in print as sites of low performance and outsider spectatorship. The texts, although only occasionally explicitly moralizing, nevertheless perform an imaginative discipline upon the urban poor as they stage the underclasses as textual entertainment. Many of these books aimed at juvenile audiences; urban scenes infiltrated everyday lives and subtly trained the class sensibilities of early American audiences. By the early nineteenth century, depictions of the city's laborers, beggars, and street vendors had become entertainment for young Americans.

In addition, an increasing number of published guides to the city appeared in the early 1800s. Samuel Mitchill's 1807 *The Picture of New-York, or, The Traveller's Guide, Through the Commercial Metropolis of the United States*, for example, advertised itself as a product of a "gentleman residing in this city"; well before *Tom and Jerry*, the privileged had already begun to mediate insider knowledge of urban areas. Knowledge of the city conferred privileged status, or perhaps such expertise required the theatrical pose of privilege in order to avoid contamination by its own urban scenes. Allowing that New York City, America's center of commerce and population growth, is "not as well known to its own inhabitants as it deserves to be," Mitchill's text stages a homegrown version of the London rambles.[46] Compared to later scenes, the *Picture of New-York* paid scant attention to the city's underclasses. Mitchill's laboring types or urban poor appear only in brief discussions of almshouses and prisons, institutions designed to alleviate misfortune and control misbehavior. Ultimately, those sites discipline and even erase the less fortunate from the imagined city. Only after fascination with the entertainment value of urban scenes escalated in the 1820s would depictions such as Anne Newport Royall's 1826 *Sketches of History, Life, and Manners, in the United States* note the "gay cook and merry chamber-maid, with some scores of honest tars, hucksters, rude boys, and chimney sweeps" that "may give some idea of this life-inspiring city."[47] Although with none of the edginess of *Tom and Jerry*'s voyeurism, Royall's sanguine account participates in the emergent practices of voyeuristic observation and description. *Tom and Jerry* thus appeared in the midst of changing material conditions and shifts in encountering and perceiving class division in urban sites.[48]

Early American culture parlayed its focus on observation into participatory forms, mimicking and reenacting the low. An "alternative public sphere" developed "for white-collar men who rejected the values, tastes, and lifestyles of middle-class moralists," as James W. Cook observes.[49] Often newcomers to the city, these young men self-consciously inserted themselves into urban underworlds.[50] A "flash press" even developed in the 1820s—not coincidentally on the heels of *Tom and Jerry*—and peaked in the 1840s, providing a reading public with additional forms of access to the "inverted world of easy male sexuality" and crime that "co-existed (for many of them) with simultaneous participation in respectable society."[51] Such behavior made conventional the habits of watching and mimicking underclasses—rather than inhabiting the position in any real way, the pose involved a theatrical rejection of developing middle-class mores from an ultimate position of privilege. Charles Dickens entered this world and replayed *Tom and Jerry*'s flâneurie when he visited a Five Points dance hall and chronicled the event in his 1842 *American Notes*. Later texts, such as E. Z. C. Judson's 1848 *The Mysteries and Miseries of New York* and George Foster's 1850 *New York by Gas-Light* shaded the distinctions between urban observation and an increasingly titillating and participatory entertainment. These texts (more sensational than, for example, Henry Mayhew's encyclopedic mid-century *London Labor and the London Poor*) continued to position readers as the knowing audiences of urban underclass acts. Readers replaced *Tom and Jerry*, textually reenacting their flâneurie, patronage, and mimicry. Eventually, American antiquarians such as Thomas De Voe in the 1860s conjured retrospective, even nostalgic, textual performances of the urban poor. Conditioned indirectly by the influence of *Tom and Jerry* and its theatrical forms, De Voe could remember compelling vernacular performances in terms provided by Moncrieff's play and the new styles of spectatorship that arose in its wake.

Ultimately, the newly reimagined urban landscape produced new modes of spectatorship. Publics formed, their group identity defined by their ability to slum, to access the sites of low performance, and to help produce and reproduce those acts. Such representations compact a complex mix of reality and perception, of experiences, conventions, and projections by the observer onto the underclass subject. The early American city certainly hosted low street performances, but the realities remain problematic, shaped and produced by outsider perceptions. Rather than sorting "reality" from "perception," immediate experience from nostalgic reconstruction, the theatre offers a chance to examine how acts of the urban undersiders were always already scripted, shaped by acts of spectatorship and forms of presentation.

Moreover, the theatre (especially in the case of *Tom and Jerry*) contributes forms that defined public understanding of classes and their interrelationships. As *Tom and Jerry* produces its underclasses as performers, it also foregrounds the possibilities of upper-class performance enacted by the play's stars. The dramatis personae from Philadelphia's 1824 performance lists Francis C. Wemyss, Joseph Jefferson, and Henry Wallack in the lead roles of Tom, Logic, and Jerry.[52] Wemyss, like Joe Cowell in New York City, had just arrived from London. Even more so than Cowell, he struck audiences as appropriate to the play in which he starred. He appeared a "*fac simile* of the Bond street man of fashion—the Corinthian Tom" as Durang remembers; his act convinces even a theatre and fashion-savvy crowd:

> we verdantly concluded that he was one of old England's sprigs of nobility or gentry on a tour of pleasure of our wilds. The sight was certainly novel in Chesnut street, where then a fashionable *ne plus ultra* was seldom seen. If we remember truly, he was in this wise costumed:—A green Bond street promenading coat, rounded at the skirts; red waistcoat, with pocket flaps and bell buttons—these upper garments beautifully cut; salmon-colored stockinet pantaloons, fitting to the skin; Cossack polished boots; lace ruffles at the breast and similar wrist ruffles; and a Bob Logic hat, with correspondent equipments of the ornamental kind.[53]

Wemyss impresses the local fashionables, who immediately queried his performance; Durang recounts that they "wished to know the legitimacy of Wemyss' costume, and if it was authorized in Europe."[54] The performance of gentility convincingly reenacts *Tom and Jerry*'s fashionable lead. Recently arrived, Wemyss, still speaking in an English accent noticeably different from the American "nasal twang," brought coveted performances of fashion, on- and offstage, from the distant metropole.[55]

We get brief and intriguing glimpses at the ways in which such roles played out offstage; Wemyss's cohort imagined themselves offstage as Tom, Jerry, and Logic. A playful note from visiting English actor Edmund Kean to Wemyss, for example, imagines the actors playing out Tom and Jerry offstage. In the play's flash language, Kean invites Wemyss to "visit our crib on the banks of the Hudson," where the "honorable" gang would welcome the actors to a carousal.[56] The note is a playfully doubled performance that has the marks of low scenes from *The Beggar's Opera* to *Tom and Jerry*. Kean invites Wemyss's participation as a slumming observer, an intruder into scenes of life in the same style as Moncrieff's voyeurs. As the onstage characters do, Kean also imagines the actors as a "swindling fraternity" reminiscent of Macheath's gang. The note (and Wemyss's pleased reproduction of it in his memoir) shows the bohemian slumming and performing propensities of sporting culture. It also shows the extent to which the stage could inflect everyday acts. By the 1820s, actors reenacted and reproduced Atlantic sporting cultures very much like Tom and Jerry's rakish London circle. The rakish crew of actors, moreover, drew inspiration from the roguish low amongst whom they imagined slumming.

PRIVILEGED REENACTMENTS

On stage, *Tom and Jerry*'s elites gain access to low dives through their own theatrical acts of disguise. The three ramblers masquerade as beggars and street characters to gain access to the urban dives where authentic underclass types performed. In *Tom and*

Jerry, such performance represents deceptive artifice and admired skill, secret crime and public display. It also provides the elites with access, and Tom, Jerry, and Logic don disguises to penetrate the sites of low performance. Some of Moncrieff's most striking scenes—scenes that American theatres retained—depict the privileged observers dressing up as underclass characters in order to access the secret locales of low culture. The men, in "rags," enter the Beggar's Hall in St. Giles to observe the performing mendicants; playing the beggar themselves, they hope to observe hidden versions of street performance.[57] In the hall, however, they encounter the play's women (Kate, Sue, and Jane, also in disguise), whom the men thought remained at home. The female disguise is perhaps more intriguing than the male acts, trading openly in the appeal of doubled identity and cross-class and cross-gender performance. A Baltimore advertisement, for example, lists the female characters in all their roles, laying bare the mechanisms of their theatricality. Kate appears as "otherwise the Hon. Miss Trifle, otherwise Sir Jeremy Brag, other wise Nan, the Match Girl."[58] Likewise, Sue and Jane take on multiple identities (transforming into a ballad singer and a beggar, respectively) in their pursuit of the men. The women thus perform the tripled roles of genteel ladies, young rakes, and performing underclasses. At times, the play layers those identities in a dizzying compound of multiple performances. The women's cross-class and cross-gender disguises outperform and satirize the masculine sporting culture in which female participation seems most often in the forms of sexual objectification. As we might expect, the plot of *Tom and Jerry* limits such satire by its patronizing toleration of such acts and the final reinstatement of gender and class norms.

The mimicry of the play's flâneurs models audience responses as well. The play inspired disorderly imitation and fears of further rowdy reenactment as audiences mimicked both the upper-class ramblers and the low objects of their voyeurism. Charles Durang, remembering the season in his theatrical memoirs, saw the effects of the play's popularity as alarmingly pervasive:

> We had a racy group of Toms, Logics, and Jerrys, which sprang from its vicious froth instanter, in every city where it was represented. Tavern haunts and mixed drinks were created, and christened "Tom and Jerry," to suit the vitiated new-born taste. Boxing schools, and the elegant auxiliaries of genteel rowdy education, arose and became prevalent. After the inauguration of this dramatic monstrosity (which for the time being drew good houses) at the Park theatre and at the Chesnut street house, we thought that we saw a visible decline from the original positions of those dramatic temples. The reflective and respectable heads of families gradually dropped away from their once hearty support, and other intellectual amusements and changes intervening, radically altered the character of our theatrical audiences.[59]

An 1824 review of the play in the *Columbian Centinel* decries a horde of "rude, unlettered barbarians, all running after 'Life in London.'"[60] "The truth is not to be disguised," the account continues dramatically; those "who have not seen Tom and Jerry, have at least come within wind of it; and the sound is still ringing in our ears from the noddle of every boy in the street."[61] Worse, the performances encouraged rowdy crowd behavior both inside the playhouse (where audiences were "let loose, like so many mad dogs") and in the streets afterward, their "ungovernable yells" accompanying the more genteel patrons on their way home.[62]

The review evocatively shows the kinds of admiration and paranoia that slumming imitation and imitation of slumming creates. The male youth sporting culture, with its mixed drinks and boxing schools, does seem to transgress expectations of class performance. The real threat posed by *Tom and Jerry*, however, was not the creation of a new class of carefree young elites. The play's alarmed detractors seem most perturbed by imitation of the low and the consequential inversion of social hierarchies. Similar kinds of impersonations, of course, formed the very core of the play. As George Daniel observed in an introduction to Cumberland's London edition of the play, the "moral *intended* is to show the necessity of a young gentleman being carefully initiated into the trickeries of the town."[63] The *actual* moral, however, seems closer to a celebration of the inversion or erasure of social distinctions and the promotion of a new ethos that privileged certain kinds of access and imitation. The charismatic appeal of (and the possibility of manipulating) underclass performance had long been *Tom and Jerry*'s draw as well as its threat. Audience members, like the play's own slumming elites, might act as the low.

URBAN UNDERCLASSES AT THE GROVE THEATRE

Tom and Jerry changed hands multiple times, emerging in unlikely venues. On June 7, 1823, a company of black performers led by William A. Brown presented *Tom & Jerry, or Life in London* at an African American theatre on New York City's Mercer Street.[64] Brown's theatre adapted the play only three months after the Park's production, transforming a script of white flâneurie and class voyeurism into a more complex cross-racial and cross-class act. Brown, a former steward and proprietor of a pleasure garden on Thomas Street, had opened New York City's first known African American theatre in the summer of 1821. As Shane White has shown, African Americans maintained an increasingly active, independent, and visible public presence in New York; Brown's theatre represents independent black cultural production and public participation in one of its most prominent guises.[65] Insisting upon their right to the same variety of theatricals as other venues, Brown's theatre offered a conventional repertoire of Shakespeare and contemporary English comedies, even competing with the Park Theatre. Brown's garden and later his theatre entertained and profited from African American patrons. The venue reveals the desire to occupy not merely the margins, but, as Marvin McAllister has argued, the center of New York City's social landscape.[66] In response, their opponents (especially Park Theatre manager Stephen Price and Mordecai Manuel Noah, playwright and editor of the *National Advocate*) portrayed the black actors as the same riotous and lawless scoundrels that existed in the imagined world of Tom and Jerry's rambles.

The African theatre's production of *Tom and Jerry* seems largely based on the sequence of Moncrieff's play. A playbill—all the information that has survived—advertises the characters' progression from a country manor ("Life in the Country") through bustling London mercantile scenes, finally descending into the haunts of London's urban underworld. After traversing "Life on Foot" and "on Horseback," Tom and Jerry treat audiences to scenes of "Life in Fancy" among London's sporting society. The trio visit Bond Street, a bustling business and shopping district since the eighteenth century. The play also follows its urban tourists into shadier locations. Scenes vicariously take audiences to Wapping, London's dock area and a site famous for its "execution dock" where pirates were hanged. Such scenes seem derivative of the Euro-American adaptations, but the simple and material fact of African American actors would have changed them radically. *Tom and Jerry*'s production by an interracial cast layers the play's preexisting blackface with other acts of masking; Brown's troupe presented a heady compound of whiteface, cross-dress, posing as English and disguising as the low. Brown's production improvises radically on *Tom and Jerry*'s themes of underclass voyeurism, displaying London's dives and blackface performances for a city fascinated by black and blackface performance.

The play also stages cross-racial transgression and spectatorship. "Life in the Dark," for example, reveals the black character Dusty Bob performing urban misbehavior opposite a watchman. "Life in Rags," likewise, reproduces the All-Max dance scene of African Sal and Dusty Bob. Moreover, the African theatre led the trend of "naturalizing" *Tom and Jerry*'s "Life in London," introducing local scenes and distinctly American forms. The playbill boasts an "additional scene," as the playbill notes, entitled "Life in Limbo—Life in Love," set at the "Vago Range," in Charleston, South Carolina (probably a corruption, according to George Thompson, of the Vendue Range, a noted site of slave auctions and other commercial transactions).[67] The black troupe had moved beyond a strictly local scope and from comic to apparently sentimental modes to perform southern U.S. chattel slavery. The next week, the same playbill promises, the play would transplant the play to New York City and include "an Additional Scene, Life in Fulton Market!!!"[68] This last scene most likely adapts the scenes of market performance that Thomas De Voe remembers forty years later in his *Market Book*, and that spectators had long patronized in the mixed and mobile crush of bodies flooding American urban markets.[69]

Those exclamation points trace the excitement such naturalized, even reclaimed, performances of black New York must have generated. It is appealing to think of such scenes as culturally autonomous black theatre. For perhaps the first time on the formal American stage, black Atlantic actors re-crafted the black roles bequeathed upon them and shaped by the conventions of blackface. Indeed, scholars rightly admire Brown's theatre for its public performances of blackness, and the important stories of black actors playing black roles have retaken their place in early American theatre history. Their acts also trouble both the received stereotypes of blackface and the elusive ideal of authentic black performance. *Tom and Jerry* seems a particularly canny way of doing so, allowing Brown's theatre to send up slumming playgoers, imitating the imitators, inverting and mocking the privileged views of London's highs and lows. The African theatre's actors in 1823 further complicated an already-rich script. Notably, evidence suggests that the Mr. Smith, who played the auctioneer in the Charleston slave auction and the watchman in "Life in Rags," was a veteran white New York City actor. If so, then his appearance, as Marvin McAllister argues, makes Brown's troupe "the first integrated company in Manhattan, if not the United States."[70]

The African theatre's productions thus torque conventional theatrical forms of cross-racial and cross-class representation. They newly ironize scenes' performance and patronage, reconfiguring the relationships between authenticity and imitation.

Their innovative roles and scenes did not just reclaim performance, but radically rewrote the play's script. The fictive template of privileged white spectatorship could not withstand the African theatre's theatrical reinventions. *Tom and Jerry* appeared alongside other radically renovated black acts. For example, Brown's theatre presented *Obi; or, Three-Finger'd Jack* and a now-lost production entitled *Shotaway; or, the Insurrection of the Caribs, of St. Domingo* in the same season as *Tom and Jerry*.[71] Those plays both represent strong, nonwhite, and resistant circumatlantic figures, and *Tom and Jerry* seems part of a concerted institutional attempt to elaborate upon the thematics of independent black theatricality. The troupe even claimed Shakespeare, as Shane White has compellingly argued. The diminutive, disfigured, but cunning character of Richard III, like *Tom and Jerry*'s black actors, embodied aspects of the conflicted social and theatrical roles available to black New Yorkers in the 1820s.[72] Responding to an entertainment industry that normally reserved the right to stage race in blackface, Brown's theatre performed what Marvin McAllister has called "whiteface minstrelsy," reenacting "white privilege" in masked but potentially emancipatory theatricals.[73]

Although white observers would have been quick to dismiss such performances as demeaning and imitative "aping" (as M. M. Noah's *National Advocate* did in 1821), the African theatre's adaptation of *Tom and Jerry* represents a more complex cross-racial and cross-class performance.[74] Brown's troupe appeared before white spectators who may have perceived themselves as slumming at the black theatre, enacting the sort of class voyeurism that Moncrieff's script called for. *Tom and Jerry* seems rather more complex on the African American stage than its author probably intended, presenting a mixed-race cast of Americans in black- and white-face, posing as elite Englishmen slumming among the low. To complicate matters, white audience members who may have imagined themselves as privileged voyeurs also engaged in spectacular "hooliganism" at the theatre. That sort of behavior led to the apocryphal notice that Marvin McAllister takes as an emblematic expression of Brown's dilemmas, which proclaimed that "white people do not know how to behave at entertainments designed for ladies & gentlemen of color."[75] The events that inspired the supposed sign include series of disturbances culminating in a riot in the summer of 1822. A "gang of fifteen or twenty ruffians, among whom was arrested and recognized one or more of the Circus riders," assaulted Brown and disrupted the black troupe's performance.[76] Records of the legal proceedings confirm that James West's neighboring and competing circus employed the rioters, who performed equestrian and possibly blackface acts.[77]

The circus performers committed a multivalent act of (perhaps) blackface-on-black violence. As an essentially intramural mêlée among performers, the disturbance reveals a professional rivalry. It reveals common histrionic pursuits despite the differences of race and social affiliation. As West's established business sought to defend its turf against interlopers (much as Price's Park Theatre had), those professional mutualities receded in the face of the insecurities endemic to economic and social competition. The circus performers activate race and cultural positioning as a defining categorical difference, justifying their violence with disdain and translating economic competition into social antagonism. Ultimately, the rioters operated by crossing boundaries, revealing as permeable the very distinctions they sought to maintain. White men enter a black space, established venues reveal their tenuous economic position, and representatives of lowbrow entertainments assault aspiring highbrow performers. Ultimately, the rioters attempt to establish their racial and cultural power by behaving in the very disdained, low, and unruly ways they accused black performers of acting.

LIFE IN NEW ORLEANS AND AUTHENTIC IMITATION

Its unique combination of elements—play, actors, venue, audiences, and competition—make *Tom and Jerry*'s appearance at the African theatre a distinctive event in the staging of the Atlantic underclasses. As one of the first plays to focus on urban black underclass life in detail, *Tom and Jerry*'s New York City appearance would alone constitute a notable event. But when a racially integrated troupe of actors adapted it to its American contexts and reenacted its racial masking and elite patronage, the play becomes a landmark moment in the Atlantic performance of lowness and blackness. In its articulation of the relationships among audience and actors, black and white, high and low, the play signifies in radically self-conscious ways upon American theatre's dominant practices. Brown's *Tom and Jerry* also reveals a key chapter in the history of the forms that would later emerge as American blackface minstrelsy. The play stands, as it were, between the eighteenth-century popularity of *The Beggar's Opera* and the blackface minstrelsy popular in the 1830s and 1840s. In the 1820s and 1830s, local performances of "Life in" various American cities met and helped produce the emergent forms of urban blackface performance. Although Brown's theatre offers particularly rich instances of such acts, their inherent theatrical reflexivity shaped other stagings of low black performance. American theatre would continue to explore the possibilities of authenticity, imitation, in acts of the racialized underclasses.

Life in New Orleans, for example, appeared at the St. Charles Theatre in New Orleans on May 13, 1837.[78] Developing a local version of *Tom and Jerry*'s scenes of "Life in London," the play featured Old Corn Meal, a locally famous black figure. Newspaper notices referred to him as the "celebrated sable satellite" and "the well known vendor." The songs of the "celebrated amateur" had already enjoyed popularity in the market when he burst upon the theatre scene.[79] His songs were on the tips of everyone's tongues, and his stage appearances caused a minor sensation in the theatre world. Driving his cart on stage and singing his well-known street tunes, Old Corn Meal reveals the prominence of local scenes that had come to dominate the format of *Tom and Jerry* in its Atlantic travels. Although Old Corn Meal had other scheduled performances soon after his 1837 debut, the vendor's horse fell and was killed in a stage mishap—putting a damper, one assumes, on both his day and night jobs. For the next three years or so, however, the street performer's vocal talents and comic acts continued to make regular appearances on the New Orleans stage and in local newspapers.

"Signor Cornmealiola," as newspaper commentators labeled him, performed as himself just as earlier lore reputed Billy Waters to have done in London.[80] The black vendor on his horse and cart became the focal point of *Life in New Orleans*, adding authentic street performance to a play that emerged from *Tom and Jerry*'s tradition of reenacting the peculiarities of the urban underclasses. Commentators celebrated his acts as "originals," construing Old Corn Meal as a black theatrical embodiment of the Crescent City's common street scenes. Multiple and mutual imitations, however, vex those declarations of originality. Old Corn Meal performed his own songs, but he also adapted the acts of white blackface performers who themselves trafficked in stylized imitations of blackness. The vendor-performer's acts participate in complex dynamics of authenticity and imitation, copying the copiers who had ripped off the acts in the first place. Moreover, our view of Old Corn Meal comes entirely from an archive produced by local white commentators and relatively privileged visitors such as English diplomat Francis Cynric Sheridan, who recounted their observations in New Orleans newspapers and travel accounts.[81] Old Corn Meal thus appears in the historical record through the same techniques that *Tom and Jerry* used to produce figures such as Billy Waters, Dusty Bob, and Sal.

Such acts drew on and profited from their connections to the wider worlds of Atlantic performance, and observers noticed the connections. Old Corn Meal's act engaged outsiders in an interplay of local and regional or national forms. A notice in the Macon, *Georgia Telegraph* on February 16, 1841, for example, describes a "grand trial of skill" between Old Corn Meal and some probably forever-untraceable "Dutch Broom Sellers." According to the notice, Old Corn Meal sang his minstrel standards "Old Rosin the Bow" and "Sitch a Gitting Up Stairs," and the Dutch Broom Sellers pitted "Buy a Broom" against his black (but also blackface) ditties.[82] Although the Dutch broom sellers could have been associated with or perhaps even from New York, they also show the traveling diversity of ethnic identities classed as low in early American performance culture. Brooms, as Dale Cockrell has pointed out, played an important material and symbolic role in popular performances; they emblematized labor, purity, and pollution, clearing "a symbolic space for public performance."[83] Brooms also featured in blackface performance and other modes of low theatricality. The competitive performance ritual pits "original" New Orleans urban blackness against Dutch styles perhaps associated with mid-Atlantic cities—but in the exchange, both become routed, transacted, mobile underclass performances. Moreover, the news item appealed to and appeared before rural residents in the landlocked South, far away from whatever urban authenticity in which those performances may have traded.

In what Meredith McGill identifies as a culture of reprinting, compiling, and excerpting, regional culture jumped fences energetically.[84] Old Corn Meal's fame outran his actual range, his obituary appearing in New England newspapers.[85] Even as the *Picayune* had enthusiastically claimed him as a "New Orleans Original," the performer performed his local authenticity on a much broader Atlantic stage—an arena also occupied by the residual acts of Macheath and his gang, by the still-popular scenes in *Tom and Jerry*, and by the embryonic moves of T. D. Rice and other blackface performers. Observers detected echoes of other Atlantic underclass performances in Old Corn Meal's songs, revealing tenuous connections among apparently far-flung acts. Francis Sheridan, for example, observed the vendor's wild oscillations between falsetto and bass voice, judging the effect as "precisely the same effect as one of our street duets where the Man & Boy alternately sing a line."[86]

Likewise, a commentator in the *New Orleans Picayune* wrote, "[w]e never heard a vocalist who could make his voice 'wheel about and turn about' so quick from tenor to bass and from bass to falsetto."[87] The New Orleans journalist deliberately references the catch phrase from T. D. Rice's well-known song "Jump Jim Crow," which had already proven wildly successful on both sides of the Atlantic. Different observers associate Old Corn Meal's act with familiar performance contexts—one with London street performance and the other with Rice's urban blackface. A "New Orleans original" reenacts the vocal techniques of London street singers, and transatlantic performances tap into residual forms of wheeling and turning about to generate new ones. Extended chains of influence and connotation connect the acts, outlasting traceable links.

T. D. Rice's Jim Crow act itself emerged at the end of the 1820s from the urban culture shaped by the same processes of class voyeurism and reenactment that *Tom and Jerry* recruited and spectacularized. Although blackface minstrelsy would later perform its elaborately mythologized origins in rural southern plantation culture, Rice's early acts acknowledged his own training in New York City's racially mixed and economically diverse Five Points neighborhoods. Rice's performances revised the dynamics of *Tom and Jerry* once again, offering knowing spectators access to its scenes of underclass and black performance. It thus seems no surprise, then, that Rice's performances cross-pollinated with Old Corn Meal's version of "life in New Orleans." Sheridan's 1840 account of the black performer cites his performance of "Old Rosin the Bow," and "Such a Gittin' Up Stairs"; Old Corn Meal's audiences knew those songs as minstrel-show standards. According to Henry Kmen, Rice visited New Orleans three times during the height of Old Corn Meal's popularity. In 1836, the blackface performer even appears to have incorporated the performing vendor into his burletta *Bone Squash Diavolo*, a farcical send-up of popular genres. Old Corn Meal, the "original," was thus acting out the complex interchange of authenticity and imitation in the Atlantic theatre world: he generated acts that others then copied, copied other acts himself, and (perhaps most intriguingly) turned out the original—and better—versions of minstrelsy's own imitations. Old Corn Meal's act, brought onstage by New Orleans theatrical manager James Caldwell in a cagey act of cultural arbitrage, continued the theatrical trends energized by Moncrieff's *Tom and Jerry*.

T. D. Rice's early blackface acts produced underground scenes of urban low around the same time. Rice had even exported his urban scenes to Caldwell's American Theatre in New Orleans on March 27, 1835.[88] In New Orleans, Rice had performed a version of *Life in Philadelphia* that a "Dr. Burns, of Philadelphia" had written for him.[89] Just as Old Corn Meal was gaining in popularity, Rice helped constitute a public and perhaps even modeled urban black performance for the New Orleans act. Local urban scenes had truly gone on the move, thoroughly confounding any expectations of local authenticity and originality. Although the script no longer exists, that play almost certainly adapted *Tom and Jerry*'s earlier urban voyeurism to Rice's increasingly popular blackface act. The play probably also restaged scenes depicted in Edward Williams Clay's "Life in Philadelphia" prints, which disdained underclass pretensions in a circulating graphics imagining black dandyism on American city sidewalks.[90]

As Lhamon explains, *Life in Philadelphia* also contributed to Rice's blackface burletta *Bone Squash Diavolo*, which Rice performed in New York City and Philadelphia in the autumn of 1835.[91] In a crescendo of convergences, Rice brought *Bone Squash Diavolo* to New Orleans in 1836, where he either shared the stage with Old Corn Meal or played him in an entr'acte skit.[92] *Bone Squash* ratchets up the stakes in urban blackface by casting a class-crossing vagrant as the hero and making its antagonist a white Yankee devil. It seems edgy humor for New Orleans, its satire capable of playing multiple ways. Revealing its debt to *Tom and Jerry*, the burletta's opening scene displays a performing working-class cohort, streets populated with black laborers and their implements. One of the characters, Caesar, sleeps in a wheelbarrow as the curtain and the sun rise together. In a conventional stage fantasy of urban low performance, the black characters transform props (a violin, a pole, brushes, scrapers, brooms, and whitewashing tools) from implements of labor into tools of entertainment. The scene lays the physical burden of performance upon the laborers; the "business in this scene," the stage directions read, "is very particular, as there is no symphony between the verses of the solos."[93] The burletta's song, dance, gesture, and physical comedy present city life and work as entertainment. Such scenes offer counter-performances to the plantation labor depicted in earlier blackface theatricals such as John Fawcett's 1800 pantomime *Three-Finger'd Jack*, in which contented slaves sing of kind "buckra man" and merry slaves. Rice's protagonist Bone Squash emerges from the newly urbanized cohort of singing and dancing workers. Tapping into and parodying the conventional humor of its day, the play centers on Bone's attempt to rise above the constraints of his social standing as a black laborer. He ultimately, if ambivalently, succeeds by escaping amidst an increasingly kinetic fracas of stage business, exaggerated pantomimes, and comic pratfalls.

Rice's play reenacts the motley conventions of Atlantic urban street scenes, including *Tom and Jerry*. Notably, the urban voyeurs no longer frame the entertainments—that role has shifted to minstrelsy's audiences, perhaps in a deliberate send-up of the audience's participation in elite voyeurism.[94] Ranging from New York City to New Orleans, Rice's spectators probably encountered blackface with mixed motivations, variations on the pattern enacted by *Tom and Jerry*. They could admire the hidden cachet of underclass acts; at the same time, by the 1840s, those same audiences patronized the performances, putting themselves in the same positions that Tom and Jerry had performed for over a decade. Rice's blackface and Old Corn Meal's New Orleans act do not collide accidentally, nor do they perform a competition between original and imitator. Instead, Rice's act and Old Corn Meal represent two of the many replications of Atlantic urban lowness, performances that enact mimicry as authenticity based on well-established theatrical practices of observation and self-conscious patronage. These connections reveal lateral and oblique transmissions in the genealogies of Atlantic performance; Atlantic culture reproduces and cross-pollinates in fractal patterns that follow the chaotic order of markets and human movements. Distinct techniques and gestures find their way through hidden channels of cultural transmission, and observers notice the similarities. Wheeling and turning about unpredictably, whether articulating lumpen cross-dressing or African American culture, whether at the end of the eighteenth century or the middle of the nineteenth, had become an embedded and energizing feature of performing the Atlantic underclasses.

DICKENS AND THE NEW YORK LOW

I close this chapter with another noteworthy if unconscious improvisation on *Tom and Jerry*'s scenes, which occurred when Charles Dickens toured the United States as a celebrity author in 1842. By his own account in *American Notes*, Dickens had tired of the genteel fêtes and celebrations and wished to observe American urban underclasses.[95] Accordingly, Dickens departed a ball held at the Park Theatre to observe the performing low firsthand. Escorted by police, Dickens "descended" to a dance hall in the Five Points neighborhood known as Almack's. There, he witnessed an energized and compelling dance by the black performer William Henry Lane, better known as Master Juba. His vivid account of Lane was not simply a vaguely theatrical narrative; it actually helped launch the dancer's transatlantic career—his London press releases, as Marian Hannah Winter observes, used the description from Dickens's text.[96] Dickens worked within and perpetuated the Atlantic practices of patronizing and producing class performance.

Dickens's path is just as significant as the scene he witnesses. He leaves the site where *Tom and Jerry* had first enacted their voyeurism and enabled American audiences to access the purported authenticity of low performance. Perhaps unconsciously, even inevitably, he reenacts *Tom and Jerry*'s rambles in a topography shaped by two decades of Moncrieff's play in performance. The dive's name, of course, commemorates or satirizes the elite London club Almack's, but in the wake of *Tom and Jerry*'s satirical presentation of the East End "All-Max," the name may even have self-consciously referenced the play's den of underclass performance. Such sites had clearly become conventions in the Atlantic world of urban stage scenes. The name signals an indecipherable mix of earnest and burlesque imitations of London's elites and the staged low, and its multivalent referentiality seems just right, considering the cross-class transactions of performance and entertainment that it hosts.

As narrated in Dickens's *American Notes*, Master Juba's dance reveals underclass performance conditioned by the practices that *Tom and Jerry* had condensed on stage. The club's proprietors appear in a display of finery that contrasts sharply with the "world of vice and misery" Dickens had just left behind. The characters he meets dissemble and perform, recalling the deceptive acts of the Atlantic low from *The Beggar's Opera* to *Tom and Jerry*. It is a scene of potentially titillating transracial sexual danger as well. Two "young mulatto girls," for example, appear to Dickens to be "as shy, or feign to be, as though they never danced before."[97] Following the play's formula, acting covers up a more authentic sort of performance that emerges in the dance that occurs next. The dance itself is a vexed performance of authenticity and imitation, conjured into existence by outsider patronage just as Dusty Bob and African Sal's had been. The upper-class observers call for a dance and witness a "regular break-down," as Dickens writes, an updated version of Dusty Bob and African Sal's comic minuet in *Tom and Jerry*. The dance, led by Lane, offers Dickens a virtuoso performance of urban black style, which he famously described:

> Single shuffle, double shuffle, cut and cross-cut; snapping his fingers, rolling his eyes, turning in his knees, presenting the backs of his legs in front, spinning about on his toes and heels like nothing but the man's fingers on the tambourine. Dancing with two left

legs, two right legs, two wooded legs, two wire legs, two spring legs—all sorts of legs and no legs.[98]

In the end, the description reveals its debt to Atlantic theatre's complex performances of authenticity and imitation, listening to the "inimitable sound" of Lane's laugh, the "chuckle of a million of counterfeit Jim Crows."[99] Like Old Corn Meal, like Billy Waters, like Dusty Bob and African Sal, Lane performs within the structures becoming in the eyes of his audience (and perhaps by default) a genuine sort of "counterfeit," an authentic and inimitable mimicry of Rice's Jim Crow—an act that was itself a conventional imitation of the performance values that Lane reenacted in Almack's.

The English author of fictions of English low culture wheels about to star in a literary performance as a Tom or a Jerry, observing, patronizing, and re-presenting underclass acts. Dickens did not indicate that he was consciously reenacting *Tom and Jerry*'s scenes, nor did he need to. His actions trace routes made possible by Atlantic culture industries that had transmitted low performance across decades and oceans. Earlier acts had already produced forms that render the sites of underclass performance visible; theatre had practiced the procedures by which outsiders accessed those scenes for decades. Creating a stage genre, *Tom and Jerry* rehearsed the moves of urban voyeurism for two decades before Dickens arrived. Dickens's descent into Almack's adroitly follows the paths of least resistance through a landscape produced and marked by the signs of those earlier performances. This cultural geography populated itself with characters such as Lane; it called forth actors who staged counterfeit versions of themselves before outsiders, acts compelling for their authentic mimicry. The Atlantic culture industries continue to produce and reproduce act and reenactment in dizzying dynamics of imitation and authenticity, of twice-removed self-imitation and performatively genuine inauthenticity.

The point, of course, is not to sort original from mimic. Instead, these scenes foreground the self-replicating genealogies of Atlantic performance that shaped such acts and powered their proliferation. Dickens account reveals the dissemination and continual reinvention of *Tom and Jerry*'s Atlantic underclasses in new locations. Moncrieff's scenes and their multiple adaptations make visible the complex transmissions of such acts and the privileged access to low and racialized performance. Those scenes code valued insider experiences and specialized local knowledge, delivering pleasures similar to those produced in the dens of criminal gangs, the caves of Jamaican Maroon rebels, and among early America's runaway slaves and unruly mobs. In the 1820s and later, that insider knowledge shifted to city streets and slums, growing increasingly self-conscious and deliberately theatrical, performing the practices that enabled its own reproduction.

Staged encounters with the underbellies of Atlantic cities show the social and economic hierarchies that enable mobile voyeurism and class mimicry. They question the shifty theatricality of imagined underclasses and what these performances amount to when watched self-consciously. In rendering Juba visible in Almack's dance hall, Dickens inherits and reenacts *Tom and Jerry*'s interlinked rituals of mobility, spectatorship, patronage, and performance. The formula is portable, but it does more than simply conjure an act like Juba's into visibility. *Tom and Jerry* shows what William Brown, Stephen Price, Francis Wemyss, James Caldwell, and Old Corn Meal knew at their respective theatres—that charismatic lowness produces the privileges and profits of further reenactment. Dickens had obviously learned that lesson as well. Ironically, access and mimicry has the potential to transform performance spaces, remaking them into rowdy offstage imitations of the hidden spaces appearing on stage. Such scenes locked Atlantic underclass performance in insolvable conundrums of obscurity and revelation, of originality and imitation, of self-conscious performance and immensely productive and profitable reenactment.

7. Slave Revolt and Classical Blackness in *The Gladiator*

As Moncreiff's play tutors audiences in the dangerous appeal of watching and mimicking the urban underworld, other plays demonstrate equally vexed relationships to the forms of stage blackness. Robert Montgomery Bird's *The Gladiator*, a melodrama of Roman slave revolt, is one such play. Transforming the antebellum low into neoclassical rebels at the Park Theatre, it enacts the problems of redeeming and fitting plebeian types into the selective traditions of America's neoclassical imagination. *The Gladiator* also presents a story of the costs of theatrical success, revealing the ways that celebrity and popularity can effectively smooth over troubled underclass performance genealogies.

Audiences, crowding the playhouse despite torrential downpours, seemed eager to applaud the play's September 26, 1831, premiere. It became a hit despite the fact that it was "miserably got up," as Bird later wrote.[1] As one reviewer wrote, "the whole last speech of Spartacus was utterly inaudible by reason of the clamorous applause, and when the curtain fell, the theatre was literally shaken with the energetic demonstrations of pleasure given by the spectators."[2] Audience members responded to iconic moments in the theatre of underclass revolt, such as when Forrest strikes a defiant pose and proclaims:

> Death to the Roman Fiends, that make their mirth
> Out of the groans of bleeding misery!
> Ho, slaves, arise! It is your hour to kill!
> Kill and spare not—For wrath and liberty!—
> Freedom for bondmen—freedom and revenge!

At his cry, battle commences; "Shouts and trumpets" sound and enslaved gladiators "rush and engage in combat," as the curtain falls on the first act.[3] Punctuating moments of noble rhetoric with such chaos, *The Gladiator* found immediate popularity. By November of that year, the play had successful runs in New York City, Philadelphia, and Boston, and it held the stage for the next three decades, cementing Forrest's reputation as perhaps the preeminently popular American actor. By 1854, according to Curtis Dahl, the play had seen over one thousand performances.[4]

The class-conscious rewriting of Roman history presents yet another take on the Atlantic low. Bird's play stages charismatic rogues in rebellion against the powerful, drawing upon reserves popularized by plays such as *The Beggar's Opera*. The playwright's own notes reveal a fascination with the class conflict between Roman "patricians" and "plebeians," interests that show in the script as Romans repeatedly call their gladiatorial slaves "rogues," "knaves," "villains," and "miscreants."[5]

Although the play takes mythologized neoclassical history as its most obvious frame of reference, its characters speak in the class accents of antebellum America. The play restages the American mob that had appeared in *Slaves in Algiers* and *The Glory of Columbia*, embodying again American culture's variegated constituencies. Moreover, *The Gladiator*'s mythologized rebels appeared in venues accustomed to the cross-class encounters of plays such as *Tom and Jerry*, elaborating upon the forms the low could take.

Bird based his dramatization of *The Gladiator*'s revolt on the well-known story of the first-century BCE Thracian gladiator Spartacus, who led a slave revolt against the Roman authorities. The play tells an ambivalent parable of class and political action, first celebrating low solidarity in Spartacus's attempts to forge rebellious unity out of a fragmented and diverse polity. Attracting oppressed masses, Spartacus's revolt achieves its initial goal of escape from Italy. The rebels, however, hubristically return for more plunder, and Roman forces eventually killed their leader and executed the rebels in mass crucifixions. Bird's script reshapes this history to present a heroic leader's inspiring but doomed revolt against Roman power. The plot of failed revolt enacts the rabble's reliance on, but ultimate betrayal of, their heroic leader, as Bruce McConachie argues.[6] Even in failure, however, *The Gladiator*'s revolt presents democratic and violent direct action, elevating Spartacus and his rabble to heroic, if problematic status before audiences who remained capable of their own melodramatic acts of riotous violence.

The play presents two competing performances: a class plot manifest at the level of content (which I will discuss first), and another, more subtle scene of racial embodiment. The play represents a scene with deeply vexed racial overtones, and the play shapes its class commitments in the masked blackness of Bird's slave revolt. Interpreting blackness presents a problem in the play's revolt, and the play seems particularly unwilling to divulge its relationship to African American slavery. Certainly we can say, with Jeffrey Richards, that the play "articulates the very longings we now recognize that African-American slaves themselves maintained."[7] Audiences (southern and northern) seemed content to understand the play as primarily concerned with nonracial revolt. They could even identify the play's uprising with Poland's emancipatory struggles. The New York *Evening Post* remarked upon the "peal of warm applause" responding to a "happy allusion" to Poland in the play's prologue.[8] Such happy allusions had their limits, apparently. As McConachie concludes, few audience members ever seem to have sensed that a "drama centering on a white slave revolt was meant to apply to Nat Turner or the threats of William Lloyd Garrison."[9]

In script and staging, however, Bird's Roman revolt subtly presents the stage forms of blackness, not least through the body of its star actor. Especially when embodied by Forrest, *The Gladiator*'s classical acts inscribe subtle racial associations on working-class American bodies. Forrest performed blackness during his early career, debuting at the Park Theatre in 1826 as Othello after playing in blackface in America's provincial theatres.[10] The stage transformation of "black" into "neoclassical" seems rare but not impossible. *Obi; or, Three-Finger'd Jack*, for example, compares its black banditry to Spartacus's revolt. William H. Murray's 1830 melodramatic rewrite of *Three-Finger'd Jack* gave the linkage of blackness and neoclassical forms antebellum popularity. Perhaps half-consciously, Forrest's Spartacus would put such lessons into action beginning in 1831. At the same time, *The Gladiator* clearly departs from earlier versions of transracial cultural forms. Rather than presenting low-black characters openly, it permits observers to experiment with the spectacles of underclass and interracial mutualities masked behind the dominating presence of Forrest's heroic, muscular performance.

THE THEATRICAL STAR OF UNDERCLASS HEROICS

Bird's play, driven by Forrest's starring performance, seems a signal moment in early American theatre history. Its success apparently confirms American theatre's maturation, proving the workability of the emergent star system and the rise of "native" authorship. It also offered a vehicle for the promotion of melodrama's new styles and for self-consciously defined new audiences. The play's textual genesis itself represents a spectacular change in the way American theatre did business. Robert Montgomery Bird had submitted the winning entry in Forrest's second play competition, but his classically themed *Pelopidas* remained unperformed, perhaps because it did not promote the central character enough, as Jeffrey Richards observes.[11] Bird, however, immediately produced *The Gladiator,* another neoclassical drama, for Forrest's consideration, better highlighting the doomed heroism and muscular style that appealed to popular audiences.[12]

More than just a sign of jingoistic Jacksonian attitudes about American culture, Forrest's prize play competitions mark the emergence of new institutional and performance practices in American theatre. Forrest's ability to commission plays catering directly to his acting style was neither an accidental result of forceful histrionic personality, nor a sign of the inevitable maturation of native American theatre. Instead, wide-ranging and drastic institutional and economic changes enabled the star-driven prize play. Forrest, luckless as a novice on the Philadelphia stage, began his career during a time of near-crisis among the established American theatres. Returning from stints in the small-town circuits of the Midwest and New Orleans, Forrest found established Philadelphia playhouses facing increasing competition from a variety of upstart theatres and other popular diversions.[13] New York's Park Theatre, rebuilt after an 1820 fire, faced competition from the upstart Chatham and Bowery theatres in the early part of the decade.[14] New theatres and emergent forms of entertainment also increasingly competed with established playhouses, cutting into profits and forcing managers to reevaluate conventional repertoires. Both enterprising actors and playwrights could take some advantage of the situation. As historian Reese James wrote, "now that the theatres were in trouble, the managers were more open-minded in the matter of experiments."[15]

Bird's play thus emerged from a period of institutional crisis, a series of complex realignments resulting from far-reaching changes in the American entertainment industry's commercial practices. The changes had material effects on actors. The existing routine of season-long engagements in regional circuits gave way to a system in which prominent actors toured in shorter and more spectacular engagements. Stars could become cosmopolitan celebrities, contrasting with the local stock companies who supported their efforts. As McConachie argues, the practices "constructed the unique individuality of the itinerant 'genius' in part by contrasting him or her to the bumbling anonymity of the actor's support."[16] In the preceding decade, for example, the well-publicized and sometimes controversial tours of English stars such as Charles Mathews and Edmund Kean had accustomed American audiences to expect some novelty. As early as 1808, Thomas Abthorpe Cooper (an influence on Forrest, and one of the earliest actors whom we might identify as a "star") had shuttled back and forth between New York City and Philadelphia, performing in two different cities four nights a week.[17] While surely a publicity stunt, the exploit also demonstrates the shifting status and changing mobility of leading actors within the system. Under the new dispensation, Cooper showed, an actor could offer a service so valued that he could dominate multiple markets simultaneously. Forrest's plays manifested these new entertainment market practices, and his career traded on the unique and portable aura of the individual performer in ways that actors rarely could in the older forms of American theatre's circuit system. The system certainly had its costs; local actors became a caste of undertrained, overworked, and underappreciated laborers who found themselves forced to "play or starve," as provincial actor John Banvard later wrote.[18] The emergent practices of the new entertainment industry also devalued playwrights, as Bird would find when he received paltry compensation for his contributions to American dramatic authorship. The system itself depended upon its in-house theatrical underclasses.

In the process of adapting to the changing market conditions, Forrest's prize play provided a sort of cultural ritual around which new audiences coalesced in the 1830s. *The Gladiator* rehearses emergent class identities in ways that spectators could have construed as relevant to themselves. As scholars have argued, the play rehearsed Jacksonian ideologies, opposed British-style imperialism, and valorized the common man.[19] The play's script certainly permits this sort of interpretation, with its frequent references to freedom and liberty, and elitist Roman antagonists. The Centurion Jovius, for example, sneers at Rome's republican citizenry, calling them "something bauble-brained, and like to children."[20] They exhibit no better judgment politically, as Jovius comments:

> What do they
> In their elections? Faith, I have observed,
> They ask not if their candidate have honour,
> Or honest, or proper qualities;
> But, with an eager grin, *What is his wealth?*
> If thus and thus—*Then he can give us shows
> And feasts; and therefore is the proper man*.[21]

Voiced by a Roman imperialist, the derision baldly vocalizes suspicions of Jacksonian democracy, especially the political effects of broadening white male enfranchisement and fears of the rabble's susceptibility to demagoguery. The lines probably sounded elitist to Forrest's audiences, and almost certainly insulting to an audience enjoying the same sorts of "shows" that the Roman derides. The speech marks the play's sympathies as leaning toward the populist, and *The Gladiator* reclaims classical models for use by the common person. Slave revolt thus becomes a performed possession of nineteenth-century audiences. As McConachie asserts, Forrest advocated revolt "not only to the enslaved gladiators on stage, but directly to the audience as well."[22] In a theatrical culture prone to rioting audiences could identify and even participate vicariously in Spartacus's rebellion.

Who revolted, then? Given the play's rhetorical encouragement of audience identification with revolt, it seems important to try to define Forrest's spectators—or at least to trouble some of the scholarly mythology shaping our perceptions of that audience. Observers began early to imagine Forrest's performances as expressions of a simple, monolithic populist masculinity. Walt Whitman, for example, assessed the play's enduring appeal in 1846, arguing that the play was "calculated to make the hearts of the masses well responsively to all those nobler manlier aspirations in behalf of mortal freedom!"[23] Such assessments, however, conflate the star and his act with the later emergent cultural formations of New York's Bowery working classes—as one reviewer wrote in 1845, audiences, apparently including Whitman, wished to "[take] him at the start, as a Bowery Boy."[24]

The Bowery B'hoy's cultural style presented extravagant street performances of working-class identity, which peaked around mid-century. Mose and his "g'hal" Lize, the popular working-class stars of Benjamin Baker's 1848 *A Glance at New York*, further energized urban street performance on the stage.[25] Forrest's performances in the 1830s, however, preceded and perhaps even contributed to the rise of self-conscious urban working classes. *The Gladiator* did not necessarily have such ready-made audiences in 1831. Forrest's audience remained under construction—arguably, the star's performances helped to construct that cohort, and *The Gladiator* appeared at an early moment in that process. Certainly, the play met early audience receptions that would resemble later Bowery behavior. Forrest supposedly received nine cheers, for example, at the end of *The Gladiator*'s second act at Philadelphia's Arch Street theatre on October 24, 1831.[26] As those New York reviews noted, energetic and vocal applause drowned out Spartacus's final speech. However, such playgoers did not yet adhere to the later and well-defined neighborhood identities, and *The Gladiator* met equally enthusiastic receptions at the working-class theatres and the relatively upscale Park Theatre. The class bifurcation that divided antebellum New York City, according to theatre historian Peter Buckley, had not yet fully determined the sites of class performance.[27] The Bowery district was "swiftly becoming" the "workingmen's counterpart to fashionable Broadway," as Sean Wilentz writes, but when Forrest began collecting his plebeian audiences in the late 1820s, the district was only a few years old, and its residents were one-quarter immigrant and 15 percent African American.[28] In the 1830s, at the emergence of Bowery culture, *The Gladiator* most likely reckons with economically and socially diverse audiences.

Audiences and their spectatorship pose significant problems for historical interpretation; even with the best of evidence, acts of reception only become visible through further cultural productions. However, theatre (and especially the rowdy early American playhouse) does not stage the simple one-way transmission of performance. Audiences do not sit silently and simply "receive" acts. In the early 1830s, Forrest and his audiences instead participated in mutual performative acts of self-construction. They cooperatively staged class difference and affiliation through scripted performance and offstage act, working out the mutually constitutive roles of star and audience, hero and followers, even demagogue and mob. The collective identities those performances rehearsed by no means represent foregone conclusions. As David Roediger has argued, racial and class attitudes and identities seem always complex and often self-contradictory, and the performances they patronized reflect such contradictory desires and modes of self-awareness.[29] Stabilized identity, if it ever exists, emerges only in conventions repeatedly rehearsed and always in need of reinforcement. In the 1830s, Forrest's public may have self-identified as Jacksonian democrat or working-class, but they also entered the playhouse as theatregoers attracted by the playhouse's social opportunities, the cachet of performance, and lowered prices. Certainly, class and racial politics entered the playhouse, but it accompanied the processes of learning how to be spectators for the emergent forms of melodrama and star celebrity culture.

THE MUSHROOM WARRIORS OF ROMAN DUNGHILLS

Some assessments find *The Gladiator* a conservative and pessimistic reenactment of American class dynamics. Bruce McConachie, for example, reads Forrest's heroic plays as allegories, even predictions, of failed working-class emancipation. Invariably, the epic hero inspires a violent revolution against an oppressive and much more powerful foe, only to be undone by his own followers. The plays thus only superficially assert democratic ideals, elevating the hero and further victimizing the low. The interpretation, though pessimistic, is believable. Forrest, playing the part with vigor and action, dominates his supporting cast and his audience just as his heroic characters govern their common followers. The play's lower-class characters, divided and confused, seem to abandon their best chance at freedom, undermining their own rebellion. Worse, the people do not simply fail to emancipate themselves; they actively turn on and defeat the rebellion, working against their own best interests and making such plays at best hero-worshiping and at worst simply antidemocratic. As McConachie concludes, such productions in the end declare the "fickle, childlike nature of the people."[30]

My understanding of the play both accepts and diverges from this argument. Bird's script develops a more intricate vision of class dynamics than McConachie's interpretation suggests. However, he and Forrest apparently collaborated to shorten and simplify the script for performance. Four extant manuscripts transcribe Bird's revisions of the play and his writerly processes, from taking notes on historical sources to negotiating the material staging in the playhouse.[31] Those manuscripts reveal complex processes of negotiating over how to ventriloquize and verbally describe the play's rebellious gladiators. They also reflect a central struggle between ways of shaping the underclass hero: Bird's version scripts a tragic figure whose flaws lead to his downfall, and the cuts resulting from Forrest's input seem to have produced a more dynamic and perhaps melodramatic hero. The substantial cuts almost certainly diminish the complexity of Bird's characterization, producing a more simplified view of revolt and class dynamics; indeed, as McConachie suggests, the cuts reinforce the antidemocratic scene of heroic leader betrayed by unreliable rebels. In the end, a script merely provides suggestions, inscribing the textual wishes of the performance event. At the same time, it can also reveal the processes and contingencies of performance.

Bird's script, with or without its elisions, imagines complex negotiations preceding its pessimistic result. Its rebels demonstrate talents for constructing and disrupting structures of authority. The play first consolidates a multinational and multiethnic cohort of gladiators, setting the roguish Spartacus at their head. The motley enslaved cohorts provide the raw material—the manpower and the social conditions—for Spartacus's rebellion, and revolutionary solidarity emerges from a motley collection of sub-working-class characters. The play also plots piecemeal productions and realignments of class coalitions. In the processes, *The Gladiator* experiments with transnational and cross-racial characters and their effect on collective affiliations. Even their failure seems due to the complexities of class formation as rogue factions betray the rebellion when Spartacus begins to aspire to aristocratic status. By the play's end, the rabble behaves something like Marx's lumpenproletariat, the outcast subclasses that betrayed the European revolutions of 1848. The underclasses, continually disrupting existing class relations, embody both cause and effect of shifting class formations. Eventually, they exercise their own cagey agency, breaking away and undercutting Spartacus's revolt. The upwardly mobile classes find themselves haunted by the capricious and unpredictable outlaw cohort that performs the grunt work of class formation.

Bird's script imagines a hodgepodge collective built on the contingencies of captivity and servitude. Class appears constructed rather than stable, inevitable, and naturalized. Spartacus's rebellion gathers the "remnants of those tribes by Rome destroyed," a multinational and multi-tribal assortment of characters:

> Arm'd with retributive and murderous hate,
> The sons of Fiery Afric,—Carthaginians
> Out of their caves, Numidians from their deserts;
> The Gaul, the Spaniard, the Sardinian;
> The hordes of Thessaly, Thrace, and Macedon,
> And swarming Asia;—all at last assembled
> In vengeful union 'gainst this hell of Rome.[32]

United by a common foe, the rebellion of "slavish, ragged scum" recalls *Slaves in Algiers* and perhaps even *Three-Finger'd* Jack.[33] Disdained cohorts affiliate across tribal and national lines. At least one newspaper reviewer found that passage, and the transnational underclasses it describes, compellingly poetical.[34] *The Gladiator*'s transnational cohort stages more than the awareness of American culture as a multiethnic nation of immigrants, and more than a fantasy of the erasure of ethnic and national differences. They stage the many-headed hydra of transnational collectivity, the low forms of mutuality that the more respectable classes had long feared.

Spartacus, for example, asserts a primitive sort of anti-imperial and transnational class solidarity when his master tries to force him to fight a fellow slave from Spain. The scene provides the hero with an opportunity to assert transnational underclass solidarity against Rome, "our common foe," as the Thracian declares.[35] Bird seems to have carefully weighed the dramatic possibilities of slaves from Gaul and Carthage, making notes to himself about the historical likelihood of such transnational and intertribal encounters.³⁶ Spartacus's revolutionary class builds on the slaves' collective sense of rootlessness, of disconnection, of dispossession—traits that had consistently defined outcast low characters in American theatre. The play wields this understanding of class deliberately, and eloquently. Spartacus refers to his men as "the mushroom warriors of Roman dunghills," the dirtiest, least respected, and most disdained cohort in Roman society.³⁷ Audiences recognized the play's transnational cohort, remarking, for example, on the collection of "great numbers of peasants, slaves, pirates, &c." who joined Spartacus's rebellion.³⁸

Those underclasses become a characteristically stagey group, and *The Gladiator* traffics openly in the self-conscious display of theatrical low violence. Bird's script imagines gladiatorial fight scenes as popular theatre, a "spectacle" before the mobs of Rome. Their theatricality marks them as the troubling subjects of economic and cultural exploitation. They are slaves, but they fight rather than work, laboring at (literally) unproductive performance. As a gladiator, Spartacus himself is an actor, fighting in public performances. Those scenes, of course, allowed Forrest to show off his muscular physique and appealed to audiences; they also present a metatheatrical critique for those inclined to sense it. The gladiatorial show presents a miniature play-within-a-play that subtly indicts the popular American melodramatic theatre. As Crassus, a Roman Praetor, remarks, "my flesh always creeps, to see these cold-blood slaughters."³⁹ The line reveals, perhaps, a latent distrust of such performances and for the laborers who produced them. The assessment seems motivated by the traditional American anti-theatrical prejudice, but also distaste for the unproductive labor of theatrical entertainments.

Such acts remain, as they had long been in Anglo-Atlantic culture, troublingly appealing. Julia, the niece of Crassus, models a feminized response to such rogue performances: "I swear I ne'er will look again. But when / They battle boldly, and the people shout, / And the poor creatures look so fearless, frowning, / No groaning, when they are hurt.—Indeed, 'tis noble!"⁴⁰ Theatrical display renders rogues noble. The lines recall the forms of spectatorship made famous in *The Beggar's Opera*, whose women find an outlaw never more handsome than when he is going to die. In a line marked for deletion, Julia also describes Phasarius as "a handsome rogue," who "kills a man the prettiest in the world."⁴¹ Feminized spectatorship responds, as it did in *The Beggar's Opera* and later in Stoker's circus hanging act, in conflicted ways to the dangerously attractive scenes of rogue (self) execution.

Spartacus himself stands at the head of this cohort, embodying the instability of his roguish cohort. The script plots Spartacus's roundabout journey from alien outsider to monumentalized hero of the urban working classes. Before his capture, he is a shepherd, but also a sort of rogue leader. Spartacus in fact has a Thracian rap sheet, a prehistory of outlaw activities. The Roman gladiator master Lentulus describes him as a "leader of a horde of his savage countrymen," appealing as "the most desperate, unconquerable, and, indeed, skillful barbarian in the province."⁴² Crassus too points out Spartacus's rogue position, both outlaw and charismatic, both slave and natural leader, when he comments that the "rogue is not a common one."⁴³ This last passage, marked for deletion, may not have made it to the stage, but such language litters Bird's script, imagining Spartacus (following typical stage formulae) as admirable and uncommon, heroically outcast. The script produces Spartacus as the same sort of captive rogue that American theatre earlier applauded in Macheath, Three-Finger'd Jack, and other antiheroes.

Spartacus's ascendancy to a leadership role during the course of the revolt occurs effortlessly, as if to confirm his natural claim to pastoral leadership. *The Gladiator* endows its Thracian hero and his supporters with a class mobility that construes his society—and by implication American society—as moldable, manipulable. Spartacus's dreams of upward mobility, however, begin to crumble as his emancipatory revolt turns to attack Rome itself. After this hubristic turn, underclass disorganization and reorganization begin to thwart the rebel leader's growing authority. Competing and unreliable intertribal affiliations help speed its defeat, and the gladiatorial revolt experiences its own sub-rebellion. The "mad mutiny," as Bird's script labels it, which splits his coalition at a crucial moment and gives victory to the Roman forces.[44] This plot point provides commentators such as McConachie with the evidence that the play takes a conservative and pessimistic view of low agency. However, Spartacus's response deliberately troubles that assessment. Faced with insubordination by German confederates and by his own brother Phasarius, Spartacus indignantly abandons his erstwhile followers to defeat, exclaiming, "did I not lead them ever on to victory? And did they not forsake me? Wretched fools, this was my vengeance, yea, my best of vengeance."[45]

The script calls for an even more telling and provokingly hubristic reaction from Spartacus, one in which he assumes his oppressors' identity. "Even as the Romans punish, so I'll punish," Spartacus declares in a line that Forrest's performances most likely omitted.[46] In the final act, Spartacus begins to echo the language of the Roman oppressors; he uses the same scornful "Sirrah!" with which the Roman masters head earlier addressed their gladiator slaves.[47] As Spartacus achieves military success, his revolt generates its own authority structures, and power begins to corrupt him. The underclasses seem allergic to success. More importantly, this turn of events repositions Spartacus as a tragically flawed hero, opening up generic possibilities for understanding his character that Forrest's proposed cuts would foreclose. The tragic turn has a double effect, rendering Spartacus noble even as it dooms him to defeat for betraying his own principles. The events of the play's ending have a quality of inevitability about them reminiscent of earlier defeats in Atlantic theatre. The stage conventionally rewards rebellion with death; like Morano or Three-Finger'd Jack, Spartacus seems "born to die." The rebellious rogue cannot last long in the face of organized authority and power.

With things falling apart around him, Spartacus finds he must rely for escape upon a gang of pirates, who "are most treacherous hounds, and may set sail without us," as a lieutenant says.[48] This subclass troubles once more any assessment of the play's conservative class politics. *The Gladiator*'s rebellion eventually devolves onto pirates, an outsider cohort that offers the play's last hope for democratic rebellion. The magical appearance of the pirates reveals the play's structural reliance on such outcast groups—if they do not already exist, the dramatist will invent them. The drawing of class lines in such scenes seems to rely upon the presence of a social surplus, an excess on the underside of the more stable classes. Pirates seem a particularly resistant and theatrical cohort, much like Spartacus's rebellion in its early stages. They also represent a logical embodiment of this excess in 1831, when nautical melodramas saturated Atlantic stages with the mythologies of outlaw mobility, piratical democracy, and mutinous self-determination. Thus, as Spartacus begins tragically to act out the rhetoric of the Romans, the play turns elsewhere for its rebellious energies. Although only a minor presence in the play, the deserting pirates act on a dissenting mode of collective agency that opposes Spartacus's upward mobility and ominous tyranny. As Spartacus fears, the pirates set sail without the gladiatorial rebels, assuring both the hero's defeat and their own breakaway survival in a cyclical return to Spartacus's initial uprising.

As they must in Forrest's starring performance, the play's attention ultimately returns to the hero rather than the breakaway low. After a rather predictable sequence of heroic and doomed stands, Spartacus sinks to the stage, a heroically overcome character. Spartacus's revolt proves its valor and nobility most in failure. In an ending that Bird worked and reworked in his notes (and that he ultimately marked for deletion), his Roman archenemy Crassus declares that Spartacus should be regarded "not as a base bondman, but as a chief enfranchised and ennobled."[49] Although the Romans "denied him honour while he lived, / Justice shall carve it on his monument," Crassus declares.[50] In its final scene, the script re-enfranchises and honors characters that began as rogue rebels. It also sentimentalizes the low in scenes whose way had been paved by scenes from the likes of *The Beggar's Opera*, *Three-Finger'd Jack*, and *Slaves in Algiers*. At the same time, the play seems compelled to reenact its rejection of Spartacus's status, both evoking and erasing the memory of Spartacus as "base bondsman." In perhaps the last irony of such elaborate processes Forest and Bird seem to have agreed that they could do without it on stage. The probable nonperformance of Crassus's lines, however, only gives his emotional response increased immediacy. Rather than relying on Roman observers to verbalize the scene's affective values, Forrest himself acts out the sentiments in his self-monumentalizing death. The revolutionary slave, emotionally if not verbally memorialized, finally attains full though posthumous membership in the body politic.

His achievement depicts a political impossibility for the literal (and most likely African American) "base bondsmen" in American political culture, but it enacts a scene of great value to working-class audiences who might imagine their own membership in the body politic. Although its final heroic apotheosis does not actually seem optimistic about the possibility of liberating revolt, *The Gladiator* at least acknowledges the volatility of class relationships. The story acts out the upward urges of the low, energies that Spartacus harnesses, ordering the disparate parts of a fragmented underclass. Ultimately and tragically, he replicates the social order against which he rebels. The achievement of status and respectability (the lack of which has defined his revolt) becomes his defeat, leaving his revolt undone at play's end. Against that heroic downfall, the play's return to chaos and the piratical cohort subtly poses the potential of rebellious mobility, slipperiness, and opportunism. The insubordinate horizontal itinerary of unreliable outcasts thwarts the hero's upwardly mobile ambitions.

THE BLACKNESS OF CLASSICAL ANTIQUITY

The specter of blackness haunts its performance, working within the scenes of the transnational rabble. Although avoiding explicit reference, Bird's play cannily recruits the forms of stage blackness. Spartacus himself, in the opening scene, insists upon his status as a slave—he was bought, sold, and scourged in ways that could hardly fail to invoke images of American chattel slavery. However, the play apparently *did* fail to invoke African American slavery openly, but its staged Roman slave revolt never completely escapes the shadow of American race-based slavery. Behind the play's monolithic white masculinity lies a fascinated and covert (and perhaps more involuntary than calculated) flirtation with the forms of blackness. As Ginger Strand has argued of Forrest's performances, traces remain of the very categories (including the homosocial and homosexual) that the play seems to exclude or repress in its construction of populist masculinity.[51] In similar ways, Bird's play invokes and exorcises blackness in the service of performing working-class identity in antebellum American theatre culture. Blackness, though performed with a certain plausible deniability, centrally constitutes the urban white working-class identity performed by *The Gladiator*. The act reveals not merely the conscription of blackness, but the interlinked construction of group identities on the categories of class and racial identities.

Bird's play more obviously uses the forms of classical antiquity to navigate among the marked and unmarked political positions of nineteenth-century New York City. The play shapes classical history into scenes that critics such as Arthur Hobson Quinn could take for granted as representing "just as truly the spirit of America in 1830 as it did that of ancient Thebes."[52] Bird's play, in this line of thinking, demonstrates the antebellum democratic inclinations that allowed antebellum Americans to identify with Roman republicanism. The shaping of domestic issues in neoclassical forms was hardly unusual in early American theatre. As Caroline Winterer has shown, early Americans continued a long tradition of appropriating the relics of Greek and Roman antiquity as political and social models.[53] Americans came by their neoclassicism by way of English republicanism and revolutionary France's radical emulation of Roman models, but it extended well beyond the realm of politics.[54] Although many neoclassical forms primarily circulated by elite modes of education, they also became common currency, inevitably diluting and transforming in multiple transmissions. Appearing in architecture, literature, visual arts, and more, neoclassical and especially neo-Roman forms loom large in the American collective imagination. At the same time, such forms seem to inhabit the public consciousness covertly.

It is worth pointing out the perhaps obvious fact that such cultural recycling had long separated "classical" culture from its original contexts. American neoclassicism represents the continued renovation and redeployment of older forms in ever-changing contexts. That, of course, is precisely why slave revolt could appear in classical garb, and why such a play could become Forrest's plebeian star vehicle. As Edwin Miles points out, American culture's relationships with classical relics were always on the move.[55] By the 1830s, the classical symbols that had once dramatized political and cultural elitism had become popular (and frequently romanticized and sentimentalized), and the democratic connotations of Greek neoclassicism were replacing Roman republican forms. Older and more exclusivist modes of classical aesthetics had fallen into disfavor at the end of the eighteenth century. Neoclassical imitation appeared "servile" next to emergent Romantic individualism and invention, as Eric Slauter has pointed out.[56] Despite the connotations of neoclassical imitation, the stage literally provided a place where Americans mimicked Romans. Addison's 1713 *Cato*, for example, was reputedly a favorite of George Washington, one of the few plays to evade the official ban on theatricals when the Continental Army performed it in the winter of 1778 in Valley Forge. Neoclassical tragedies such as *Cato* constitute a dominant genre among Revolutionary-period patriotic plays; their models of "transactional death, mourning, and memorial" bolstered American political identity, as Jason Shaffer argues.[57] Later, Forrest made John Howard Payne's 1818 *Brutus, or the Fall of Tarquin* and Sheridan Knowles's 1820 *Virginius* two staples of his early career. Such popular dramas suggest a public obsession with reenacting classical cultural references, creating cultural genealogies with the found materials of ancient Rome.

Americans had long imagined themselves as the "Roman republic reborn," even if such poses did only amount to political "window dressing," as Bernard Bailyn argues.[58] Such posing and self-naming acts, however distant from any political reality, seem precisely the kind of cultural performances that shaped the classical posing of Forrest's hero. *The Gladiator* stages multiple typologies. Its neoclassical scene required audiences to understand Forrest as simultaneously Thracian gladiator, rising Roman, and American working-class hero. Likewise, audiences must see Rome as once republican but also imperial, both populist and oppressive, the source and the antithesis of American egalitarianism. As theatre tends to do, the play allows audiences to identify simultaneously with different aspects of its symbolic logic.

As *The Gladiator* shows, the templates and conventions of neoclassical borrowing remained powerful and persuasive, but also reshaped to appeal to Romantic and popular sensibilities. *The Gladiator*'s classicism also depends crucially upon the near-simultaneous invocation and erasure of slave revolt's blackness. The convergence of those two seemingly disparate discourses, classical aesthetics and black cultural forms, produces a distinct "classical blackness." Classical blackness involves the various, complex, and frequently ironic application of recycled antiquity to New World blackness. Those mythic histories in turn experience an associated blackening. As Eric Slauter has argued persuasively, both neoclassical culture and representations of blackness can raise issues of subjectivity and humanness, originality and imitation. For example, Anglo-Atlantic audiences can understand Phillis Wheatley's neoclassical forms and her embodied blackness as linked, and both relevant to questions of subjectivity and agency. Who gets to produce original culture, what models they ought to follow, and what that implies about their humanity seem deeply at issue. Forrest's classical slave revolt likewise works out questions of social status, race, and agency, while masking its investments in the imitation of nonwhite forms.

The script's alterations reveal some sensitivity about the connotations of blackness, as when Bird replaced the label "lying African" with the more neutral "man of the Punic stock."[59] On the page facing a line in the first scene of the fourth act, Bird penned a note to himself opposite the word "slavish": "Shall I substitute vile, odious, degrading, or some other word?"[60] Certainly, the revision could simply signal Bird's writerly concern with verbal style. At the same time, the note suggests that he contemplated carefully the connotations of the word slavery in his drama. Even questions of aesthetics, of "mere" verbal style, have social implications. If only obliquely and tentatively, *The Gladiator* negotiated broader discourses of race, slavery, and violence in an antebellum culture charged with fears of slave insurrections. Early commentators on the play occasionally betray a forthright awareness of race that seems surprising in view of later critical focus on a whitened working-class identity. Whitman, inclined to celebrate *The Gladiator*'s labor ethos and idealize laboring solidarity, famously observed that Bird's melodrama was "as full of 'abolitionism' as an egg is of meat.'"[61] Whitman's later review suggests that *The Gladiator*'s audiences could hear the resonances between Roman and American slave revolt, even if the text reveals absence where repressed or discounted knowledge might reside.

Bird's script, however, does not simply whitewash the blackness of antebellum slave revolt. Instead, the play continually foregrounds echoes of American chattel slavery. *The Gladiator* seems intent on reminding audiences that it enacts a *slave revolt*. As Spartacus declares, he is "a slave. I was bought, I say, I was bought."[62] Reviews note the "deeply affecting" scene of Spartacus's enslaved arrival in Rome, his "whispering agony" evoking the "sad music of an exile."[63] Spartacus also endures an auction that forcibly and emotionally separates his family members. "They are human," Spartacus protests when Roman owners "put them up for sale, like beasts."[64] The scene affectively anticipates the abolitionist theatre that would become immensely popular at mid-century as stage versions of Harriet Beecher Stowe's *Uncle Tom's Cabin*, for example, embodied the pathos of the slave trade. At the same time, Bird's script manifestly avoids the emotionally fraught scenes of *Uncle Tom's Cabin*, perhaps simply because its slaves are white. *The Gladiator* strategically embodies slavery in ways similar, for example, to Hiram Powers's 1844 sculpture *The Greek Slave*. Hiram's sculptural materialization of slavery, as Joy Kasson argues, invokes African American slavery while deflecting the emotional impact into moral outrage over the captivity of a white female body.[65] Dion Boucicault's 1859 play *The Octoroon; or Life in Louisiana* also staged the problem of the white slave. Perhaps, amidst the increasing antebellum national tensions over slavery, it seems increasingly possible to imagine racialized slavery. Boucicault's play poses white slavery as not simply the problem of a Greek woman, but the dilemmas of mixed-race characters in America's stratified systems of racial identification. Such scenes undoubtedly owe part of their effectiveness to earlier scenes of neoclassical slavery such as those in *The Gladiator*. Spectacular-yet-veiled representations stage forms of blackness strategically (if unconsciously) evacuated of their black content. Forrest's success, and the eventual enshrinement of *The Gladiator* in American dramatic history, depended upon the ability to simultaneously embody and erase blackness in a performative deniability.

The Gladiator participates in the theatre's habitual merging of blackness and classical culture. Addison's 1714 *Cato*, for example, better known now for its Roman patricians, also features the black character Juba. The North African prince simultaneously embodies colonial dilemmas and sentimental masculinity. Tellingly, American spectators sometimes identified Juba as a more memorable or important figure than the stoic republican patriarch Cato. In one of the more famous and improbable examples, George Washington contemplated amateur theatrical performances during the Seven Years' War, imagining himself playing *Cato*'s black character. Writing to Mrs. George William Fairfax, he envisions himself "doubly happy in being the Juba to such a Marcia, as you must make."[66] Washington, who seems more likely to be identified with the patriarchal Cato (and later Cincinnatus), imagines himself as a character at least nominally black. The letter reveals the perhaps-unconscious depth of early American investments in a classically shaped blackness.

Washington's imagined scene suggests that blackness provided a sort of doping agent for the "optic white" (to borrow from Ralph Ellison's *Invisible Man*[67]) staged by Cato's neoclassical decorum. Juba makes Cato that much whiter. *Cato*'s reception, however, also suggests that audiences noticed and appreciated the very blackness structuring the play. Washington's loving theft of the role of Juba is only the tip of a cultural iceberg. Julie Ellison has identified a counter-tradition of scenes that repress the "cultural memory of Addison's hero" Cato in favor of Juba; the North African prince emerges as a "quasi-ethnographic" marker of African American slave performance.[68] By 1801, for example, when Anglo-Irish writer Maria Edgeworth published her novel *Belinda*, Juba had become a portable type, a contemporary black character pursuing a cross-racial relationship. By the turn of the century, Juba evokes African American slave performance more immediately than patrician republicanism. Those processes of reconstructing Juba appear in higher resolution in 1842, when Charles Dickens toured the Five Points and extolled the skills of the African American dancer William Henry Lane, who performed under the name "Master Juba."[69] Juba seems to have come unmoored from its historical antecedents, and particularly from the patriarchal, republican drama *Cato*. He had become instead a sort of real-life minstrel show character. Dickens's account of Juba acknowledges in print the ongoing subterranean processes of association and transformation that produced an intertwining antebellum classical blackness.

Changing associations of blackness and classical culture seem partly due to apparently random or indiscriminate use of symbols. Such recycling, however, is never completely random, and it can never entirely erase residual patterns of meaning. The redeployments result in a mix of preexisting symbolic meanings and emergent new patterns. In the 1830s, for example, the young and fast New York City sporting set knew Cato as a "gentleman of colour," the proprietor of a tavern four miles outside New York City on a popular riding route. A travel account by Irish actor Tyrone Power comments upon Cato, revealing the knowing and often ironic production of classical blackness in American popular culture: "And what is Cato's? and who is Cato? Shade of Rome's patriot and sage, anger not! for Cato is a great man, foremost amongst cullers of mint, whether for *julep* or *hail storm*; second to no man as a compounder of *cock-tail*, and such a hand at the *gin-sling!*"[70] Rome's "patriot and sage," a symbol of neoclassical patrician power, becomes an ironically humorous symbolic element of popular culture. Political prestige silently becomes bartending greatness. The position also conferred symbolic membership in the urban underclasses. As Michael Kaplan has shown, Cato Alexander's tavern hosted both plebeian unruliness and elite slumming.[71] The same kinds of people that imitate *Tom and Jerry* patronize Cato's tavern. Tyrone Power himself follows the model of Tom and Jerry's flâneurie, touring the interlinked sites of low entertainment. The actor's jesting irony relies on the dissonance generated by classical symbols and black referents; perhaps more importantly, it shows the spread of such symbols into popular culture and semiconscious common usage.

Cato, attaching blackness to his tavern via his classical name (a name perhaps bestowed by a former owner), noiselessly parlays a classical symbol into an iconic brand name of popular sporting culture. The Irish actor touring the American theatrical provinces in turn reuses classical blackness, a compounded and ironic structure of feeling, to his own cultural advantage. Cato's original naming or self-naming recedes behind Power's narrative encounter, and the real event here is the actor's fluent transcription and transmission of Cato's symbolic presence. The actor's delight at finding an African American man named after a legendary Roman statesman suggests Cato's appeal as a richly evocative and heavily laden symbol of classical blackness. Classical mythoi drape the antebellum period's low, black, and unruly cohorts, the new figures of cross-class cultural charisma. By mid-century, markers of such classical borrowings had circulated and adapted in ways that naturalized and blackened ancient borrowings—as various Catos and Jubas show, elements of classical culture had grown figuratively blackened in the American popular imagination.

Classical blackness certainly did not always (if ever) involve equitable transfers of cultural power. Roman drapery did not consistently confer dignity and nobility on blackness, for example. As Tyrone Power shows, classical blackness more frequently inspires satire and mockery, responses to the disjunctions of cultural power perceived between signifier and signified. The meteoric rise of blackface minstrelsy produced another, if unlikely, convergence of blackness and classical culture. Blackface acts featured a battery of characters with names derived from Greek and Roman history and legend. Along with the predictable Caesars and Pompeys, the black dandy "Zip Coon" reveals the various deployment of classical blackness. Popularized in George Washington Dixon's blackface acts, Zip Coon satirized class pretension by way of mock Roman political and military heroism. Edward Williams Clay's *Life in Philadelphia* prints satirized his type, and he became a regular on the blackface stage. The name also has neoclassical associations; "Zip," a corruption of "Scipio," masks a reference to Scipio Africanus, a Roman general in the Punic wars who had entered the western imagination as the conqueror of Carthage and an opponent of Cato the Elder. Scipio was also known for his legendary refusal to dishonor a captive female slave. Petrarch's fourteenth-century epic *Africa* as well as Handel's 1726 opera *Scipio* had celebrated the nobility, moderation, and magnanimity of the character.

The persistent presence of Scipio reveals more than an orientalist-style construction of alterity, however. It embodies the repeated resuscitations performed by Europe's elitist literary traditions. In the figure of Scipio, western culture conscripts symbolic African identity, first to define nobility and then to stake their claim to the same values. At least some members of antebellum American audiences would have known such characters and their uses in western culture. Of course, the demeaning irony of Zip's neoclassical blackface seems apparent even without neoclassical cultural literacy. The blackface application of the noble name fails to ennoble. It instead performs demeaning satire, pointing out the lack of the epic, heroic and self-restrained qualities of minstrelsy's Scipio. Scipio Africanus's act of self-control evokes its opposite in a blackface dandy's threatening sexuality. At the same time, the act can mock the class pretensions of America's classically educated elite, taking away one of their sacralized icons. Minstrelsy's "Zip Coon" (a demeaning epithet replacing "Africanus") subverts Scipio's cultural legacy. Its satire shows the elastic availability of classical antiquity to articulate blackness, if only to use it to produce new social orders of racial difference.

Travesties and caricatures, despite their ambivalence, suggest that classical blackness was a widespread and well-known practice. Audiences had long encountered similar acts of classical naming offstage. Generations of slaveholders frequently applied classical names to slaves of African descent, publicizing the names in runaway slave notices. Later, free African Americans such as the New York tavern keeper Cato inherited or perpetuated such naming practices. David Douglass, one of the major shapers of early American theatre, had a pair of slaves named Apollo and Bacchus. Those "two great gods of the theatre," as Odai Johnson observes, perhaps worked backstage, nearly hidden reminders of the ironies of presenting blackness in classical forms.[72] As Trevor Burnard explains of Jamaican practices, owners "ransacked classical literature to come up with Apollo, Jupiter, Adonis, Ajax, Philander, Hercules, Hannibal, Mercury, Neptune, Daphne, Dido, and Juno."[73] Although the naming practiced by slave owners certainly differed in different times and places, the major themes seem to persist in North America as well. Crucially, many of the names invoke heroic characters—the sorts of types antebellum Americans would have recognized in *The Gladiator's* Spartacus. Certainly, this might simply replicate biases in the way that Americans transmitted and received Greek and Roman cultural survivals—the selection, availability, and popularity of heroic texts (including dramatic texts) supplied models to the name-giver. At the same time, owners could have applied (and frequently did) exotic African, biblical, or traditional English names. Naming involved choices, perhaps conscious and even ironic, to apply labels in the classical and heroic registers.

Such neoclassical names articulate imposed representations rather than self-representations, reflecting owner's power rather than a slave's self-constructed sense of identity. Above all, such labels show the owners' desires to perform their gentility through the dual possession of classical learning and African slaves. Applied to the enslaved, classical naming becomes inherently theatrical, acts of renaming rather than simply naming, labels that apply pre-existing character types to a person. The named person (in these cases the slave), inevitably, in some fashion, performs a role relative to the assigned character. Such acts do not flaunt their motivations. It seems nearly impossible to know what motivated these ritual invocations of classical culture, what kind of script a slaveholder follows and rewrites when naming a slave after a heroic character. On the charitable end of the spectrum, we might speculate that owners meant to elevate the name's black bearer, designating nobility or perhaps noble savagery. Just as likely, if not more, the name performs pity or even something of a parody or farce, mocking the bearer's inability to live up to the model scripted by such nobility. As a public event, the naming ultimately becomes a matter of collective interpretation, to some extent unmoored from "authorial" intention. In any case, such acts show a thoroughly theatrical sense of identity, casting actors in roles that could turn with equal ease to tragedy or travesty.

Self-naming practices followed entirely different cultural codes, and slaves could contest their masters' attempts to reshape slave identity. Intriguingly, advertisements sometimes tell of slaves discarding such appellations. Richard Stillwell of Middletown, New Jersey, publicized a particularly vigorous example of slaves challenging the assigned role. Stillwell advertised more than once in the 1750s for the apprehension and return of his slave "named *Cato*, alias Toby." The "sly artful fellow," Stillwell's notice informs readers, "pretends to be free"; and had "since his Elopement changed his Name several times."[74] Toby perhaps renamed himself simply for practical reasons, to facilitate his escape; even so, his self-renaming rejects the symbolic orders imposed by slaveholders' naming practices. Perhaps with self-conscious irony, Toby invents a new role. His act strategically takes advantage of the very theatricality of identity presupposed and even produced by his owner's neoclassical name. The classical names imposed by owners, as Cato/Toby (and many other escaped Caesars, Pompeys, and Scipios) shows, are entirely superficial, disposable. Names imposed from without do not exercise uncontested powers to articulate the reality of the named. Names articulate but do not fix some version of identity, connected to but also strategically divorceable from their bearer. That is precisely why owners could apply classical labels to African American bodies, and, conversely, why embodying slave revolt as classical can articulate a compelling rendition of low blackness.

FORREST'S PHYSICAL EMBODIMENT OF CLASSICAL BLACKNESS

Forrest's own embodied performance infused *The Gladiator* with forms of blackness. As contemporary audiences understood, Forrest used the prize plays to focus attention on himself rather than on the script. Whatever it says about Forrest as a person, the actor's legendary contractual disregard for his celebrity-producing playwrights confirms that his performances aimed at the physical display of the spectacular acting body. A script's verbal virtuosity seems merely incidental to the process. As one review observed, *The Gladiator*'s "claim for approbation rests principally on what is technically termed *stage-effect*, and also, upon its happy adaptation to the peculiar powers which characterize the acting of Mr. Forrest."[75] In terms that Bird found less than complimentary, the reviewer describes the play as "a drama of startling incident—of intense passion—of hurried action; and its progress is never retarded by philosophical soliloquies, nor by strains of sentiment merely poetical."[76] Audiences knew the actor's performances primarily as intensely physical experiences, and reviewers sensed the importance of his embodied style. Forrest's stardom centered on the raw, dramatic forcefulness of his physicality. Later reviewers remarked upon his "commanding" presence and the "most perfect symmetry" of his physique, traits well suited to the demands of neoclassical roles.[77] Audiences found Forrest's face "handsome and expressive," his black eyes "capable of exhibiting, with equal truth, every gradation of feeling, from the meltings of love, to the lightning glances of revenge."[78]

Audiences could interpret Forrest's body as affective, but also as physically forceful in ways that invoke the laboring underclasses. One reviewer evocatively described Forrest's performance as labor, working "sturdily, and with great outlay of muscular power."[79] Bird himself wrote in his notes that "no other man could sustain the labour" of the role of Spartacus.[80] Forrest's performance evokes toil, both African American labor and the work of white, urban working classes tenuously clinging to claims of whiteness. The interpretation of Forrest's body as simultaneously playing and working is doubly ironic. He certainly did not perform on stage the kind of labors that many in his audiences had to carry out on a daily basis, and the review attempts to transform the privileged leisure of theatre into productive labor. At the same time, Forrest's act presented the forms, and potentially the values, of work and physical toil on the stage. The focus on Forrest's muscular physicality suggests that audiences valued him and his performance style for some of the same reasons (perhaps unconscious ones) that slave buyers and dealers valued slave African bodies in marketplaces—their ability to perform work, brawnily, with "great outlay of muscular power."

The review also defines Forrest's performance as unpainted and genuine. "It is no painted shadow you see in Mr. Forrest, no piece of costume," the reviewer wrote, "but a man, there to do his four hours of work."[81] Forrest evidently embodies the obverse of blackface's increasingly popular imitation. The remark shows a dim and perhaps repressed awareness of the force of blackface, the literal "painted shadows" on antebellum American stages. It also represents a perhaps inadvertent irony of theatrical interpretation. The reviewer probably knew that Forrest had built his early career with performances of black characters. After the actor first made a hit at the Park Theatre in 1826 as *Othello*, he drew praise for his embodiment of the "intricate and heart-rending perplexities of the noble and credulous Moor."[82] In 1829, two years before he would embody Spartacus, another reviewer declared that the actor "always elicits continual admiration, by the force, beauty and discrimination of his conception and execution" of Othello.[83] Although Othello (like other stage representations of Moors and North Africans) could appear on stage with varying skin tones, antebellum audiences came increasingly to read the character in terms of New World blackness. Audiences and critics, praising the beauty and skill of Forrest's Shakespearean blackface, seem capable of completely ignoring Othello's relationship to literal blackness even while increasingly reading the lesson of the play as explicitly racial.

Moreover, Forrest's biographers assert that he built his success not on tragic performances like *Othello*, but on the very different, more physical, and disdained genre of blackface comedy. Biographer Montrose Moses writes in passing that the actor "joined the ranks of the cork-faced players" while playing in Cincinnati, and Richard Moody describes Forrest's blackface and even brief stint with the circus in 1823.[84] Antebellum players commonly served a sort of apprenticeship playing popular genres to less numerous audiences on underpaying rural circuits. Most of Forrest's contemporary reviewers, however, seem less eager to report that part of his history in their celebrations of the novice player's untrained genius in tragedy. The apprenticeship in blackface does not fit so well with the career path Forrest envisioned for himself. Compared to the myth of the untrained genius emerging as Othello on New York City's Park Theatre, the blackface prehistory demotes Forrest doubly from metropolitan to rural and from tragic to comic.

The connection between minstrelsy and Forrest's later melodramatic style eventually entered the body of popular lore audiences used to interpret Forrest's performances. After mid-century, for example, Artemus Ward claimed the working-class actor for the minstrel tradition in a satirical description of the theatre. Ward's spectators discuss the actor's career, concluding that he "'does a fair champion jig...but his Big Thing is the Essence of Old Virginny,'" a blackface dance popularized, according to Hans Nathan, by minstrel performers in the 1840s and 1850s.[85] Ward's account is a satire; Forrest's specialty by the end of his career was certainly not minstrelsy. But the audience's "big thing" was indeed blackface, and to some extent, Forrest's act worked and appeared through those same practices of spectatorship. In Ward's satire, Forrest's tragic and melodramatic thespian skills recede behind the real draw—imagined here as the physical comedy of blackface. Audiences reconstruct the age's greatest emblem of working-class whiteness on stage in burnt cork makeup. Despite their muddled chronology of Forrest's career, such tales restore a lingering cultural memory of the theatrical blackness that had long shaped Forrest's performances. They also provide alternate narratives of Forrest's mid-century ascendancy at the head of a white and sometimes nativist working-class Bowery cohort.

The reactions code a subtly persisting sense that Forrest had trained his acting style in the provincial theatrical laboratories specializing in masked imitations of blackness. The brawny physicality that articulates his performance of Spartacus in fact did emerge from blackface in a tenuous linkage of thespian practice. Even as T. D. Rice and other blackface comedians worked out the relationships between whiteness and blackness in American theatre's regional circuits, Forrest too enacted blackness, first in minstrelsy, then as Othello, and later in surrogate form as Spartacus in *The Gladiator*. Forrest's 1831 embodiment of a Thracian slave improvises on the forms of early blackface, masking blackface tenuously in neoclassical whiteness. With Artemus Ward's satirical and satirized literary theatregoers, audiences could understand Forrest's physicality as contiguous with the energetic and agile maneuvers of blackface performance. The tales implicitly acknowledge that the forms of blackness can indeed shape whiteness. Minstrelsy does shape melodrama, even if observers only mock and satirize that recognition.

Blackface and melodrama appeared in closely connected performances sites. *The Gladiator* competed with the blackface acts it drew upon. As Reese James has noted, the Walnut Street Theatre had brought in T. D. Rice to counter Forrest's new popularity.[86] In 1834, Rice held the Bowery stage until the trumpeted arrival of Forrest from New Orleans, acting a version of *Life in New-York* with scenes written expressly for his popular blackface acts.[87] Later images, like an 1848 illustration of the Bowery favorite *New York as It Is*, reveal the working-class cohort's persistent spectatorship of performing blackness (figure 7.1).[88] The scene, centering on the interplay among observers and performers, reveals the centrality of black performance to the working-class theatrical imagination. It imagines access to that performance style as key to Mose's status.

Mose, the Bowery B'hoy at the center of the scene, does more than simply observe; he becomes an active part of the depicted performance. Lounging against a barrel in the foreground, he visually marks the New York City working-class audience's role in such scenes. The B'hoy's limbs, though at rest, faintly mimic the dancer's moves; both exhibit the hand on the hip, the bent knee, and the turned head that had all become emblematic of blackface by the middle of the nineteenth century. Mose's patronage merges with the performance, making him the kind of insider that Tom, Jerry, and Logic had casually attempted to become. As his relaxed, authoritative pose suggests, the play positions Mose firmly "in the know," a participant-observer

Figure 7.1 James and Eliphalet Brown, "Dancing for Eels; A scene from the new play of *New-York As It Is*, as played at the Chatham Theatre, N.Y." (1848); Library of Congress Prints and Photographs Division [LC-USZC4-632].

(or spectator-actor) moving through the subcultural sites of low performance at the heart of the city's theatre culture. While certainly shaped by outside intrusions of voyeurism and patronage, the scene as importantly demonstrates the theatrical co-production of black and white bodies. Mose and the black(face) dancer both embody cross-racial constructions. Their co-performing bodies appear over and against each other, posed within the converging practices of street culture and the Atlantic entertainment industries. The scene's audience, a mixed-race crowd of onlookers including ragged beggars and well-dressed gentlemen, imagines heterogeneous mixing in mid-century spectatorship. The image suggests the complexities of coding racial spectatorship in the antebellum period. As when circus performances had invaded William Brown's African theatre in the 1820s, violent rivalries of working-class whiteness and minstrelized blackness hardly enact social opposites. Instead, such cohorts remained entangled in interconnected and cross-racial forms of physical theatricality.

The differences between character and actor always seem a bit vexed in such scenes. Like Mose himself, Forrest presented a performance trained in and subtly evoking the forms of stage blackness. Certainly, his act represents an alternate mode of minstrelsy's physical performances. Unlike the cagey, wheeling, disempowered but clever characters of T. D. Rice's blackface acts, Forrest offered a muscular, physically, and even politically empowered type. *The Gladiator*'s blackness comes in a rebellious and tragic package. Its form is more serious than comical, apparently the diametrical opposite of blackface minstrelsy. Only his sentimentalized death produces any hint of weakness, but the play's tragic form ultimately redeems Spartacus's failure. Forrest acts the part of an empowered slave in a script that makes multiple, if veiled, allusions to blackness; Spartacus thus represents a suggestively dangerous revision of stage blackness, a pugilistic alternative to Jim Crow's cagey evasions.

Forrest's performance of Spartacus thus enacts various forms of theatrical blackness without, so to speak, ever filling those forms with content. The same audiences who applauded Forrest as Spartacus understood to some extent both the earnest and mocking forms of classical blackness. Such mixed impulses undoubtedly characterized such performances. Audiences enthusiastically reenacted blackness even if, as Eric Lott has argued, theft tempered their love, even if audiences unconsciously conscripted blackness to build white working-class identities. More radically, perhaps, W. T. Lhamon has argued that white appropriations of blackness have deeper roots in transracial and interracial interactions at the lower levels of society.[89] Even if underclass blackness appears only obliquely in Forrest's acts, the antebellum theatrical practices of linking blackness and antiquity shaped Forrest's performances of slave rebellion. On- or offstage, antebellum Americans seemed willing to imagine blackness in various forms of classical garb. As minstrelsy and naming practices show, these performances could range from mocking to serious. In *Three-Finger'd Jack*, blackness had appeared at once heroic and dangerous, relying on classical antecedents to define Jack's slave revolt. As it had in *Cato*, blackness could supplement noble heroism, but it could also assume the independent cultural life of a minstrelized urban Juba. Stage minstrelsy and offstage naming practices ironically couple classical forms to blackness, ambivalently articulating the aspirations to culture and education that African Americans might harbor. Taken together as a complex of related theatrical and performative modes, these examples suggest that Americans understood blackness as intertwined with classical cultural relics.

THE CULTURAL WORK OF CLASSICAL BLACKNESS

The theatricality of everyday life shapes and interprets *The Gladiator*'s classical blackness in hazy and suggestive ways. Slave naming practices, minstrelsy acts, and the legendary elevation of a few North African characters surely do not force Bird's text to expose fugitive forms of blackness. They show a matrix of taken-for-granted assumptions and naturalized gestures of the relationships among blackness and whiteness, products that erase the traces of their own processes. If such tenuously connected pieces of evidence offer a "smoking gun" of sorts, it is certainly a thin trail of vaporous evidence. Hints and allegations hardly "prove" that *The Gladiator* was "about" race revolt. Instead, they show the play's participation in a complex of gestural, visual, and rhetorical practices that repeatedly experiment with dovetailed forms of classical culture and blackness. Even so, the collective critical brush-off of the play's radical slave revolt requires explanation—or at least historicizing. Forrest's performance of whiteness was neither self-sufficient nor inevitable, as its adaptation of the forms of blackness and blackened performance shows. The only way to reconstruct such historical contingencies, however, is to re-articulate discursive continuities that processes of cultural repression, both contemporary and retroactive, have dismembered and erased. Audiences did not have to acknowledge openly or self-consciously the links between classical culture and American blackness. The unconscious, tentative, and collective cultural urges that gradually transformed a memorialized Roman patrician into an act of blackface entertainment reveal much about the racialized possibilities of classical culture in America. Likewise, imagining a Jamaican bandit in terms of a Roman slave revolt suggests that theatre audiences could act on, if they did not acknowledge, the links between performances of blackness and whiteness. Even the derisive renaming of a slave "Jupiter" or "Caesar" improvises on the same imaginative associations.

Some cultural commentators seem more willing to articulate such connections than others. William Lloyd Garrison himself offered an indictment of American culture's ability to simultaneously envision and avert their gaze from blackness. Writing in the aftermath of Nat Turner's Virginia revolt, Garrison disputes the charge that white abolitionists instigated slave revolt. "The slaves," he writes, "need no incentives at our hands"; they will find them

> wherever you and your fathers have fought for liberty—in your speeches, your conversations, your celebrations, your pamphlets, your newspapers—voices in the air, sounds from across the ocean, invitations to resistance above, below, around them! What more do they need? Surrounded by such influences, and smarting under their newly made wounds, is it wonderful that they should rise to contend—as other heroes have contended—for their lost rights?[90]

Garrison was not writing about *The Gladiator*, but he seems to sense the relationships between white and black expressions of political agency and violent rebellion that responses to the play seem content to erase or ignore. Speeches, conversations, celebrations—the performed "invitations to resistance" that seem thoroughly wrapped in white, working-class, nationalistic, and frequently classical garb—can also invoke the specter of African American slave revolt. *The Gladiator*, activating radical republican rhetoric and performing those emancipatory gestures, enacts those voices. As such, Garrison implicitly recognizes, it performs an incitement to black slave revolt.

Bird himself seems to have realized his play's racially radical implications about the same time, and the archive is not completely silent on the subject. In documents titled "Secret Records," he observes that if "the Gladiator were produced in a slave state, the managers, players, [and] perhaps myself in the bargain, would be rewarded with the Penitentiary!"[91] Various historians have cited that remark, sometimes to suggest the author's radical social awareness. The rather striking passage that immediately follows Bird's remark reveals a far more conflicted understanding of his play's slave revolt. It is worth transcribing at length:

> At this present moment there are 6 or 800 armed negroes marching through Southampton County, Va. murdering, ravishing, & burning those whom the Grace of God has made

their owners—70 killed, principally women & children. If they had but a Spartacus among them—to organize the half million of Virginia, the hundreds of thousands of other states, & lead them on in the Crusade of Massacre, what a blessed example might they not give to the world of the excellence of slavery! What a field of interest to the playwriters of posterity! Some day we shall have it, and future generations will perhaps remember the horrors of Hayti as a farce compared with the tragedies of our own happy land! The *vis, et amor sceleratus habendi* will be repaid, violence with violence, & avarice with blood. I had sooner live among bedbugs than negroes.[92]

The Gladiator's premiere preceded Nat Turner's famous and bloody Southampton, Virginia, revolt by a matter of mere months, and the rebellious slave's death occurred immediately before *The Gladiator* appeared in Boston. Forrest enacted slave revolt less than a decade after Denmark Vesey's insurrection in South Carolina, and hard on the heels of the furor generated by David Walker's revolutionary *Appeal to the Coloured Citizens of the World*. The wonder, perhaps, is that *The Gladiator*'s "Four hundred arm'd slaves, that hate their masters!" did not evoke stronger fears of slave revolt.[93]

Bird's comments remained in the private "secret records," and the script of *The Gladiator* never makes public and literal reference to American chattel slavery or slave revolt. Turning from the spectacle of revolt in his Secret Record, Bird continues histrionically, "But the play, the play—Ay, the play's the thing." The playworld of *The Gladiator* and the aesthetic concerns of dramatic authorship fit uncomfortably with the reality of avarice and the horrors of slave revolt. The passage reveals a remarkable sensitivity to the play's connotations of slave revolt, an awareness that only appears in glimpses in *The Gladiator* itself. Bird's own assessment, however, seems void of the celebratory or emancipatory urges that his play performs. Only a faint sense of slave revolt's aesthetic value ("what a field of interest" to posterity!) glimmers through his unconcealed fear and disdain. Such statements demand careful interpretation. Various scholars have characterized the document as a letter or a secret diary, but it seems both and neither at the same time. It appears a sort of hybrid document aimed at an imagined and anonymous literary posterity. Self-consciously assessing his young career and recording responses to *The Gladiator*, Bird speaks from behind a professionally detached mask. Ultimately, he addresses those "playwriters of posterity" who might appreciate slave revolt's dramatic enormity better in historical perspective, as Bird's own audience does the reconstructed classical revolt of *The Gladiator*.

Both in its script and in embodied performance, *The Gladiator* shapes its figures from a formal mélange of its shadowy others. *The Gladiator* requires its audience to recover the heroic Spartacus repeatedly and imaginatively from among the ranks of a transnational rabble. The nobility of Forrest's performance relies heavily upon disdained forms, classical whiteness resting on the absent implication of enslaved blackness. As Crassus suggests in the play's cut final scene, the play produces its hero as an emblem of patriotic, white workers by simultaneously recognizing and overwriting his status as enslaved and alien. Likewise, the play simultaneously embodies and averts the racial implications of the label "slave." Audiences in turn could both recognize and turn away from the play's theatrical forms of blackness, the minstrelsy still coded into Forrest's theatrical physicality. Moreover, the play presents potentially destabilizing forms; the black heroes that Garrison imagines arising from such performances could easily mimic the antebellum performances of Roman heroes who seem already, in various small and imaginative ways, blackened in the public imagination. Garrison, for his part, seems one of the few observers willing to imagine the connections between such performances, although a subtle and perhaps repressed awareness seems to permeate American theatre culture in various ways. Classical blackness remained a complex and covert performance, invoking and erasing the blackness at the center of performances of whiteness, but also carrying forward a radical charge built up through myriad small acts.

Epilogue: Escape Artists and Spectatorial Mobs

The unruly, energized Atlantic low continued to appear in nineteenth-century American playhouses, their mobility, theatricality, and rebellion captivating audiences. On December 30, 1839, Jonas B. Phillips's *Jack Sheppard, or the Life of a Robber!* premiered at the Bowery Theatre, by that time a dominantly working-class theatre.[1] Phillips was no stranger to the stage forms of underclass life; five years earlier, he had penned the popular *Life in New York*, a local version of *Tom and Jerry* that featured blackface performer T. D. Rice at the Bowery Theatre.[2] *Jack Sheppard*, however, turns from contemporary scenes to present the fictional jailbreaks and punishment of the London thief and jailbreaker Jack Sheppard. Sheppard, the historical model for *The Beggar's Opera*'s Macheath, pursued a short but spectacular criminal career that ended on the gallows in 1724. Ever a century later, the play reenacts Jack's rakish, defiant underclass heroics, reconstructing them after more than a century lurking in the popular Atlantic memory. *Jack Sheppard* unabashedly celebrates the excarceration that *The Beggar's Opera* offers in stylized form. Instead of stylized escape, *Jack Sheppard* displays literal escapes. However, unlike Macheath, Jack dies on the gallows at play's end. As Odell writes, the thief enjoyed "all the rope he needed" at the Bowery theatre.[3] *Jack Sheppard* offers this study some closure, bringing old forms around to new contexts. Its scenes look to the past for its emancipatory energies, but they also seem entirely of the 1830s, performing a new clampdown on criminal energies.

Jack Sheppard died at Tyburn on November 16, 1724, after a career of petty crime; his repeated prison breaks had made him an improbable celebrity. Jack Sheppard's thievery and housebreaking seem relatively commonplace, especially in light of the wave of property crime that hit London in the 1720s. His breaking in seems no big deal. His breakouts, however, were much more spectacular, epitomizing the excarceral energies that Peter Linebaugh explores in *The London Hanged*.[4] Jack became legendary for his transgressive, boundary-breaking acts, and his youth (he was twenty-two at his death) made him an icon of early Atlantic youth culture. His mechanical skills, honed in his apprenticeship, enabled his breakouts and made them symbolic demonstrations of laboring self-emancipation. His four escapes between April and October of 1724 led to the kinds of profoundly theatrical scenes that Levi Ames and Thomas Mount reenacted across an ocean and fifty years later.

Jack energized the forms of criminal celebrity that shape rogue performances for over a century, and that seem still active today. His crimes and escapes produced a body of popular lore almost immediately; London newspaper reports and popular narratives retold his exploits and disseminated his popularity. He became the subject of theatrical forms such as the ballads, flash songs, and even graphic depictions that celebrated his actions. A number of competing histories of Sheppard's exploits appeared; as the *Narrative of all the Robberies, Escapes, &c. of John Sheppard* succinctly states, his life "afforded matter of much Amusement to the World."[5] The anonymous *Narrative* presents the escape artist as the author and narrator of his own adventures, imagining for Jack the chance to set the record straight. Various "Pamphlets, Papers, and Pictures relating thereunto are gone abroad, most or all of them misrepresenting my Affairs," Jack claims in the *Narrative*, making it "necessary that I say something for my self, and set certain intricate Matters in a true Light."[6] The jailbreaks produce imagined and performed social critique. At one point, Jack condemns thief-takers such as Jonathan Wild, the man ultimately responsible for his execution. They "deserve the Gallows as richly as any one of the thieves," Sheppard argues; such characters "hang by proxy, while we do it fairly in person."[7] At least in the imaginative form of his criminal narrative, Jack Sheppard becomes a self-representing character, walking to center stage and stating his own case through the narrative's literary treatment.

Jack's exploits made him famous as a charismatically stagey and even dandyish character. Accounts depict him as small but strong and agile. He appears stoic and clever, and prone to nonchalant satire and public displays of extravagant criminality. In one of his most strikingly theatrical acts, he ironically arrayed himself as a gentleman after a breakout. He "furnished himself with a Black Suit of Cloaths, a Silver-hilted Sword, and a light Tye-Wig, as proper ornaments for his Person, and a few Watches, Diamond Rings, Snuff-Boxes, and several other Pretty little Toys, besides a Small Quantity of Ready Cash for the lining of his Pocket."[8] The scene of costuming culminates Jack's career of chronic textile thieving. As Linebaugh shows, the act constitutes a logical form of mutiny in London of the 1720s. The labor discipline of London's textile industries governed the lives of the working classes, and cloth fashioned into costuming differentiated gentility and lowness. The costs of conspicuous consumption and the informal codes of sumptuary display excluded the underclasses from enjoying the goods that they produced, and Jack's act has the ring of deliberate transgression. His final insouciant costuming also reveals outsiders' fear of and fascination with such potential transgressions of class standards. Jack commits the crimes of assuming a gentleman's "proper" adornments with none of the substance. He carries little more than trinkets and some ready cash, the better to perform his role. In the end, his accoutrements fail to hide his true nature, and Sheppard ultimately is taken up again, guilty of piratically "hoisting up *false Colours*," as one narrative phrases it.[9]

Jack's final escape in 1724 became an extravagant demonstration of his rogue theatricality. He ended his excarceral career on a high note, hiring a hackney coach, carrying his sweethearts about town, and carousing near Newgate. The rogue and his cohort even audaciously passed by the prison in the coach, with the "Windows drawn up."[10] He spent the remainder of the evening sharing drinks with comrades, women, even his mother. He seems to have made no effort to hide his identity, converting his fugitive roguery into further public celebrity. Authorities finally apprehended him in a stupor at the end of the evening. Jack's final tour transforms

London into what Linebaugh refers to as the "theatre of his climactic and defiant last act."[11] It is thus not surprising that the Theatre Royal at Drury Lane quickly transformed his exploits into a pantomime, *Harlequin Sheppard*, bringing the housebreaker's exploits in song, dance, and gesture to theatre audiences.[12] Sheppard also provided John Gay with a model for Macheath, supposedly inspiring Gay's "Newgate pastoral."

Even when apprehended, Jack generated a troubling popular appeal; as one 1724 account put it:

> His escape and his being so suddenly retaken made such a noise in the town, that it was thought all the common people would have gone mad about him, there being not a porter to be had for love nor money, nor getting into an ale-house, for butchers, shoemakers, and barbers, all engaged in controversies and wagers about Sheppard.[13]

Common people crowd public spaces, gamble, drink, and publicly debate the chances of resistance to the law. It seems an alternative and threatening version of the emergent public sphere of London's coffeehouses, where aspiring mercantile gentry could do business and place their own wagers on business ventures. The broadsides, criminal pamphlets, and ballads appeared alongside and troubled the sanctioned print-performances of newspapers, royal decrees, and shipping reports. "In short," the account continues, Jack's celebrity produced "a week of the greatest noise and idleness among mechanics that has been known in London."[14]

Jack Sheppard's incarceration and execution became a theatre unto itself. Spectators paid to visit Sheppard's prison cell and hear affecting tales of the thief's criminal adventures and escapes.[15] His death provided the most striking theatricals. Upon his conviction, "the curious from St Giles's and Rag-Fair" awaited his execution, the "rabble" watching the gallows by night "lest he should be hanged incog."[16] Even at his execution, the rogue seems subject to the demands of a spectatorial audience. Two decades before actor David Garrick proclaimed from the London stage that "we that live to please must please to live," Sheppard's final procession embodied a more cutting assessment of a performer's obligations.[17] Sheppard indeed had to submit to the laws of drama's patrons, but performance could not rescue or sustain the rogue. Ultimately, a crowd of people perhaps numbering in the tens of thousands saw his hanging in 1724.[18] The lower sorts were a dominant presence among the large and almost certainly socially diverse audience for Jack's execution. Newspaper accounts suggest their empathy for Sheppard, describing "uncommon Pity from all the Spectators."[19] The crowd even seems to have rioted, their ire aroused by the possibility that Sheppard's body was going to be given to the surgeons for dissection; the same account describes guards alert to "prevent the Violence of the Populace, who had been very tumultuous all Day."[20]

Sheppard continued to circulate in the eighteenth-century cultural unconscious, collective memories of his escapes surviving in Newgate calendars, popular songs, and theatrical reenactments. Jack's story became popular literary fare in 1839 when William Harrison Ainsworth's novel *Jack Sheppard* revived the character. Dickens's 1838 *Oliver Twist* also seems to owe something to Jack Sheppard's youthful criminality.[21] Jack's charismatic lawlessness even produced a moral panic after a

London valet named B. F. Courvoisier, admittedly inspired by the novel, murdered his elderly employer.[22] On the stage, John Baldwin Buckstone's *Jack Sheppard*, a direct adaptation of Ainsworth's novel, premiered at London's Adelphi Theatre on October 28, 1839, creating a mania for Jack Sheppard among London playgoers.[23] Buckstone's play refurbishes Sheppard's scenes and character types, extending the theatrical roguery that Macheath and his gang had made famous a century before. It was, like most Anglo-Atlantic rogues, a mobile act, and American theatres quickly adapted Buckstone's *Jack Sheppard*. The play thus became another Atlantic rendition of rogue performance, condensing once again the traveling scenes of theatrical underclasses.

COSTUMING THE UNDERCLASSES

The Adelphi Theatre's 1839 stagings of Buckstone's *Jack Sheppard* featured popular London actress Mary Ann Keeley in cross-dress as the youthful felon. Cross-dressing as Jack was not entirely new. Early histories of Sheppard's escapes dress him in a nightgown and bonnet in one of his escapes, and his dandyism has the faint ring of sartorial gender transgression. Keeley's performances, however, enact gender crossing with a new intensity. Cross-dress, or more properly breeches performance, became a conventional underclass theatrical technique in the 1830s. Keeley, for example, had also starred as Macheath in "Burlesque" productions of *The Beggar's Opera*. Keeley's *Jack Sheppard* also revises the breeches mobility that Polly acts out in the sequel to *The Beggar's Opera*. Although her adventures end in a restoration of gender identities and proper costuming, cross-dress allows the mobile underclass female character to assert some forms of independence throughout much of the play. Keeley's cross-dress, far from the distracting "lark" that Martin Meisel labels it, became an important marker of rogue theatricality in the 1830s and 1840s.[24] Keeley followed precedents set by performers such as Eliza Vestris, who had garnered fame by managing London's Olympic Theatre while performing Macheath in breeches. Keeley's cross-dressed thief links housebreaking and the sartorial extravagance of cross-dressing.[25] It thus performs a stagey version of the low's simultaneous transgression of property and gender norms.

Keeley's act trades on the comedic appeal of travesty, enduring since the Haymarket's popular productions of the 1780s. It also enacts a sexualized role, showing off the female figure; that bodily display also, however, exhibits a small, supple, sinewy criminal body. The 1840s cross-dressing of *Jack Sheppard* thus codes the gender of lowness. Keeley's act turns on a feminized presentation of small size and physical weakness that also signals a cagey and flexible mobility. Keeley's small, seemingly powerless body fashions the role, and the Adelphi consciously capitalized on her ambivalent and multiple appeals. As a woman playing the youthful rogue, Keeley re-poses the problems of gender, power, and costuming in a low part. The inversive act highlights paradoxical underclass strength in weakness, generating audience sympathy for a resourceful but doomed character. Her stagey character embodies the slippery strength, the suppleness, the resilience of the low. This, too, echoes one of the long-standing appeals of such characters—the strength of weakness, of slippery resistance, the appeal of Macheath dodging the gallows, of Three-Finger'd Jack lurking in his obeah cave, of beggars gammoning a maim to fleece elite voyeurs. The stagey underclasses thrived on such poses, and Keeley's act continued the tradition. Notably, Buckstone's script does not require cross-dressed roguery; the gender reversal represents a theatrical supplement to the story even as it fulfills the play's thematics of sartorial transgression. The material and sartorial practices of the playhouse converge to costume an acting body in ways that only enhance Jack Sheppard's symbolic social value.

The act became popular to the point of transatlantic conventionality; American theatres, including the Bowery, followed the Adelphi's lead by frequently casting a female actor as Jack while all the rest of the characters appeared conventionally cast. Jonas B. Phillips's Bowery *Jack Sheppard*, casting a Mrs. Shaw as Jack, shares the London productions' thematic emphases on costumed transgressive.[26] The script indicates multiple outfit changes, culminating in Jack's trademark garb, a "Scarlet hunting coat, trimmed with gold lace, high boots, and feathered hat."[27] Early productions of *The Beggar's Opera* dressed Macheath similarly. Jack's audacious sumptuary display signals once again resistance to cultural authority through appropriation of elite status symbols. In the play's second act, for example, he flaunts his freedom in a "rich scarlet riding suit, a broad belt, with a hanger attached, high boots and laced hat in his hand."[28] Later, in the third act, he enters the dining room of Kneebone, one of his archenemy Jonathan Wild's partners. It is an act of housebreaking, but also of theatrical display; Jack throws off his "handsome roquelaure," revealing a "coat of brown flowered velvet laced with silver, richly embroidered white satin waistcoat, shoes with red heels and large diamond buckles, pearl coloured silk stockings a muslin cravat edged with lace, ruffles of the same material and handsome sword."[29] The clothing and fashion accessories that Jack steals punctuate his brazen entrances, claiming a place at fashionable tables where he is not welcome. The sartorial displays marks Jack as having transgressed class-associated roles.

Like Macheath's singing in *The Beggar's Opera*, such acts are not merely incidental to the drama; they are integral parts of the rogue process. Eighteenth-century narratives of Jack's life had consistently commented upon his tendency to pilfer cloth and other decorative fashion accessories between stays in Newgate. London thieves targeted those sorts of items quite commonly in the early eighteenth century, revealing another side of what Linebaugh evocatively calls the "shadow economy of clothing."[30] Clothing, trinkets, handkerchiefs, trimmings, and the like—items signaling luxury and prestige, for the most part—constantly recur in the records of London petty crime. In *Jack Sheppard* a long-standing, concerted plebeian assault on the costuming prerogatives of the affluent returns with charismatic staginess. It is impossible to tell whether the act turns primarily on the fear of class transgression or on the admiration of class mobility and self-possessed display—but perhaps that is precisely the point. On stage, Jack's escapes lead him inevitably to a cloth dealer or to a tailor's shop for a new costume. Having acquired his new clothing, the appearance of upper-class status permits access to London's fashionably decadent nightlife and enables his class-transgressive displays. Costuming displays both contribute to and become proof of Jack's theatrical criminality.

On stage, Jack's escapes produce other extravagantly theatrical displays, including singing and demonstrative unruliness. Having finished his escape offstage, Sheppard rejoins his crew in the "Flash Ken," the literal underworld site of gang activity. The site is an important one in the history of rogue performances; it had appeared before as the site of Thomas Mount's criminal career, for example, and the hidden mountain lair of Three-Finger'd Jack and his gang of Maroon bandits. The ken provides a sanctioned space for transgressive self-presentation, a site where audiences, as *Tom and Jerry* had showed, can safely observe low acts. In the flash ken, Jack appears in an elaborate costume that he has just stolen. Buckstone's version describes a spectacular appearance that conventionally appeared in the various versions of *Jack Sheppard*:

> Jack...advances with Poll Maggot on one arm, and Edgeworth Bess on the other. They all shout, jump up, and surround him. Jack stands laughing in the centre—his coat is of brown flowered velvet, laced with silver—a waistcoat of white satin, richly embroidered—smart boots with red heels—a muslin cravat, or steenkirk, edged with point lace—a hat smartly cocked and edged with feathers.[31]

Jack's triumphant entrance (following a "call at our tailor's and milliner's") spurs a performance of the famous "Nix My Dolly Pals, Fake Away," a song that became enormously popular.[32] Drinking, singing, and dancing to their flash song, the criminal crew celebrates the contiguity of escape, crime, sumptuous display, and low theatricality. With a girl on each arm, Jack also acts out the polygamous performance that titillated audiences and frustrated authorities ever since Macheath's multiple wives first surrounded him at the end of *The Beggar's Opera*.

As a cross-dressed lead performer suggests, the play maintains a thematic focus on the interplay of costume, gender, and class. Jack's dandyish clothing, always stolen, provides material evidence of his roguishness, and Buckstone's play shows a strikingly self-reflexive interest in costuming. The stage low had long worked through the implications of various forms of masking and disguise. *Polly*, for example, had envisioned cross-dressing as a counterpart to racial masking. *Three-Finger'd Jack* likewise sets Jack's blackface next to Rosa's breeches performance. The theatrical roguery of disguise and deception remained live issues in Atlantic theatre. Pirates, highwaymen, slumming gentlemen, and beggars alternately hide and expose their identities in conventional scenes of masking and unmasking. Such acts, of course, reveal in the very act of concealing. Masked identity may hide specific identity, but it tells an audience precisely what kind of type a character represents. Since the open disguise of Morano, masking had become the open marker of disdained and rebellious outsider status. Likewise, the heavily advertised cross-gender costuming of *Jack Sheppard* hardly attempts to deceive. Instead, the fascination with clothing and sumptuary display, the thematic expression of Keeley's cross-dressing, marks open transgression of social boundaries. *Jack Sheppard* in breeches thus taps into persistent aspects of acting outcast.

The 1839 resurgence of *Jack Sheppard* thus carries the earlier social critique of Macheath and his gang to its logical conclusion, explicitly linking thievery to deceptive performances of false gentility. Jack's excarceration becomes not merely the act of survival but the flaunting of resistance, the show of disregard for authority. The play enacts the associations that structured theatricals of the Atlantic underclasses for over a century. It stages a self-consciously repetitive cycle—although Jack's jailbreak enables the celebration, he eventually returns to Newgate Prison.

JACK SHEPPARD'S SCAFFOLD STAND

The 1840s saw multiple versions of Jack Sheppard in Atlantic theatres, and American audiences saw more than one ending to Jack's story in the 1840s. Although the plays were usually associated with specific theatres in England, they crossed the Atlantic in a promiscuous proliferation of textual reproductions. It remains difficult to track which version American audiences might have seen, since many newspaper notices simply give the play's title. Alongside Phillips's script, marked and pinned promptbooks in the Harvard Theatre Collection show that American audiences probably saw Buckstone's script and a competing version of the play by T. L. Greenwood.[33] There are longer and shorter editions; the most detailed, and probably the most frequently performed, are the London acting editions based on Buckstone's script. American theatres, however, most likely improvised upon those London templates to produce locally compelling variations of rogue performance. Ainsworth's novel or Buckstone's script (the two highest-profile versions of the story) thus provide loose guidelines rather than fixed determinants of performance.

One of the key differences among the different versions is the final scene of Jack's execution. In it, various playwrights work out the dynamics of rogue performance and mob spectatorship. Buckstone's script, for example, avoids the display of public execution altogether, ending with Jack's capture by the thief-takers. Its finale, however, does present an intriguing scene of violent popular disorder. Calling for riotous mobs to attack Wild's house, the play stages a "war of the populace" that might have seemed dangerously evocative in performances occurring after the 1849 Astor Place riots.[34] Most versions of *Jack Sheppard* hint at such unruly collective energies, although the scripts repeatedly thwart Jack's final escape. Jack invariably dies, proving once again that radically resistant underclass acts have their onstage limits.

Phillips's revision changes the process (if not the ending) of rogue performance, uniquely staging Jack's public execution. Those New York City productions of *Jack Sheppard* centered on the gallows, presenting a scene that *The Beggar's Opera*, allowing Macheath's reprieve, had long refused to stage. Jack delivers a scaffold speech not found in any other version of the play, a version of the "last words" conventionally delivered by condemned criminals. As other condemned criminals have shown, Jack's speech could take various conventional directions. He could take the opportunity, as some of his narratives had done in 1724, to set the record straight. He could voice defiance or penitence, or even a gallant nonchalance, as the slave bandit Tom had in Norfolk in 1818. Even if Jack performed an abjectly remorseful speech, it would have enacted Jack's agency, and especially his centrality as the star of the scene. Such scenes customarily center on the acts, the words, the poses, and style of the condemned. Even if he only repeated moral lessons as Levi Ames or Thomas

Mount had years before, his words would have connected with audiences, gathering a public around the spectatorship of criminal display.

In Phillips's script, Jack delivers what audiences might have heard as an inspiring speech. "Might have heard" because the last leaf of the only known archived manuscript is torn. It seems one of the unpredictable but not precisely random accidents of historical preservation and cultural transmission. The rip raggedly excises most of Jack's speech, along with the top quarter or so of the page, from the only known script of Phillips's *Jack Sheppard*. In the end, an unknown hand has silenced Jack, and the script will not tell what Jack said. The tear, of course, almost certainly happened inadvertently. In all likelihood, it expresses no personal stake in muting the underclasses. The torn page shows the uneven edge of unintentional dismemberment, as if someone had simply tried to turn the page and the age-weakened paper gave way.

The disappearing scene leaves in its stead something just as intriguing: proof of the delicate materiality of the archive. The tear shows the ways texts can deconstruct, even when well preserved and cared for. Those processes of decomposition are many, and it seems worth noting what did and did not happen to the *Jack Sheppard* manuscript in the Harvard Theatre Collection. *Jack Sheppard* does not seem deliberately dismembered; the ragged rip suggests inadvertent tearing. Nor have we found the page elsewhere, recycled for a new purpose. The evidence does not suggest that someone textually reconstituted the fragment in a chapbook or an extra-illustrated volume, and the remainder of Phillips's manuscript remained intact. It does not reveal the deliberate action of unbinding and rebinding, of reconstructing a text's presentation for better or worse. Although inadvertent, the rip does not seem precisely an act of nature, the simple age-related deterioration or the action of insects, dampness, or mildew. Texts can fall apart at the binding or along folds, the result of wear patterns and the particular value and emphasis placed on durability of paper and binding. Phillips's manuscript, however, had no binding to dismember, and its pages remain flat and unfolded.

No evidence remains to suggest when it happened, nor does the scene of the archival crime reveal a motive. The documentary dismemberment could have happened from either neglect or use. It could have been the result of enthusiastic overuse just as likely as abuse. It seems indicative of the all-too-frequent fate of popular culture's physical remains. Although inadvertently, the rip suggests *Jack Sheppard*'s implicit exclusion from the pantheon of monumentalized high culture. Its manuscript, the physical remains of the performance in the archive, follows suit. Such objects survive marked by indiscriminate combinations of casual disregard, elite disdain, and loving (ab)use. Excluded from the collection of sacralized culture, at least through some stages of their life span, such documents do not merit multiple replications and careful preservation. *Jack Sheppard* seems no national treasure, no Declaration of Independence with multiple contemporary copies. Nor does Phillips's script reside in case built with cutting-edge material technology, breathing a carefully climate-controlled inert atmosphere. *Jack Sheppard* inspires no television documentaries or websites on its creation and preservation.

Of course, at some point, the collecting and preserving energies of the archive did rescue the text. Nevertheless, the fellow travels of manuscript and character through the less-valued sites of cultural production produced a wild popularity that left material devaluation in its wake. In the end, the manuscript's partial dismemberment presents a provocative moment of non-knowledge, reminding us of the difficulties of ever knowing for certain how the resistant low act in their own defense. It also reminds us of the disappearance that constitutes performance. The self-annihilation that happens in the very act of performance infiltrates theatre's textual evidence, showing the material fragility of the print-performance archive.

INSANE DREAMS AND RIOTOUS ACTS

At the Bowery Theatre performances, Jack spoke, though his words have disappeared. Nevertheless, his theatrical oratory does not free him from the noose in Phillips's play. The only escape in Phillips's Bowery performance occurs in a fantastic dream sequence narrated by Jack's mother. It reveals a striking kernel of collective Atlantic gallows memories:

> Listen. I'll tell you of a dream I had last night. I was at Tyburn. There was a gallows erected and a great mob around it. Thousands of people, with corse-like faces. In the midst of them, there was a cart with a man in it, and it was Jack, my son Jack. They were going to hang him. And opposite to him, with a book in his hand, sat Jonathan Wild in a parson's cassock and band. And when they came to the gallows, Jack leaped from the cart. And in his stead they hung Jonathan Wild. How the mob shouted and so did I. Ha! Ha! Ha![35]

Jack's mother relates scenes that had become in a way antiquarian by the 1830s; the procession to Tyburn had long since ceased and the charismatic criminal's scaffold had moved indoors. Mobs still acted, but in far different ways than Jack's mother imagines. The rogue's mother, perhaps delusional but quite coherent, dreams an alternate dispensation of popular justice from the madhouse. The mob turns on the thief-taker, freeing the thief. Dreaming the crowd's approval, she hints at the potential for collective plebeian disorder and mayhem. Her insane laughing brings to audible life the power of the rioting mob; it also signals that by 1839, such fantasies of collective antiauthoritarian action had become less believable, less sane. As Phillips suggests, such sentiments only issue from the madhouse. Ultimately, the dream is as socially dead as Jack's mother. The "corse-like" dream crowd, however much they should, ultimately reveals the underclass's excarceral potential as an insane dream, a remote fantasy of the dead past.

The irony of the insane dream of Jack's mother is that American riot action grew increasingly visible and apparently chaotic in the 1830s. *Jack Sheppard*'s 1839 emergence stands at the midpoint of perhaps the most important two decades in the formation of an American culture of collective violence and mob identity. In the antebellum American theatre, violent forms of direct popular action had become increasingly common in the antebellum period. Riots, as Paul Gilje has shown, "became a fixed feature of the American stage in the first half of the nineteenth century."[36] Peter Buckley describes the period from 1833 to 1837—the years directly preceding the appearance of *Jack Sheppard*—as the "Anni Mirabili" of collective violence, a period remarkable for the fifty-two riots recorded in New York City newspapers. The frequency of these events points to a social logic of rioting; the mob's habitual success, Buckley argues, testifies "to the fact that the legitimacy of audience sovereignty remained in effect."[37] Even as Phillips's script forecloses the dream of riotous mobs onstage, ritual modes of low behavior became effective forms of social expression. Rioting represents an assertion of violently scripted conventional prerogatives that temporarily claims power from the powerful. The specter of those rioting crowds haunts *Jack Sheppard*, enacting the power of radical "mobocracy" in ways that the play itself could only dream.

The audiences, actors, and managers of antebellum New York participated in an extended period of cultural reshuffling of theatrical orders. In the years before Phillips's *Jack Sheppard*, the Bowery entertained the working classes while the Park Theatre gradually became the upper-class venue. The Bowery Theatre, with Thomas Hamblin as manager after 1830, embarked on a deliberate campaign of innovative offerings, drawing working-class crowds with "performing animals, scenic effects, including dioramas, and afterpieces such as living statuary and impersonations."[38] Although New York City had very different institutions of theatrical authority, the Bowery resembles the illegitimate theatres of London of the same period in its variety and its refusal to limit itself to "legitimate" forms. As one might expect from a theatrical venue in the middle of public negotiations over the relationships between class and culture, the Bowery hosted significant tumult. It saw frequent riots, which shape Phillips's production of *Jack Sheppard*. Offstage violence shoves up against onstage rioting, intermingling modes of performance that each construct the contexts for the other. The staging of popular violence in and along Phillips's *Jack Sheppard* impels underclass performances in significant new directions.

Although the period immediately following Jack's debut seems quieter than the riot-prone 1830s, the cessation of rioting was not permanent. Instead, it led up to the Astor Place Riots of May 1849, which mark a dramatic shift in American public culture and the role of the unruly low. Just a few years before Marx names the lumpenproletariat in the *Eighteenth Brumaire*, the Astor Place riots of May 1849 theatricalized mob action and class relationships in the New York streets. For three days in early May, mobs rioted outside the Astor Place Opera House, which had scheduled performances by English actor William Macready, a fierce competitor of Edwin Forrest in the 1840s. The riots gather force on the evening of May 10, when thousands of working-class New Yorkers gathered outside the Astor Place Opera House to protest Macready's presence. Amidst showers of stone and shattered windowpanes, members of the vastly outnumbered militia fired into the crowd. Twenty-one people died and over a hundred received serious injuries. The following night, mobs again gathered to hear inflammatory speeches in City Hall Park, and some five thousand of them then marched toward Astor Place, where others had already gathered.[39] This crowd, however, dispersed without violence or loss of life, and the riots drew to their denouement on their third night.

The simple version of the story is that an actor's quarrel caused the riots. Edwin Forrest, by 1849 an icon of melodramatic theatrical style, working-class Bowery masculinity, and American nationalistic sentiment, had developed a public and contagious rivalry with William Macready. Their careers first converged in 1826, when Macready, on his first American tour, competed with Forrest for publicity and audiences. The feud simmered for nearly two decades, until 1845 and 1846, when audiences hissed during Forrest's London tour and Forrest returned the insult at one of Macready's Edinburgh performances. Those events returned to haunt Macready's American tour in 1848 as audiences protested his appearance and rioted repeatedly at his performances. Although later demeaned as the result of a "paltry quarrel of two actors," the riots represent more than an absurdly disproportional response to temporary cultural frictions.[40] Instead, they represent the ludic turned agonistic, playing out a street-level version of the Bowery's acts. The press had followed the actors' quarrel that led up to the riots with the "interest usually reserved for the careers of notorious pugilists, criminals and politicians," as Buckley notes.[41] Audiences watched the events, one might argue, with the enthusiasms they usually reserved for theatricals of Atlantic rogues. The riots represent a logical response to cultural developments in New York City. The Astor Place riots enacted the growing tensions in a city characterized by an "Upper Ten" and a "Lower Million," as George Lippard sensationally described the city's class divisions in 1853.[42] The riots, in addition, responded to the self-proclaimed elite association with English culture. Theatrical style fed class antagonism in the riots. As Lawrence Levine has argued, the riots performed a "struggle for power and cultural authority within cultural space."[43] The riots mark a watershed moment on the American side of looming class struggles. Before the Astor Place events, theatre rioting had held an accepted and conventional place in American theatres. Before those troops fired on the mob, authorities made infrequent use of force against rioters, and the "mobocracy" exercised informally legitimated forms of control in the theatre. The Astor Place riots, however, presented the more insidious transformation of the working class into an uncontrollable and violent mob.

The events outside the Astor Place Theatre in 1849 are important because they make obvious—physically, locally, and painfully evident—the trends that had long been at work shaping Atlantic culture. The city, with its densely layered topographies of theatre culture, condenses the relationships constantly reproduced and riffed upon all across the Atlantic world. The violent disturbances evoked broader contexts of riot and rebellion, even connecting to the European revolutions of 1848. Indeed, the American city streets had apparently become a battlefield. Among New York's elite, observers such as the lawyer George Templeton Strong noticed that everything "looked much in earnest there—guns loaded and matches lighted—everything ready to sweep the streets with grape at a minute's notice, and the police and troops very well disposed to do it whenever they should be told."[44] The Astor Place riots, as much as the European revolutions of the year before, marked in the minds of contemporaries an irreconcilable split between high and low culture. More dangerously, it marked the possibility that New York City's less advantaged ten million could collectively transform from disciplined workers into violent insurrectionists such as Morano, Three-Finger'd Jack, and Jack Sheppard and his riotous mob.

LORE CYCLES OF THE ATLANTIC UNDERCLASSES

The offstage events follow patterns long rehearsed by onstage rogues and their low cohorts. Marx blames the failure of Europe's 1848 revolutions on the lumpenproletariat, echoing the onstage plot of Bird's *Gladiator*. In America, less revolutionary-minded observers seem intent on identifying and isolating the low elements who seemed most guilty of causing the riots. Post-riot legal proceedings and the incarceration of some rioters at Blackwell's Island expel mob violence "once again outside the normal social relations of the city," as Buckley observes.[45] The act resonates with onstage rogue performances. In playhouses, courts sentence thieves to transportation in *The Beggar's Opera*; Rowson imagines the revolutionary mob in the Mediterranean, and *The Glory of Columbia* marches its riotous yeomen offstage. *Three-Finger'd Jack* uses Caribbean exoticism and *The Gladiator* classical forms to distance the contagious threat of slave revolt from its audiences. Everywhere, the chaotic and violent, perplexing and even counterrevolutionary underclasses seem destined for the dustbin of history.

Participating in the circumatlantic transmissions of such acts, early American theatre's rogue performances suggest several important lessons. The low, especially in their criminalized forms, enact the recurring tensions between discipline and unruliness. The impulse to disorderliness, revolt, mobbing, violence, and criminalized acts seems endemic to the stage underclasses. As *The Beggar's Opera* and early American gallows performances show, various parties could deploy rogue acts to enact various forms of cultural work. The low represents a flexible and contested resource in Atlantic culture. It also provides a useful template for the co-production of racial and ethnic identities. Outsider and "undersider" qualities lent themselves to performances of other identities, for example, Algerian, Jewish, Irish, and African identities in the Atlantic world. The convergences seem consistently less than deliberate, and more than accidental. The class dimension of such performances remains important. The production of race and ethnicity frequently requires imagined collective boundaries and hierarchies of cultural value and class distinction. As *Slaves in Algiers* and *The Glory of Columbia* demonstrate, such performances helped construct the national imaginary, even as rogues and rioters troubled the celebratory national ideal. Such performances reveal the extent to which early American culture, imagining itself alongside and through its ever-present low, recruited them for patriotic performances. Rogue performances demonstrate varying levels of self-consciousness about the mechanisms of staging and spectatorship. When plays lay bare their own processes, as *Tom and Jerry* does, they reveal underclass acts as not simply unconscious resources for Atlantic theatre culture, but instead as part of uniquely self-aware processes. The entangled dialectics of mimicry and authenticity that drive performance come to the fore when acting across the apparent boundaries of class division.

As much as anything, this account hopes to foreground the forward transmissions and continuing transformations of rogue performances. The Astor Place riots of 1849 do not end the Atlantic world's cultural representations of popular outcasts and class struggles. Performances of the unruly, rebellious, disconnected classes continued to appear on stage. The performances of the underclasses passed along in what W. T. Lhamon has called "lore cycles," staging further improvisations on low themes.[46] Harriet Beecher Stowe's *Uncle Tom's Cabin*, for example, translates excarceration into abolitionist scenes. The novel and its theatrical adaptations present Tom's sentimentalized racial body (even performed by T. D. Rice at the end of his career) against the heroics of the mixed-race slave George Harris and his runaway wife Eliza. Even more troubling and popular were the insouciant antics of Topsy, who "jes' grew" from the soil of imagined slavery and sprang into fully formed transatlantic popularity. Those scenes reached audiences in America and England just as Marx began publishing his views on the lumpenproletariat. Henry David Thoreau spends a night in jail and recalls the secret histories and songs of Concord's imprisoned population just as Melville examines the implications of slave revolts in his 1855 *Benito Cereno*. Martin Delany's *Blake, or the Huts of America*, serialized at the end of the 1850s, applies a Latinate name to its slave hero, producing another classically black Caribbean insurgency. The same issues continued to shape cultural productions in other parts of the Anglo-Atlantic world; Charles Dickens's 1861 *Great Expectations*, for example, revolves around the relationship between Pip and the transported convict, a less celebrated version of Jack Sheppard who nevertheless becomes a significant cultural resource for the novel's hero. In a less openly fictional vein, Henry Mayhew's encyclopedic project of the 1850s, *London Labour and the London Poor*, feeds audiences' appetites for depictions of London's seamy underbelly. It also reproduces outcast and disdained people as a subject of knowledge, identifying the genres of dislocated, unproductive, but ever-theatrical underclasses.

These circumatlantic literary productions share and shape theatre's takes on the low. Dickens's novels, for example, were popularly adapted, and increasingly professionalized playwrights such as Dion Boucicault produced plays such as the 1857 *The Poor of New York*, which staged the urban impoverished as pitiable but compelling objects of the spectatorial gaze. Long after the black tragedian Ira Aldridge made *Obi; or, Three-Finger'd Jack* one of his staple roles, it finally reached the stage in Jamaica in 1862. The rogue figures propelled onstage in the eighteenth century continue to capture the imagination of Atlantic audiences more than a century later. Charismatic rogues, always on the verge of breaking out, continued to act out the notorious theatricality, mobility, and class resentment that had brought them to popular audiences for the past century. Those acts continue to reveal the transmissions and transactions of underclass lore cycles. The Atlantic world's less privileged types repeatedly appear on stage to trouble class hierarchies and the interplay across social boundaries.

The Atlantic world's cultural currents continue exchanging elements of the lore cycle, staging theatrical revivals of the low. Such acts represent, as *Jack Sheppard* suggests, dreams of desire and disavowal. This account of the stage underclasses hopes to foreground those processes of cultural transmission. The classic paradigms of cultural production and consumption only tell part of the story, for the "goods" made and handled are gestural and performative. Stage acts emerge from a heterogeneous mix of on- and offstage influences; they do not wear out, and their audiences do not precisely "consume" them. In part, that explains their continued circulations. Moreover, the lore cycles do not create clean narratives. Their acts revive lowness in intermittently visible poses and snatches of overheard song. Their resonances appear far removed, as flashes of metaphorical resemblance, but the lore cycles leave their traces at ground level. They connect far-flung structures of feeling in vast, decentered networks of metonymical relationships. Theatre shapes ever-changing acts of collective memory and creative reinvention. The fragmented collective trajectories of the low do not ever reach full closure. This becomes at once a liability and an advantage, preventing the easy transmission of coherent messages while ensuring their survival in fragmented bits. The histories of cultural transmissions, however, do not need or allow smoothing over. This account, though it traces patterns, does not propose to order them into a sequential historical narrative. The Astor Place riots and Marx's naming do not produce historical closure. Outlaw repertoires overleap porous boundaries of nationality, racial identity, and class affiliation. Their hurtling, hurdling acts lead quickly past the arbitrary geographic and chronological edges of the view from here. The charismatic rogue performances of Atlantic culture carry on, insisting that we not contain or ignore them.

Notes

1 ATLANTIC UNDERCLASSES AND EARLY AMERICAN THEATRE CULTURE

1. *London Morning Post*, January 5, 1776.
2. Philip H. Highfill, Kalman A. Burnim, and Edward A. Langhans, eds., *A Biographical Dictionary of Actors, Actresses, Musicians, Dancers, Managers & Other Stage Personnel in London, 1660–1800* (Carbondale: Southern Illinois University Press, 1973), 3: 65.
3. *London Morning Chronicle and London Advertiser*, October 7, 1776.
4. Ibid.
5. Terry Castle, *Masquerade and Civilization: The Carnivalesque in Eighteenth-Century English Culture and Fiction* (Stanford: Stanford University Press, 1986), 5.
6. See E. J. Hobsbawm, *Bandits* (New York: New Press, distributed by W.W. Norton, 2000); E. P. Thompson, *Customs in Common: Studies in Traditional Popular Culture* (New York: New Press, 1991); and *Whigs and Hunters: The Origin of the Black Act* (New York: Pantheon Books, 1975).
7. Joseph R. Roach, *Cities of the Dead: Circum-Atlantic Performance* (New York: Columbia University Press, 1996).
8. Richard Schechner, *Between Theater and Anthropology* (Philadelphia: University of Pennsylvania Press, 1985), 35–37.
9. Erving Goffman, *The Presentation of Self in Everyday Life* (Garden City, NY: Doubleday, 1959), 17–76; Judith Butler, *Gender Trouble: Feminism and the Subversion of Identity* (New York: Routledge, 1990).
10. Indeed, only *The Gladiator* appears in Jeffrey H. Richards, *Early American Drama* (New York: Penguin Books, 1997).
11. This is one of the recurring and valuable themes in Jeffrey H. Richards, *Drama, Theatre, and Identity in the American New Republic* (Cambridge: Cambridge University Press, 2005).
12. Mikhail M. Bakhtin, *Rabelais and His World* (Cambridge, MA: MIT Press, 1968) and Jean-Christophe Agnew, *Worlds Apart: The Market and the Theater in Anglo-American Thought, 1550–1750* (Cambridge: Cambridge University Press, 1986) explore the relationships between markets and performance.
13. Peter Stallybrass and Allon White, *The Politics and Poetics of Transgression* (Ithaca, NY: Cornell University Press, 1986), 94–96.
14. Karl Marx, *The Eighteenth Brumaire of Louis Bonaparte: With Explanatory Notes* (New York: International Publishers, 1963), 75; emphasis in the original.
15. Marx had addressed the concept of a "lumpenproletariat" before publishing the *Eighteenth Brumaire*, although it is perhaps his first thoroughgoing exploration of the idea. Peter Stallybrass, "Marx and Heterogeneity: Thinking the Lumpenproletariat," *Representations* 31 (1990): 70, notes that Marx and Engels used the term in various conflicted ways, ranging from sartorial to racial. See also Robert L. Bussard, "The 'Dangerous Class' of Marx and Engels: The Rise of the Idea of the Lumpenproletariat," *History of European Ideas* 8.6 (1987): 675–92.

16. Marx, *Eighteenth Brumaire*, 124.
17. Ibid., 15.
18. Simon P. Newman, *Embodied History: The Lives of the Poor in Early Philadelphia* (Philadelphia: University of Pennsylvania Press, 2003), 11; see also Gary B. Nash, "Poverty and Politics in Early American History," *Down and out in Early America*, ed. Billy G. Smith (University Park: Pennsylvania State University Press, 2004), 1–37.
19. A few examples of the excellent recent scholarship on early American and Atlantic low include Christine Stansell, *City of Women: Sex and Class in New York, 1789–1860* (New York: Knopf, distributed by Random House, 1986); Timothy J. Gilfoyle, *A Pickpocket's Tale: The Underworld of Nineteenth-Century New York* (New York: W.W. Norton, 2006); Shane White, *Somewhat More Independent: The End of Slavery in New York City, 1770–1810* (Athens: University of Georgia Press, 1991); Tyler Anbinder, *Five Points: The 19th-Century New York City Neighborhood That Invented Tap Dance, Stole Elections, and Became the World's Most Notorious Slum* (New York: Free Press, 2001); Marcus Rediker, *Between the Devil and the Deep Blue Sea: Merchant Seamen, Pirates, and the Anglo-American Maritime World, 1700–1750* (Cambridge, New York: Cambridge University Press, 1987); Ira Berlin, *Many Thousands Gone: The First Two Centuries of Slavery in North America* (Cambridge, MA: Belknap Press of Harvard University Press, 1998); and A. Roger Ekirch, *Bound for America: The Transportation of British Convicts to the Colonies, 1718–1775* (Oxford: Oxford University Press, 1987).
20. Sean Wilentz, *Chants Democratic: New York City and the Rise of the American Working Class, 1788–1850* (New York: Oxford University Press, 1984); Alexander Saxton, *The Rise and Fall of the White Republic: Class Politics and Mass Culture in Nineteenth-Century America* (London: Verso, 2003); David R. Roediger, *The Wages of Whiteness: Race and the Making of the American Working Class* (New York: Verso, 1991).
21. Gordon S. Wood, *The Radicalism of the American Revolution* (New York: A.A. Knopf, 1992); Alfred F. Young, *The Shoemaker and the Tea Party: Memory and the American Revolution* (Boston: Beacon Press, 1999); Eric Foner, *Free Soil, Free Labor, Free Men: The Ideology of the Republican Party before the Civil War* (New York: Oxford University Press, 1970).
22. Peter Linebaugh, *The London Hanged: Crime and Civil Society in the Eighteenth Century* (Cambridge: Cambridge University Press, 1992); Peter Linebaugh and Marcus Rediker, *The Many-Headed Hydra: Sailors, Slaves, Commoners, and the Hidden History of the Revolutionary Atlantic* (Boston: Beacon Press, 2000).
23. Irving's commentary appears in the *New York Morning Chronicle*, December 4, 1802.
24. Paul A. Gilje, *The Road to Mobocracy: Popular Disorder in New York City, 1763–1834* (Chapel Hill, NC: Published for the Institute of Early American History and Culture by the University of North Carolina Press, 1987), 5.
25. Elaine Hadley, "The Old Price Wars: Melodramatizing the Public Sphere in Early-Nineteenth-Century England," *PMLA* 107.3 (1992): 524–37; Marc Baer, *Theatre and Disorder in Late Georgian London* (Oxford: Oxford University Press, 1992); Bruce A. McConachie, *Melodramatic Formations: American Theatre and Society, 1820–1870* (Iowa City: University of Iowa Press, 1992), 9, 144–54; Leonard L. Richards, *Gentlemen of Property and Standing: Anti-Abolition Mobs in Jacksonian America* (New York: Oxford University Press, 1970).
26. James Fennell, *An Apology for the Life of James Fennell, Written by Himself* (Philadelphia: Published by Moses Thomas, No. 52, Chesnut-Street, J. Maxwell, printer, 1814), 406.
27. Michael Denning, "Beggars and Thieves: *The Beggar's Opera* and the Ideology of the Gang," *Literature and History* 8.1 (1982): 41–55.

28. Linebaugh, *The London Hanged*, 38; see also George F. E. Rudé, *The Crowd in History: A Study of Popular Disturbances in France and England, 1730–1848* (New York: Wiley, 1964).
29. W. T. Lhamon, Jr., *Jump Jim Crow: Lost Plays, Lyrics, and Street Prose of the First Atlantic Popular Culture* (Cambridge, MA: Harvard University Press, 2003), 15.
30. See also the first chapter of Marcus Rediker, *Villains of All Nations: Atlantic Pirates in the Golden Age* (Boston: Beacon Press, 2004); Linebaugh, *The London Hanged*, broadens the possibilities of scaffold display in the wake of the foundational work of Michel Foucault, *Discipline and Punish: The Birth of the Prison* (New York: Pantheon Books, 1977).
31. George Clinton Densmore Odell, *Annals of the New York Stage* (New York: Columbia University Press, 1927), 3: 58; Odell cites the *New York Evening Post*, February 19, 1823.
32. Shakespeare, *Hamlet* 3.2.20, in Stephen Greenblatt, Walter Cohen, Jean E. Howard, Katharine Eisaman Maus, and Andrew Gurr, eds., *The Norton Shakespeare*, 2nd ed. (New York: W.W. Norton, 2008).
33. I of course borrow the immensely useful formulation of cultural work from Jane P. Tompkins, *Sensational Designs: The Cultural Work of American Fiction, 1790–1860* (New York: Oxford University Press, 1985), xv.
34. Jonas A. Barish, *The Antitheatrical Prejudice* (Berkeley: University of California Press, 1981); see also Claudia D. Johnson, *Church and Stage: The Theatre as Target of Religious Condemnation in Nineteenth Century America* (Jefferson, NC: McFarland & Co., 2008).
35. Odai Johnson, *Absence and Memory in Colonial American Theatre: Fiorelli's Plaster* (New York: Palgrave Macmillan, 2006), 4.
36. Jeffrey H. Richards, *Theater Enough: American Culture and the Metaphor of the World Stage, 1607–1789* (Durham: Duke University Press, 1991), 294.
37. Heather S. Nathans, *Early American Theatre from the Revolution to Thomas Jefferson: Into the Hands of the People* (Cambridge: Cambridge University Press, 2003).
38. Richards, *Drama, Theatre, and Identity*, 17–33.
39. Johnson, *Absence and Memory*, 31. Odai Johnson, William J. Burling, and James A. Coombs, eds., *The Colonial American Stage, 1665–1774: A Documentary Calendar* (Madison, NJ: Fairleigh Dickinson University Press, 2001) details the documentary evidence of these changes.
40. See Jason Shaffer, *Performing Patriotism: National Identity in the Colonial and Revolutionary American Theater* (Philadelphia: University of Pennsylvania Press, 2007); Jared Brown, *The Theatre in America During the Revolution* (Cambridge: Cambridge University Press, 1995).
41. Nathans, *Early American Theatre*, 2.
42. Suzanne Sherman, *Comedies Useful: History of Southern Theater, 1775–1812* (Williamsburg, VA: Celest Press, 1998); Charles S. Watson, *Antebellum Charleston Dramatists* (Tuscaloosa, AL: University of Alabama Press, 1976); William Stanley Hoole, *The Ante-Bellum Charleston Theatre* (Tuscaloosa, AL: University of Alabama Press, 1946).
43. Johnson, *Absence and Memory*, 24.
44. Thomas Clark Pollock, *The Philadelphia Theatre in the Eighteenth Century, Together with the Day Book of the Same Period* (Philadelphia: University of Pennsylvania Press, 1933), 143.
45. Lawrence W. Levine, *Highbrow/Lowbrow: The Emergence of Cultural Hierarchy in America* (Cambridge, MA: Harvard University Press, 1988); David S. Reynolds,

Beneath the American Renaissance: The Subversive Imagination in the Age of Emerson and Melville (New York: Knopf, 1988); see also Patricia Cline Cohen, Timothy J. Gilfoyle, and Helen Lefkowitz Horowitz, eds., *The Flash Press: Sporting Male Weeklies in 1840s New York* (Chicago: University of Chicago Press, 2008).

46. See Susan L. Porter, *With an Air Debonair: Musical Theatre in America, 1785–1815* (Washington: Smithsonian Institution Press, 1991); and Karen Ahlquist, *Democracy at the Opera: Music, Theater, and Culture in New York City, 1815–60* (Urbana: University of Illinois Press, 1997).
47. Heather S. Nathans, "A Much Maligned People: Jews on and Off the Stage in the Early American Republic," *Early American Studies* 2.2 (2007): 310–42; Lucy Rinehart, "'Manly Exercises': Post-Revolutionary Performances of Authority in the Theatrical Career of William Dunlap," *Early American Literature* 36.2 (2001): 263–93.
48. The collective urge to narrate recollections appears, e.g., in John Durang, *The Memoir of John Durang, American Actor, 1785–1816*, ed. Alan Seymour Downer (Pittsburgh: University of Pittsburgh Press, 1966); William Dunlap, *A History of the American Theatre* (New York: J. & J. Harper, 1832); and Joe Cowell, *Thirty Years Passed among the Players in England and America* (New York: Harper & Brothers, 1844).
49. Nathans, *Early American Theatre*, 149.
50. Rosemarie K. Bank, *Theatre Culture in America, 1825–1860* (Cambridge: Cambridge University Press, 1997).
51. Errol Hill, *The Jamaican Stage, 1655–1900: Profile of a Colonial Theatre* (Amherst: University of Massachusetts Press, 1992), e.g., builds upon Richardson Wright, *Revels in Jamaica, 1682–1838* (New York: Dodd Mead, 1937). Sean X. Goudie, *Creole America: The West Indies and the Formation of Literature and Culture in the New Republic* (Philadelphia: University of Pennsylvania Press, 2006) critically examines North America's cultural interconnections with the Caribbean.
52. Isabelle Lehuu, *Carnival on the Page: Popular Print Media in Antebellum America* (Chapel Hill: University of North Carolina Press, 2000).
53. Dale Cockrell, *Demons of Disorder: Early Blackface Minstrels and Their World* (New York: Cambridge University Press, 1997), 78–82.
54. Lhamon, *Jump Jim Crow*, 1–30; W. T. Lhamon, Jr., *Raising Cain: Blackface Performance from Jim Crow to Hip Hop* (Cambridge, MA: Harvard University Press, 1998), chapter 1, "Dancing for Eels at Catherine Market."
55. Eric Lott, *Love and Theft: Blackface Minstrelsy and the American Working Class* (New York: Oxford University Press, 1993), 29.
56. See Jane Moody, *Illegitimate Theatre in London, 1770–1840* (Cambridge: Cambridge University Press, 2000).
57. See Levine, *Highbrow/Lowbrow*, 84–168, part 2, "The Sacralization of Culture."
58. Janelle G. Reinelt, "The Politics of Discourse: Performativity Meets Theatricality," *SubStance* 31.2 (2002): 208.
59. Richards, *Theater Enough*, xiii; numerous other figures of the world as a stage circulated alongside Shakespeare's metaphor, as Richards notes.
60. McConachie, *Melodramatic Formations*, 79–80.
61. Faye E. Dudden, *Women in the American Theatre: Actresses and Audiences, 1790–1870* (New Haven: Yale University Press, 1994); Lesley Ferris, *Acting Women: Images of Women in Theatre* (New York: New York University Press, 1989).
62. See Gareth Stedman Jones, *Languages of Class: Studies in English Working Class History, 1832–1982* (New York: Cambridge University Press, 1983).
63. Peggy Phelan, *Unmarked: The Politics of Performance* (New York: Routledge, 1993), 148–66.

64. Johnson, *Absence and Memory*, 9.
65. Jürgen Habermas, *The Structural Transformation of the Public Sphere: An Inquiry into a Category of Bourgeois Society* (Cambridge, MA: MIT Press, 1989); Benedict R. Anderson, *Imagined Communities: Reflections on the Origin and Spread of Nationalism* (London: Verso, 1983).
66. Nancy Armstrong, *Desire and Domestic Fiction: A Political History of the Novel* (New York: Oxford University Press, 1987), 203–50.
67. See Cathy N. Davidson, *Revolution and the Word: The Rise of the Novel in America*, expanded ed. (New York: Oxford University Press, 2004), especially 121–50. Tompkins, *Sensational Designs*.
68. Julia A. Stern, *The Plight of Feeling: Sympathy and Dissent in the Early American Novel* (Chicago: University of Chicago Press, 1997), 7; other recent work usefully engages the relationships between early American print culture and performance practices; see Christopher Looby, *Voicing America: Language, Literary Form, and the Origins of the United States* (Chicago: University of Chicago Press, 1996); Jay Fliegelman, *Declaring Independence: Jefferson, Natural Language, and the Culture of Performance* (Stanford: Stanford University Press, 1993); Sandra M. Gustafson, *Eloquence is Power: Oratory and Performance in Early America* (Chapel Hill: Published for the Omohundro Institute of Early American History and Culture, Williamsburg, Virginia, by the University of North Carolina Press, 2000).
69. Michael Warner, *The Letters of the Republic: Publication and the Public Sphere in Eighteenth-Century America* (Cambridge, MA: Harvard University Press, 1990), 42–43.
70. Diana Taylor, *The Archive and the Repertoire: Performing Cultural Memory in the Americas* (Durham: Duke University Press, 2003), 19–23.
71. See Michel de Certeau, *The Practice of Everyday Life* (Berkeley: University of California Press, 1984); Roger Chartier, *The Order of Books: Readers, Authors, and Libraries in Europe between the Fourteenth and Eighteenth Centuries* (Stanford: Stanford University Press, 1994); Peter Burke, *Popular Culture in Early Modern Europe* (New York: New York University Press, 1978).
72. Raymond Williams, *The Long Revolution* (New York: Columbia University Press, 1961), 48–71. McConachie, *Melodramatic Formations*, xii–xiii.
73. Victor Witter Turner, *From Ritual to Theatre: The Human Seriousness of Play* (New York: Performing Arts Journal Publications, 1982).
74. See Clifford Geertz, *The Interpretation of Cultures: Selected Essays* (New York: Basic Books, 1973); James Clifford, *The Predicament of Culture: Twentieth-Century Ethnography, Literature, and Art* (Cambridge, MA: Harvard University Press, 1988); James Clifford, *Routes: Travel and Translation in the Late Twentieth Century* (Cambridge, MA: Harvard University Press, 1997).

2 GALLOWS PERFORMANCE, EXCARCERATION, AND *THE BEGGAR'S OPERA*

1. Ames appears in newspapers as early as his late August capture (*Boston Post-Boy*, August 23, 1773); various newspapers announce his execution in the days leading up to October 21, and confirmations of the event appear soon after, mostly advertising the newly printed sermon texts, including Samuel Stillman, *Two Sermons: The First from Psalm CII. 19, 20. Delivered the Lords-Day before the Execution of Levi Ames, Who Was Executed at Boston, Thursday October 21, 1773*, second edn (Boston: Printed and sold by J. Kneeland, in Milk-Street; sold also by Philip Freeman, in Union-Street, 1773);

Andrew Eliot, *Christ's Promise to the Penitent Thief. A Sermon Preached the Lord's-Day before the Execution of Levi Ames, Who Suffered Death for Burglary, Oct. 21, 1773. Aet. 22* (Boston: Printed and sold by John Boyle, next door to the Three Doves in Marlborough-Street, 1773); Samuel Mather, *Christ Sent to Heal the Broken Hearted. A Sermon, Preached at the Thursday Lecture in Boston, on October, 21st. 1773. When Levi Ames, a Young Man, under a Sentence of Death for Burglary, to Be Executed on That Day, Was Present to Hear the Discourse* (Boston: Printed and sold at William M'Alpine's printing office in Marlborough-Street, 1773).

2. Elhanan Winchester, *The Execution Hymn, Composed on Levi Ames, Who Is to Be Executed for Burglary, This Day, the 21st of October, 1773, Which Was Sung to Him and a Considerable Audience, Assembled at the Prison, on Tuesday Evening, the 19th of October, and, at the Desire of the Prisoner, Will Be Sung at the Place of Execution, This Day* ([Boston]: Sold by E. Russell, next the cornfield, Union-Street., 1773); broadside, American Antiquarian Society.
3. Stillman, *Two Sermons*, t.p.
4. See Michel Foucault, *Discipline and Punish: The Birth of the Prison* (New York: Pantheon Books, 1977), 32–72.
5. See Dianne Dugaw, *Warrior Women and Popular Balladry, 1650–1850* (New York: Cambridge University Press, 1989); and Curtis Alexander Price, Judith Milhous, Robert D. Hume, and Gabriella Dideriksen, *Italian Opera in Late Eighteenth-Century London*, 2 vols (New York: Oxford University Press, 1995).
6. See William Eben Schultz, *Gay's Beggar's Opera: Its Content, History, and Influence* (New Haven: Yale University Press, 1923); Frank Kidson, *The Beggar's Opera, Its Predecessors and Successors* (Cambridge: The University press, 1922); William A. McIntosh, "Handel, Walpole, and Gay: The Aims of the Beggar's Opera," *Eighteenth Century Studies* 7.4 (1974): 415–33; John Richardson, "John Gay, the Beggar's Opera, and Forms of Resistance," *Eighteenth-Century Life* 24.3 (2000): 19–30; Cheryl Wanko, "Three Stories of Celebrity: The Beggar's Opera 'Biographies,'" *Studies in English Literature, 1500–1900* 38.3 (1998): 481–98.
7. Daniel Defoe, *A Narrative of all the Robberies, Escapes, &c. of John Sheppard* (London: Printed and sold by John Applebee, 1724).
8. Dianne Dugaw, *Deep Play: John Gay and the Invention of Modernity* (Newark, DE: University of Delaware Press, 2001).
9. Michael Denning, "Beggars and Thieves: *The Beggar's Opera* and the Ideology of the Gang," *Literature and History* 8.1 (1982): 41–55.
10. William Empson, *Some Versions of Pastoral* (London: Chatto and Windus, 1935), 202–50 incisively examines the play's ironies.
11. See, e.g., Hobsbawm's postscript discussing debates about social banditry in E. J. Hobsbawm, *Bandits* (New York: New Press, distributed by W.W. Norton, 2000), 167–99.
12. Andrea McKenzie, "The Real Macheath: Social Satire, Appropriation, and Eighteenth-Century Criminal Biography," *Huntington Library Quarterly* 69.4 (2006): 584.
13. Gillian Spraggs, *Outlaws and Highwaymen: The Cult of the Robber in England from the Middle Ages to the Nineteenth Century* (London: Pimlico, 2001), 211. McKenzie, "The Real Macheath," 582, usefully summarizes debates about the various functions of Macheath's roguery.
14. John Gay, *The Beggar's Opera*, eds. Edgar V. Roberts and Edward Smith (Lincoln: University of Nebraska Press, 1969), 1.4.41–42.
15. Ibid., 1.11.6–7.
16. See Peter Linebaugh, *The London Hanged: Crime and Civil Society in the Eighteenth Century* (Cambridge: Cambridge University Press, 1992), 7–41. See also V. A. C.

Gatrell, *The Hanging Tree: Execution and the English People 1770–1868* (Oxford: Oxford University Press, 1994); Andrea McKenzie, "Martyrs in Low Life? Dying "Game" In Augustan England," *Journal of British Studies* 42.2 (2003): 167–205.
17. Peter Linebaugh, "The Tyburn Riot against the Surgeons," *Albion's Fatal Tree: Crime and Society in Eighteenth-Century England*, eds. Douglas Hay, Peter Linebaugh, John G. Rule, E. P. Thompson, and Cal Winslow (New York: Peregrine Books, 1975), 115.
18. Gay, *Beggar's Opera*, 2.1.9–11.
19. Ibid. (Air 22).
20. William Cooke, *Memoirs of Charles Macklin, Comedian, with the Dramatic Characters, Manners, Anecdotes, &c. of the Age in Which He Lived*, second edn (London: J. Asperne, 1806), 27–28, quoted in Schultz, *Gay's Beggar's Opera*, 37.
21. James Boswell, *Boswell's London Journal, 1762–1763*, ed. Frederick Albert Pottle (New Haven: Yale University Press, 1991), 264. Entry for May 19, 1763.
22. See McKenzie, "The Real Macheath," 596.
23. Gay, *Beggar's Opera*, 1.3.
24. Ibid., 1.1.10–11.
25. See Patricia Cline Cohen, Timothy J. Gilfoyle, and Helen Lefkowitz Horowitz, eds., *The Flash Press: Sporting Male Weeklies in 1840s New York* (Chicago: University of Chicago Press, 2008); and Guy Reel, *The National Police Gazette and the Making of the Modern American Man, 1879–1906* (New York: Palgrave Macmillan, 2006).
26. See Terry Castle, *Masquerade and Civilization: The Carnivalesque in Eighteenth-Century English Culture and Fiction* (Stanford: Stanford University Press, 1986); and Dror Wahrman, *The Making of the Modern Self: Identity and Culture in Eighteenth-Century England* (New Haven: Yale University Press, 2004). On early American performance culture, see Sandra M. Gustafson, *Eloquence is Power: Oratory and Performance in Early America* (Chapel Hill: Published for the Omohundro Institute of Early American History and Culture, Williamsburg, Virginia, by the University of North Carolina Press, 2000); Jay Fliegelman, *Declaring Independence: Jefferson, Natural Language, and the Culture of Performance* (Stanford: Stanford University Press, 1993); and Jeffrey H. Richards, *Theater Enough: American Culture and the Metaphor of the World Stage, 1607–1789* (Durham: Duke University Press, 1991).
27. Gay, *Beggar's Opera*, 3.16.5–6.
28. *The Beggar's Opera* trails only *Romeo and Juliet*, the *Beaux' Stratagem*, and *Richard III* in the colonial main-piece repertoire. Odai Johnson, William J. Burling, and James A. Coombs, eds., *The Colonial American Stage, 1665–1774: A Documentary Calendar* (Madison, NJ: Fairleigh Dickinson University Press, 2001), 64.
29. Both Schultz, *Gay's Beggar's Opera*, 285–306 and Kidson, *Predecessors and Successors*, 100–104, discuss and list plays inspired by the form and content of *The Beggar's Opera*.
30. Kidson, *Predecessors and Successors*, 37–38; William Rufus Chetwood, *A General History of the Stage, from Its Origin in Greece Down to the Present Time* (London: Printed for W. Owen, 1749), 40–41.
31. Chetwood, *General History*, 40–41.
32. See George Clinton Densmore Odell, *Annals of the New York Stage* (New York: Columbia University Press, 1927), 1.37–38; *The Beggar's Opera* appeared with *The Mock Doctor* on this date. Garff B. Wilson, *Three Hundred Years of American Drama and Theatre, from Ye Bare and Ye Cubb to Hair* (Englewood Cliffs, NJ: Prentice-Hall, 1973), 10, assigns Murray and Kean the "honor of being the first known professional company to perform in America (not counting the company—if there was a company—headed by Charles and Mary Stagg in Williamsburg, Virginia)."
33. Schultz, *Gay's Beggar's Opera*, 109.

34. George Overcash Seilhamer, *History of the American Theatre* (Philadelphia: Globe Printing House, 1888), 1:261.
35. Odai Johnson, *Absence and Memory in Colonial American Theatre: Fiorelli's Plaster* (New York: Palgrave Macmillan, 2006), 137–38, cites performances by a Mr. Joan and M. A. Warwell in Boston and other parts of New England.
36. Johnson, Burling, and Coombs, eds., *Colonial American Stage*, 64.
37. Seilhamer, *History*, 1:235, 39 and 58–61.
38. Schultz, *Gay's Beggar's Opera*, 112. Although discussing the earlier effects of the 1737 English licensing act, Matthew J. Kinservik, *Disciplining Satire: The Censorship of Satiric Comedy on the Eighteenth-Century London Stage* (London: Associated University Presses, 2002), usefully argues for censorship's positive, culturally productive—rather than simply repressive—aspects.
39. Josiah Quincy, original manuscript journal of trip to England, 1774–1775, entry November 18, 1774. Quincy family papers, Massachusetts Historical Society.
40. Gwenda Morgan and Peter Rushton, *Eighteenth-Century Criminal Transportation: The Formation of the Criminal Atlantic* (New York: Palgrave Macmillan, 2003), 1.
41. Johnson, *Absence and Memory*, 226–27.
42. Alexander Graydon, *Memoirs of a Life, Chiefly Passed in Pennsylvania, within the Last Sixty Years, with Occasional Remarks Upon the General Occurrences, Character and Spirit of That Eventful Period* (Harrisburgh, PA: Printed by John Wyeth, 1811), 93.
43. Ibid., 93.
44. An account of the incident appears in a column entitled "Beggar's Opera—Midnight Hour," *The American Monthly Magazine and Critical Review*, November 1817, 62.
45. Ibid., 62
46. Daniel A. Cohen, *Pillars of Salt, Monuments of Grace: New England Crime Literature and the Origins of American Popular Culture, 1674–1860* (New York: Oxford University Press, 1993), 117.
47. Ibid., 117.
48. On early American crime and punishment, see David J. Rothman, *The Discovery of the Asylum: Social Order and Disorder in the New Republic* (Boston: Little, 1971); David H. Flaherty, "Crime and Social Control in Provincial Massachusetts," *The Historical Journal* 24.2 (1981): 339–60; Linda Kealey, "Patterns of Punishment: Massachusetts in the Eighteenth Century," *The American Journal of Legal History* 30.2 (1986): 163–86; Adam Jay Hirsch, *The Rise of the Penitentiary: Prisons and Punishment in Early America* (New Haven: Yale University Press, 1992); M. Watt Espy and John Ortiz Smykla, *Executions in the United States, 1608–1987: The Espy File* (Ann Arbor, MI: Inter-University Consortium for Political and Social Research, 1987).
49. Gay, *Beggar's Opera*, 1.4.10–12.
50. Ibid., 3.13.7, Air 60.
51. Linebaugh, *The London Hanged*, 23.
52. See Douglas Hay, "Property, Authority, and the Criminal Law," *Albion's Fatal Tree: Crime and Society in Eighteenth-Century England*, eds. Douglas Hay, Peter Linebaugh, John G. Rule, E. P. Thompson, and Cal Winslow (New York: Peregrine Books, 1975), 17–63; and Louis P. Masur, *Rites of Execution: Capital Punishment and the Transformation of American Culture, 1776–1865* (New York: Oxford University Press, 1989), 73. Ted Robert Gurr, *Violence in America*, 2 vols (Newbury Park, CA: Sage Publications, 1989), 1.36.
53. Gay, *Beggar's Opera*, 3.13. The stage directions appear at the end of scene 13 and at the beginning of scene 14; the dialogue appears in scene 13, lines 6–7.
54. Linebaugh, *The London Hanged*, 34.

55. Richardson, "Forms of Resistance," 25.
56. Gay, *Beggar's Opera*, 3.13.26, Air 67.
57. Ibid., 3.16.8–9.
58. Ibid., 3.16.11–12.
59. Ibid., 3.16.12–15.
60. Thomas Mount, *The Confession, &c. of Thomas Mount, Who Was Executed at Little-Rest, in the State of Rhode-Island, on Friday the 27th of May, 1791, for Burglary*, ed. William Smith ([Newport, Rhode Island]: Printed and sold by Peter Edes, in Newport, 1791). Masur, *Rites of Execution*, 33–39, also discusses the conventions and formulae of such confessions.
61. Cohen, *Pillars of Salt*, 22–23. Dan E. Williams, *Pillars of Salt: An Anthology of Early American Criminal Narratives* (Madison, WI: Madison House, 1993) has usefully edited a number of criminal narratives. Lincoln B. Faller, *Turned to Account: The Forms and Functions of Criminal Biography in Late Seventeenth- and Early Eighteenth-Century England* (New York: Cambridge University Press, 1987) discusses similar English texts.
62. Royall Tyler, *The Algerine Captive, or the Life and Adventures of Doctor Updike Underhill, Six Years a Prisoner among the Algerines*, 2 vols (Walpole, Newhampshire: David Carlisle, jun., and sold at his bookstore, 1797), 1.vi.
63. Kristin Boudreau, "Early American Criminal Narratives and the Problem of Public Sentiments," *Early American Literature* 32.3 (1997): 249–50.
64. Mount, *Confession*, [2].
65. See Janet Sorensen, "Vulgar Tongues: Canting Dictionaries and the Language of the People in Eighteenth-Century Britain," *Eighteenth-Century Studies* 37.3 (2004): 435–54.
66. Mount, *Confession*, 18. The generic convention of the flash dictionary appended to a criminal narrative persists for obvious reasons, but it remains a compelling supplement to the narrative; Christopher Hibbert, *The Road to Tyburn: The Story of Jack Sheppard and the Eighteenth Century Underworld* (London: Longmans, 1957), e.g., includes a "Glossary of Cant" at the end.
67. Mount, *Confession*, 19.
68. Ibid.
69. Ibid., 21.
70. Ibid.
71. Ibid., 23.
72. Ibid., 20 (italics in the original).
73. Ibid., 21.
74. Ibid.
75. Ibid., 10.
76. The Library Company of Philadelphia, the Historical Society of Pennsylvania, and the American Antiquarian Society hold an impressive array of documents representing Levi Ames's execution. The fifteen printed pieces represent a significant proportion of the surviving textual production surrounding late eighteenth century executions. Cohen, *Pillars of Salt*, 18–20, notes the survival of twenty-five different broadsides of the "dying verses" genre, and some thirty-five "last speeches" in New England between the early 1730s and the end of the century.
77. *An Exhortation to Young and Old to Be Cautious of Small Crime, Lest They Become Habitual, and Lead Them before Thay Are Aware into Those of the Most Heinous Nature. Occasioned by the Unhappy Case of Levi Ames, Executed on Boston-Neck, October 21st, 1773, for the Crime of Burglary* ([Boston]: 1773); broadside, Historical Society of

Pennsylvania. *A Solemn Farewell to Levi Ames, Being a Poem Written a Few Days before His Execution, for Burglary, Oct. 21, 1773* (Boston: Printed and sold at Draper's printing-office, in Newbury-Street, 1773); broadside, HSP.

78. *Theft and Murder! A Poem on the Execution of Levi Ames, Which Is to Be on Thursday, the 21st of October Inst.* ([Boston]: Sold [by Isaiah Thomas] near the Mill-Bridge: and at the printing office near the market, 1773); broadside, Houghton Library, Harvard University. Cohen, *Pillars of Salt*, 143.

79. *The Speech of Death to Levi Ames. Who Was Executed on Boston-Neck, October 21, 1773, for the Crime of Burglary* ([Boston]: Printed by John Boyle, 1773); broadside, AAS.

80. *The Last Words and Dying Speech of Levi Ames, Who Was Executed at Boston, on Thursday the 21st Day of October, 1773. For Burglary* (Boston: Printed and sold at the shop opposite the Court-House in Queen-Street, 1773); broadside, HSP.

81. *A Few Lines Wrote Upon the Intended Execution of Levi Ames, for Burglary, and Being Sent to Him for His Improvement, Are Now Published at His Desire* ([Boston]: 1773); broadside, HSP.

82. *The Dying Penitent; or, the Affecting Speech of Levi Ames* ([Boston]: Sold opposite the Court-House in Queen-Street, 1773); broadside, Massachusetts Historical Society.

83. *The Dying Groans of Levi Ames, Who Was Executed at Boston, the 21st of October, 1773, for Burglary* ([Boston]: 1773); broadside, AAS. See Gwenda Morgan and Peter Rushton, "Visible Bodies: Power, Subordination and Identity in the Eighteenth-Century Atlantic World," *Journal of Social History* 39.1 (2005): 39–64.

84. Boudreau, "Criminal Narratives and the Problem of Public Sentiments," 249–69.

85. Marcus Rediker, *Villains of All Nations: Atlantic Pirates in the Golden Age* (Boston: Beacon Press, 2004), 1–18.

86. Eliot, *Christ's Promise*, 27.

87. Ibid.

88. Ibid., 29.

89. Ibid., 30.

90. *Exhortation to Young and Old [...] Occasioned by the Unhappy Case of Levi Ames, Executed on Boston-Neck, October 21st, 1773, for the Crime of Burglary* ([Boston], 1773; broadside, courtesy of the Historical Society of Pennsylvania).

91. *A Dialogue between Elizabeth Smith, and John Sennet* ([Boston]: 1773); broadside, HSP.

92. Boudreau, "Criminal Narratives and the Problem of Public Sentiments," 261.

93. McKenzie, "The Real Macheath," 583.

94. Joe Cowell, *Thirty Years Passed among the Players in England and America* (New York: Harper & Brothers, 1844), 76.

95. *Aurora General Advertiser* September 14, 1824. The Boston *Commercial Gazette*, September 30, 1824, also reprinted the account. Advertisements throughout September and October 1824 tout Stoker's "wonderful performances" and "astonishing evolutions" on the slack rope. Stoker's act appeared at the Circus during the popular runs of W. T. Moncrieff's urban underclass drama *Tom and Jerry*, discussed in chapter six.

96. *Aurora General Advertiser* September 14, 1824.

97. Cowell, *Thirty Years*, 76.

98. Ibid.

99. *Aurora General Advertiser* September 14, 1824.

100. "Spectre" could have seen wax works at the Washington Museum, advertisements for which list many of the spectacles noted in the commentary. See, e.g., the *Aurora General Advertiser*, September 8, 1824.

101. Karen Ahlquist, *Democracy at the Opera: Music, Theater, and Culture in New York City, 1815–60* (Urbana: University of Illinois Press, 1997), 12–16.

3 ALGERIANS, RENEGADES, AND TRANSNATIONAL ROGUES IN *SLAVES IN ALGIERS*

1. Performance announcements first appear in the *General Advertiser* (later the *Aurora*) as well as in the *Gazette and Universal Daily Advertiser*, June 28 and 30, 1794. Later performances announced in the Baltimore *Federal Intelligencer* (November 20, 1794), the New York *Daily Advertiser* (May 9, 1796). Thomas Clark Pollock, *The Philadelphia Theatre in the Eighteenth Century, Together with the Day Book of the Same Period* (Philadelphia: University of Pennsylvania Press, 1933), 419; David Ritchey, *A Guide to the Baltimore Stage in the Eighteenth Century: A History and Day Book Calendar* (Westport, CT: Greenwood Press, 1982), 296; and George Clinton Densmore Odell, *Annals of the New York Stage* (New York: Columbia University Press, 1927), 1:411 concur.
2. *Boston Daily Advertiser*, May 27, 1816. This announcement, indicating the play's "second time these 18 years," suggests that Boston audiences had seen the play in 1798.
3. Jeffrey H. Richards, *Drama, Theatre, and Identity in the American New Republic* (Cambridge: Cambridge University Press, 2005), 145.
4. Susanna Haswell Rowson, *Slaves in Algiers; or, a Struggle for Freedom: A Play, Interspersed with Songs, in Three Acts. By Mrs. Rowson. As Performed at the New Theatres, in Philadelphia and Baltimore* (Philadelphia: Printed for the author, by Wrigley and Berriman, no. 149, Chesnut-Street, 1794), 3.7 (69).
5. Paul Michel Baepler, *White Slaves, African Masters: An Anthology of American Barbary Captivity Narratives* (Chicago: University of Chicago Press, 1999), 24; Gary E. Wilson, "American Hostages in Moslem Nations, 1784–1796: The Public Response," *Journal of the Early Republic* 2.2 (1982): 123–41. Philip Gould, *Barbaric Traffic: Commerce and Antislavery in the Eighteenth-Century Atlantic World* (Cambridge: Harvard University Press, 2003); and Paul Michel Baepler, "The Barbary Captivity Narrative in American Culture," *Early American Literature* 39.2 (2004): 217–46 examine Barbary narratives in broader contexts. Daniel J. Vitkus and N. I. Matar, eds., *Piracy, Slavery, and Redemption: Barbary Captivity Narratives from Early Modern England* (New York: Columbia University Press, 2001) shows the long Atlantic genealogy of North African captivity narratives.
6. Frank Lambert, *The Barbary Wars: American Independence in the Atlantic World* (New York: Hill and Wang, 2005), 93.
7. See Frederick C. Leiner, *The End of Barbary Terror: America's 1815 War against the Pirates of North Africa* (Oxford: Oxford University Press, 2006).
8. Elizabeth Maddock Dillon, "*Slaves in Algiers*: Race, Republican Genealogies, and the Global Stage," *American Literary History* 16.3 (2004): 407.
9. Lambert, *Barbary Wars*, 8.
10. Rowson, *Slaves*, 3.7 (72).
11. Despite the play's leanings, Rowson did not openly display partisan support in the theatre until later in the decade. Ironically, Federalists were more tolerant of some forms of female political participation in the 1790s. William Cobbett, *A Kick for a Bite, or, Review Upon Review with a Critical Essay on the Works of Mrs. S. Rowson: In a Letter to the Editor, or Editors, of the American Monthly Review* (Philadelphia: Printed by T. Bradford, 1795), 27. See Arthur Scherr, "'Sambos' and 'Black Cut-Throats': Peter Porcupine on Slavery and Race in the 1790's," *American Periodicals: A Journal of History, Criticism, and Bibliography* 13 (2004): 3–30.
12. William Cobbett, *A Little Plain English, Addressed to the People of the United States, on the Treaty, Negociated with His Britannic Majesty, and on the Conduct of the President Relative Thereto* (Philadelphia: Published by Thomas Bradford, printer, bookseller, and stationer, no. 8, South Front Street, 1795), 70.

13. Marion Rust, *Prodigal Daughters: Susanna Rowson's Early American Women* (Chapel Hill: Omohundro Institute of Early American History and Culture, University of North Carolina Press, 2008), 209–14, neatly depicts theatre's exclusive politics but inclusive form in the 1790s. See also Heather S. Nathans, *Early American Theatre from the Revolution to Thomas Jefferson: Into the Hands of the People* (Cambridge: Cambridge University Press, 2003) especially chapters three and four.
14. Rowson, *Slaves*, [73].
15. Dillon, "Race, Republican Genealogies, and the Global Stage," 421. The phrase is Dillon's but Marion Rust also reads Rowson's prodigal energies in *Prodigal Daughters* and in Marion Rust, "'Daughters of America,' *Slaves in Algiers*: Activism and Abnegation Off Rowson's Barbary Coast," *Feminist Interventions in Early American Studies*, ed. Mary Clare Carruth (Tuscaloosa: University of Alabama Press, 2006), 227–39.
16. Benilde Montgomery, "White Captives, African Slaves: A Drama of Abolition," *Eighteenth-Century Studies* 27.4 (1994): 615–30 demonstrates the ways in which the plot of Algerian enslavement did contribute to early American abolitionist sentiment.
17. As Rust, *Prodigal Daughters*, 199, note 6 argues, the female characters allow a degree of identification and theatrical free play while the "usurious Jews, drunken Spaniards, and North African Muslim potentates" forestall identification, becoming the passive objects of the viewer's gaze.
18. Rowson, *Slaves*, 1.1 (15).
19. Ibid., 1.1 (16).
20. Terry Bouton, *Taming Democracy: "The People," The Founders, and the Troubled Ending of the American Revolution* (Oxford: Oxford University Press, 2007); William Hogeland, *The Whiskey Rebellion: George Washington, Alexander Hamilton, and the Frontier Rebels Who Challenged America's Newfound Sovereignty* (New York: Scribner, 2006).
21. See William Pencak, Matthew Dennis, and Simon P. Newman, eds., *Riot and Revelry in Early America* (University Park, PA: Pennsylvania State University Press, 2002); Paul A. Gilje, *The Road to Mobocracy: Popular Disorder in New York City, 1763–1834* (Chapel Hill, NC: Published for the Institute of Early American History and Culture by the University of North Carolina Press, 1987).
22. Nathans, *Early American Theatre*, 83.
23. Rust, *Prodigal Daughters*, 202.
24. Rowson, *Slaves*, [3], [4].
25. "Dramatis Personae," Susanna Haswell Rowson, Slaves in Algiers; or, a Struggle for Freedom: A Play, Interspersed with Songs, in Three Acts. By Mrs. Rowson. As Performed at the New Theatres, in Philadelphia and Baltimore (Philadelphia: Printed for the author, by Wrigley and Berriman, no. 149, Chesnut-Street, 1794); courtesy of Historical Society of Pennsylvania.
26. Gérard Genette, *Paratexts: Thresholds of Interpretation* (Cambridge: Cambridge University Press, 1997).
27. Plays such as Isaac Bickerstaff's 1787 *A Peep into the Seraglio* demonstrate the fascination with forbidden and sexualized oriental spaces that Edward W. Said discusses in *Orientalism* (New York: Pantheon Books, 1978) and in *Culture and Imperialism* (New York: Knopf, 1993).
28. James C. Burge, *Lines of Business: Casting Practice and Policy in the American Theatre, 1752–1899* (New York: P. Lang, 1986), 97. Part II covers American stage practices from 1752 to 1839.
29. James Fennell, *An Apology for the Life of James Fennell, Written by Himself* (Philadelphia: Published by Moses Thomas, No. 52, Chesnut-Street, J. Maxwell, printer, 1814), 334.

30. Pierre Bourdieu, *Distinction: A Social Critique of the Judgement of Taste* (Cambridge, MA: Harvard University Press, 1984), 169–75.
31. See Daniel J. Vitkus, *Turning Turk: English Theater and the Multicultural Mediterranean, 1570–1630* (New York: Palgrave Macmillan, 2003), 107–62.
32. Christopher Castiglia, *Bound and Determined: Captivity, Culture-Crossing, and White Womanhood from Mary Rowlandson to Patty Hearst* (Chicago: University of Chicago Press, 1996), 4; see also Baepler, "Barbary Captivity Narrative," 239–40; Daniel E. Williams, *Liberty's Captives: Narratives of Confinement in the Print Culture of the Early Republic* (Athens: University of Georgia Press, 2006).
33. Tyler's phrase appears in a preface to the second edition of *The Algerine Captive*; quoted in Caleb Crain, introduction to Royall Tyler, *The Algerine Captive, or, the Life and Adventures of Doctor Updike Underhill, Six Years a Prisoner among the Algerines*, ed. Caleb Crain (New York: Modern Library, 2002), xxxii.
34. Rowson, *Slaves*, 62, 3.6.
35. Ibid., 20, 1.3.
36. Ibid., 23, 1.4.
37. Ibid., 27, 2.1.
38. Newburyport, Massachusetts *Political Gazette*, July 30, 1795.
39. Robert Adams, *The Narrative of Robert Adams, an American Sailor, Who Was Wrecked on the Western Coast of Africa, in the Year 1810, Was Detained Three Years in Slavery by the Arabs of the Great Desert, and Resided Several Months in the City of Tombuctoo. With a Map, Notes and an Appendix* (Boston: Wells and Lilly, 1817).
40. Ibid., xviii.
41. Ibid., xxi.
42. Today, the term properly refers to an ethnic group, the "Shluh" or "Chleuh"; the 1911 *Encyclopedia Britannica* retains the sense of vagabondage.
43. Adams, *Narrative*, xxi.
44. William Jerry MacLean, "Othello Scorned: The Racial Thought of John Quincy Adams," *Journal of the Early Republic* 4.2 (1984): 143–60; W. T. Lhamon, Jr., *Jump Jim Crow: Lost Plays, Lyrics, and Street Prose of the First Atlantic Popular Culture* (Cambridge, MA: Harvard University Press, 2003), 11–14.
45. Vitkus, *Turning Turk*, 77.
46. Heather S. Nathans, "A Much Maligned People: Jews on and Off the Stage in the Early American Republic," *Early American Studies* 2.2 (2007): 326.
47. Michael Ragussis, "Jews and Other "Outlandish Englishmen": Ethnic Performance and the Invention of British Identity under the Georges," *Critical Inquiry* 26.4 (2000): 773–97; John J. Gross, *Shylock: A Legend and Its Legacy* (New York: Simon & Schuster, 1992).
48. Nathans, "Much Maligned People," 323–24.
49. Ibid., 330.
50. See, e.g., *The Cries of London, as They Are Daily Exhibited in the Streets with an Epigram in Verse, Adapted to Each. Embellished with Sixty-Two Elegant Cuts. To Which Is Added, a Description of the Metropolis in Verse* (London: printed for E. Newbery, 1796), and published soon after for children, Samuel Wood, *The Cries of New-York* (New-York: Printed and sold by S. Wood, at the Juvenile book-store, No. 362, Pearl-Street, 1808).
51. Rowson, *Slaves*, 1.2 (17).
52. William Pencak, *Jews and Gentiles in Early America: 1654–1800* (Ann Arbor: University of Michigan Press, 2005), 232.
53. Rowson, *Slaves*, 1.2 (17).
54. Vitkus, *Turning Turk*, 37, 81–84.

55. Lambert, *Barbary Wars*, 105.
56. Mathew Carey, *A Short Account of Algiers*, second edn (Philadelphia: Printed for Mathew Carey, no. 118, Market-Street, 1794), 4.
57. Ibid., 4.
58. Ibid., 5.
59. Peter Markoe, *The Algerine Spy in Pennsylvania: Or, Letters Written by a Native of Algiers on the Affairs of the United States in America, from the Close of the Year 1783 to the Meeting of the Convention* (Philadelphia: Printed and sold by Prichard & Hall, in Market between Front and Second Streets, 1787).
60. Cathy N. Davidson, *Revolution and the Word: The Rise of the Novel in America* (New York: Oxford University Press, 1986), 85–86, briefly discusses Markoe. See also Mary Chrysostom Diebels, *Peter Markoe (1752?-1792): A Philadelphia Writer* (Washington, D.C.: The Catholic University of America Press, 1944); and Malini Johar Schueller, *U.S. Orientalisms : Race, Nation, and Gender in Literature, 1790–1890* (Ann Arbor: University of Michigan Press, 1998).
61. Federal Gazette and Philadelphia Evening Post, March 25, 1790.
62. "Profession vs. Practice," *Federal Orrery* (November 24, 1794): 42.
63. Advertisements such as the one in the *Gazette of the United States*, June 28, 1794, show that, rather than trumpeting its Algerian spectacle, the play enticed audiences with sailor frolics such as William Rowson's song "Heaving of the Lead" and the "double hornpipe" of Mr. Francis and Mrs. DeMarque.
64. John Durang, *The Memoir of John Durang, American Actor, 1785–1816*, ed. Alan Seymour Downer (Pittsburgh: University of Pittsburgh Press, 1966), 107.
65. Mythili Kaul, "Background: Black or Tawny? Stage Representations of Othello from 1604 to the Present," in Mythili Kaul, ed., *Othello: New Essays by Black Writers* (Washington, D.C.: Howard University Press, 1997), 5.
66. Rowson, *Slaves*, 1.1 (6).
67. Ibid.
68. William Dunlap, *Blue Beard: A Dramatic Romance by G. Colman, the Younger, as Altered for the New York Theatre, with Additional Songs* (New York: Printed and published by D. Longworth, at the Shakespeare gallery, 1802).
69. "Theatrics. Critique No. 19: Blue Beard," *New-York Evening Post*, March 22, 1802.
70. Gazette of the United States, April 29, 1797.
71. The performance of May 27, 1799 was first announced in *Gazette of the United States*, May 18, 1799; playbill appearing in *Gazette of the United States*, May 27, 1799, and again on January 6 and 10, 1800. The *Constitutional Diary and Philadelphia Evening Advertiser* of January 27, 1800, indicates another performance of the play.
72. Boston Constitutional Telegraphe, January 11, 1800.
73. *Salem Impartial Register*, July 20, 1801, and the *Providence Gazette*, November 14, 1801.
74. *Federal Gazette and Baltimore Daily Advertiser*, June 10, 1801, and *Charleston City Gazette And Daily Advertiser*, March 25, 1801. The Baltimore performance was a seasonal traveling performance of the Philadelphia theatre.
75. *New-York Evening Post*, March 6, 1802 and June 8, 1802.
76. John Brown, *Barbarossa; a Tragedy: With Alterations and Amendments. As Performed at the Theatre in Boston* (Printed at the Apollo press, in Boston, for David West, no. 36, Marlboro Street, and John West, no. 75, Cornhill, 1794). *Barbarossa*, originally written for David Garrick, was first performed in London in 1755. The Newport (Rhode Island) Historical Society holds a broadside announcing the Newport performance of *Barbarossa* for May 29, 1794.

77. The *Philadelphia Gazette and Universal Daily Advertiser*, April 12, 1799; Pollock, *Philadelphia Theatre*, 390 identifies this as the first known production of the pantomime.
78. Royall Tyler, *The Algerine Captive, or the Life and Adventures of Doctor Updike Underhill, Six Years a Prisoner among the Algerines*, 2 vols (Walpole, Newhampshire: David Carlisle, jun., and sold at his bookstore, 1797), 2:16–17.
79. Ibid., 2:17.
80. Ibid., 2:27–29.
81. New York *Morning Chronicle*, March 5, 1805. Playbills appear in the *Morning Chronicle* on March 6, 1805 and after, including a benefit on March 28 to supply the prisoners with clothing. See also Dillon, "Race, Republican Genealogies, and the Global Stage," 429.
82. Rowson, *Slaves*, 3.1 (42, 46).
83. See, e.g., the catalogue of northern European immigrants in Letter III of J. Hector St. John de Crèvecoeur, *Letters from an American Farmer; and, Sketches of Eighteenth-Century America*, ed. Albert E. Stone (New York: Penguin Books, 1983), 68.
84. Richards, *Drama, Theatre, and Identity*, 155–61, traces a different dramatic routing of Muley Moloc's character, whose resurrections took him on a path from high dramatic protagonist, to villain, to low buffoon by the time Rowson received and rewrote the character. See John Dryden, *Don Sebastian, King of Portugal: A Tragedy Acted at the Theatre Royal* (London: Printed for Jo. Hindmarsh, at Golden Ball in Cornhill, 1690); Isaac Bickerstaff and Charles Dibdin, *The Captive, a Comic Opera; as It Is Perform'd at the Theatre-Royal in the Hay-Market* (London: Printed for W. Griffin, 1769); Frederick Reynolds, *The Renegade; a Grand Historical Drama, in Three Acts. Interspersed with Music* (New York: Pub. by D. Longworth, at the Dramatic Repository, Shakspeare-Gallery, 1813).
85. Rowson, *Slaves*, 3.1 (48).
86. See Peter Thompson, *Rum Punch and Revolution: Taverngoing and Public Life in Eighteenth Century Philadelphia* (Philadelphia: University of Pennsylvania Press, 1999), 111–44.
87. Rowson, *Slaves*, 3.1 (48).
88. Ibid.
89. Ibid.
90. Ibid.
91. Richards, *Drama, Theatre, and Identity*, 188–210, discusses stage Irish in early American theatre; Jason Shaffer, *Performing Patriotism: National Identity in the Colonial and Revolutionary American Theater* (Philadelphia: University of Pennsylvania Press, 2007), 174–78 discusses O'Keefe specifically.
92. John O'Keefe, *Songs in the Comic Opera, Called, the Son-in-Law. By John O'Keefe, Esq. As Sung at the New Theatre, Philadelphia. Corrected and Revised by Mr. Rowson, Prompter* ([Philadelphia]: Printed for M. Carey, 1794).
93. John O'Keefe, *The Agreeable Surprise. A Comic Opera, in Two Acts* ([Boston]: Printed at the Apollo Press, in Boston, by Belknap and Hall, for William P. Blake, no. 59, Cornhill, and William T. Clap. No. 90, Newbury Street, 1794), 15.
94. Samuel Larkin, *The Columbian Songster and Freemason's Pocket Companion. A Collection of the Newest and Most Celebrated Sentimental, Convivial, Humourous, Satirical, Pastoral, Hunting, Sea and Masonic Songs, Being the Largest and Best Collection Ever Published in America* (Portsmouth, New-Hampshire: Printed by J. Melcher, for S. Larkin, at the Portsmouth book-store, 1798), 92–93.
95. *The Nightingale. A Collection of the Most Elegant Songs, Now in Vogue* (Amherst: Samuel Preston, 1797), 23.

96. See John F. Kasson, *Rudeness and Civility: Manners in Nineteenth-Century Urban America* (New York: Hill and Wang, 1990); David S. Shields, *Civil Tongues and Polite Letters in British America* (Chapel Hill: Published for the Institute of Early American History and Culture, Williamsburg, Virginia, by University of North Carolina Press, 1997); Albrecht Koschnik, *Let a Common Interest Bind Us Together: Associations, Partisanship, and Culture in Philadelphia, 1775–1840* (Charlottesville: University of Virginia Press, 2007); Simon P. Newman, *Parades and the Politics of the Street: Festive Culture in the Early American Republic* (Philadelphia: University of Pennsylvania Press, 1997); and David Waldstreicher, *In the Midst of Perpetual Fetes: The Making of American Nationalism, 1776–1820* (Chapel Hill: Omohundro Institute of Early American History and Culture, University of North Carolina Press, 1997).
97. Rowson, *Slaves*, 3.1 (42).
98. Ibid.
99. Ibid., 68–69, 3.7
100. John Foss, *A Journal, of the Captivity and Sufferings of John Foss, Several Years a Prisoner in Algiers* (Newburyport [Mass.]: Printed by A. March, Middle-Street., 1798), 25.
101. Gazette of the United States, February 9, 1797.
102. Ibid.
103. Foss, *Captivity and Sufferings*, 15.
104. Newburyport, Massachusetts *Political Gazette*, July 30, 1795.
105. Foss, *Captivity and Sufferings*, 54–55.
106. Ibid.
107. Philadelphia Gazette and Universal Daily Advertiser, February 9, 1797.
108. Ibid.
109. Ibid.

4 TREASON AND POPULAR PATRIOTISM IN *THE GLORY OF COLUMBIA*

1. See William Dunlap, *André: A Tragedy, in Five Acts: As Performed by the Old American Company, New-York, March 30, 1798. To Which Are Added Authentic Documents Respecting Major Andre; Consisting of Letters to Miss Seward, the Cow Chace, Proceedings of the Court Martial, &C* (New York: Printed by T. & J. Swords, no. 99 Pearl-street, 1798); *André* is reprinted in Jeffrey H. Richards, *Early American Drama* (New York: Penguin Books, 1997), 58–108. The revision survives in William Dunlap, *The Glory of Columbia; Her Yeomanry! A Play in Five Acts* (New York: Published by David Longworth at the Dramatic Repository, Shakspeare-Gallery, 1817); extracts of the music appear as William Dunlap, *The Glory of Columbia; Her Yeomanry. A Play in Five Acts. The Songs, Duets, and Chorusses, Intended for the Celebration of the Fourth of July, at the New-York Theatre* (New York: Printed and published by D. Longworth, at the Shakspeare-Gallery, 1803).
2. George Clinton Densmore Odell, *Annals of the New York Stage* (New York: Columbia University Press, 1927), 1:182.
3. Heather S. Nathans, *Early American Theatre from the Revolution to Thomas Jefferson: Into the Hands of the People* (Cambridge: Cambridge University Press, 2003), 47.
4. Ibid., 145.
5. A few productions of André's story also appeared in Norfolk, as Jeffrey Richards shows in *Drama, Theatre, and Identity in the American New Republic* (Cambridge: Cambridge University Press, 2005), 290–95.
6. Richards, *Early American Drama*, 61.

7. Lucy Rinehart, "'Manly Exercises': Post-Revolutionary Performances of Authority in the Theatrical Career of William Dunlap," *Early American Literature* 36.2 (2001): 263–93.
8. Odell, *Annals*, 1:182.
9. Nathans, *Early American Theatre*, 115, 210 note 118.
10. Charles Durang, *The Philadelphia Stage…By Charles Durang. Partly Compiled from the Papers of His Father, the Late John Durang; with Notes by the Editors* (Philadelphia: 1854–57), quoted in James Rees, *The Life of Edwin Forrest: With Reminiscences and Personal Recollections* (Philadelphia: Peterson and Brothers, 1874), 59.
11. William Dunlap, *A History of the American Theatre* (New York: J. & J. Harper, 1832), 223.
12. Dunlap, *Glory of Columbia*, 3.2 (29).
13. Ibid.
14. Dunlap, *André*, 5.4 (61–62).
15. Rinehart, "Manly Exercises," 276; Andrea McKenzie, "Martyrs in Low Life? Dying "Game" In Augustan England," *Journal of British Studies* 42.2 (2003): 167–205.
16. Maya Mathur, "An Attack of the Clowns: Comedy, Vagrancy, and the Elizabethan History Play," *Journal for Early Modern Cultural Studies* 7.1 (2007): 35.
17. See John Durang, *The Memoir of John Durang, American Actor, 1785–1816*, ed. Alan Seymour Downer (Pittsburgh: University of Pittsburgh Press, 1966), 5–6, 115–16.
18. Dunlap, *History*, 223.
19. Samuel Rowland Fisher, *Journal of Samuel Rowland Fisher, of Philadelphia, 1779–1781*, ed. Anna Wharton Morris (1928), 91; Historical Society of Pennsylvania.
20. Dale Cockrell, *Demons of Disorder: Early Blackface Minstrels and Their World* (New York: Cambridge University Press, 1997), 50–54.
21. Peter Benes, "Night Processions: Celebrating the Gunpowder Plot in England and New England," *New England Celebrates: Spectacle, Commemoration, and Festivity*, ed. Peter Benes (Concord, MA: Boston University, 2002), 9, 25–28, describes New England's 1821 retrospectives view of such celebrations.
22. Fisher, *Journal*, 94; Gary B. Nash, *First City: Philadelphia and the Forging of Historical Memory* (Philadelphia: University of Pennsylvania Press, 2006), 99–100.
23. Fisher, *Journal*, 94.
24. The image appears in the *Americanischer Haus- Und Wirthschafts-Calender Auf Das 1781ste Jahr Christi* (Philadelphia: 1780); Historical Society of Pennsylvania, held at the Library Company of Philadelphia. A similar but less lively image appeared as a broadside, entitled *A Representation of the Figures Exhibited and Paraded through the Streets of Philadelphia, on Saturday, the 30th of September 1780* (Philadelphia: Printed by John Dunlap [?], 1780), held at Williams College (Evans 16959). This image was also reprinted in *The Continental Almanac, for the Year of Our Lord, 1781* (Philadelphia: Printed and sold by Francis Bailey, in Market-Street, between Third and Fourth-Streets, 1780); Historical Society of Pennsylvania, held at the Library Company of Philadelphia.
25. The *Pennsylvania Packet or General Advertiser*, October 3, 1780, identifies "Joe Smith" as the coconspirator, perhaps mistakenly or using a familiar version of Joshua.
26. Ibid.
27. Ibid.
28. Fisher, *Journal*, 94.
29. *Pennsylvania Packet or General Advertiser*, October 3, 1780.
30. Ibid.
31. Fisher, *Journal*, 94.
32. *Pennsylvania Packet or General Advertiser*, October 3, 1780.

33. Alexander Graydon, *Memoirs of a Life, Chiefly Passed in Pennsylvania, within the Last Sixty Years, with Occasional Remarks Upon the General Occurrences, Character and Spirit of That Eventful Period* (Harrisburgh, PA: Printed by John Wyeth, 1811), 111.
34. Ibid., 112.
35. John Gay, *The Beggar's Opera*, eds. Edgar V. Roberts and Edward Smith (Lincoln: University of Nebraska Press, 1969), 3.16.22–24.
36. Graydon, *Memoirs*, 112.
37. Ibid.
38. S. E. Wilmer, *Theatre, Society and the Nation: Staging American Identities* (Cambridge: Cambridge University Press, 2002), 78.
39. The description appears in the list of characters, Dunlap, *Glory of Columbia: Songs, Duets, and Chorusses*, [2]. The 1817 published version of the script appears more restrained in its presentation, perhaps because of the contemporary controversies over the motivations and actions of the captors.
40. See Sean Wilentz, *Chants Democratic: New York City and the Rise of the American Working Class, 1788–1850* (New York: Oxford University Press, 1984), 61–103.
41. Robert E. Cray, Jr., "Major John Andre and the Three Captors: Class Dynamics and Revolutionary Memory Wars in the Early Republic, 1780–1831," *Journal of the Early Republic* 17.3 (1997): 375.
42. Egbert Benson, *Vindication of the Captors of Major André* (New York: Published by Kirk & Mercein, at the office of the Edinburgh and Quarterly reviews, no. 22 Wall-Street. T. & W. Mercein, printers, 1817).
43. Cray, "Andre and the Three Captors," 376; Andy Trees, "Benedict Arnold, John Andre, and His Three Yeoman Captors: A Sentimental Journey or American Virtue Defined," *Early American Literature* 35.3 (2000): 246–73.
44. See chapter two; Thomas Mount, *The Confession, &c. of Thomas Mount, Who Was Executed at Little-Rest, in the State of Rhode-Island, on Friday the 27th of May, 1791, for Burglary*, ed. William Smith ([Newport, Rhode Island]: Printed and sold by Peter Edes, in Newport, 1791).
45. Louis P. Masur, *Rites of Execution: Capital Punishment and the Transformation of American Culture, 1776–1865* (New York: Oxford University Press, 1989), 38.
46. Dunlap, *Glory of Columbia*, 1.1 (3).
47. Ibid.
48. Ibid, 1.2 (6).
49. Ibid, 1.2 (7).
50. Ibid.
51. Ibid, 2.3 (20).
52. Ibid.
53. Marjorie B. Garber, *Vested Interests: Cross-Dressing and Cultural Anxiety* (New York: Routledge, 1992), 184; Laurence Senelick, *The Changing Room: Sex, Drag and Theatre* (New York: Routledge, 2000).
54. Elizabeth Reitz Mullenix, *Wearing the Breeches: Gender on the Antebellum Stage* (New York: St. Martin's Press, 2000), 6.
55. Alfred Fabian Young, *Masquerade: The Life and Times of Deborah Sampson, Continental Soldier* (New York: Alfred A. Knopf, distributed by Random House, 2004).
56. Dianne Dugaw, *Warrior Women and Popular Balladry, 1650–1850* (New York: Cambridge University Press, 1989). See also Paul Lewis, "Attaining Masculinity: Charles Brockden Brown and Woman Warriors of the 1790s," *Early American Literature* 40.1 (2005): 37–55.
57. Herman Mann, *The Female Review, or, Memoirs of an American Young Lady, Whose Life and Character Are Peculiarly Distinguished, Being a Continental Soldier, for Nearly*

Three Years, in the Late American War (Dedham [Mass.]: Printed by Nathaniel and Benjamin Heaton, for the author, 1797); Deborah Sampson Gannett, *Address Delivered with Applause at the Federal-Street Theatre, Boston, Four Successive Nights of the Different Plays, Beginning March 22, 1802 and after, at Other Principal Towns, a Number of Nights Successively at Each Place* (Dedham [MA]: Printed and sold by H. Mann, for Mrs. Gannet, at the Minerva office, 1802); Judith R. Hiltner, "'Like a Bewildered Star': Deborah Sampson, Herman Mann, and 'Address, Delivered with Applause,'" *Rhetoric Society Quarterly* 29.2 (1999): 5–24, argues that Mann probably wrote Sampson's "Address."

58. Dunlap, *Glory of Columbia*, 4.1 (37).
59. Richards, *Drama, Theatre, and Identity*, 188.
60. Later-nineteenth-century acts have proven much more popular with scholars. Noel Ignatiev, *How the Irish Became White* (New York: Routledge, 1995); Joseph R. Roach, "Barnumizing Diaspora: The 'Irish Skylark' Does New Orleans," *Theatre Journal* 50.1 (1998): 39–51; Robert Nowatzki, "Paddy Jumps Jim Crow: Irish-Americans and Blackface Minstrelsy," *Éire-Ireland* 41.3 (2007): 162–84.
61. Carl Wittke, "The Immigrant Theme on the American Stage," *The Mississippi Valley Historical Review* 39.2 (1952): 214.
62. Paul Goring, "'John Bull, Pit, Box, and Gallery, Said No!': Charles Macklin and the Limits of Ethnic Resistance on the Eighteenth-Century London Stage," *Representations* 79 (2002): 61–81.
63. Jason Shaffer, *Performing Patriotism: National Identity in the Colonial and Revolutionary American Theater* (Philadelphia: University of Pennsylvania Press, 2007), 177–78.
64. Rinehart, "Manly Exercises," 271–72. See Hugh Gough and David Dickson, eds., *Ireland and the French Revolution* (Dublin: Irish Academic Press, 1989).
65. Dunlap, *Glory of Columbia*, 4.1 (40–41).
66. Ibid., 4.1 (41).
67. See Peter Linebaugh, *The London Hanged: Crime and Civil Society in the Eighteenth Century* (Cambridge: Cambridge University Press, 1992) for the authoritative analysis of the gallows as a site of popular memory; Brian Henry, *Dublin Hanged: Crime, Law Enforcement and Punishment in Late Eighteenth-Century Dublin* (Blackrock: Irish Academic Press, 1994) explores the Irish contexts that O'Bogg jokingly recalls.
68. Masur, *Rites of Execution*, 73.
69. Simon P. Newman, *Parades and the Politics of the Street: Festive Culture in the Early American Republic* (Philadelphia: University of Pennsylvania Press, 1997), 5–9.
70. Susan G. Davis, *Parades and Power: Street Theatre in Nineteenth-Century Philadelphia* (Philadelphia: Temple University Press, 1986); David Waldstreicher, *In the Midst of Perpetual Fetes: The Making of American Nationalism, 1776–1820* (Chapel Hill: Omohundro Institute of Early American History and Culture, University of North Carolina Press, 1997); Albrecht Koschnik, *Let a Common Interest Bind Us Together: Associations, Partisanship, and Culture in Philadelphia, 1775–1840* (Charlottesville: University of Virginia Press, 2007).
71. Rosemarie K. Bank, *Theatre Culture in America, 1825–1860* (Cambridge: Cambridge University Press, 1997).
72. Dunlap, *Glory of Columbia*, 2.4 (21).
73. Dunlap, *Glory of Columbia: Songs, Duets, and Chorusses*, 4.
74. Dunlap, *Glory of Columbia*, 2.4 (21).
75. William Dunlap, *Diary of William Dunlap, 1766–1839; the Memoirs of a Dramatist, Theatrical Manager, Painter, Critic, Novelist, and Historian*, ed. Dorothy C. Barck (New York: B. Blom, 1969), 1:318.

76. Paul A. Gilje, *The Road to Mobocracy: Popular Disorder in New York City, 1763–1834* (Chapel Hill, NC: Published for the Institute of Early American History and Culture by the University of North Carolina Press, 1987), 108.
77. *New York Spectator*, August 1, 1798.
78. Ibid.
79. Cockrell, *Demons of Disorder*, 80.
80. Graydon, *Memoirs*, 138.
81. Shane White, "'It Was a Proud Day': African Americans, Festivals, and Parades in the North, 1741–1834," *The Journal of American History* 81.1 (1994): 13–50.
82. See Eric Lott, *Love and Theft: Blackface Minstrelsy and the American Working Class* (New York: Oxford University Press, 1993), 46–47.
83. Shane White, *Somewhat More Independent: The End of Slavery in New York City, 1770–1810* (Athens: University of Georgia Press, 1991), 95–106; Pinkster appears in various retrospective antiquarian accounts, but the descriptions closest to the actual event appeared in the *Albany Centinel*, June 17, 1803, reprinted in the *New York Daily Advertiser*, June 29, 1803.
84. White, *Somewhat More Independent*, 101.
85. White, "It Was a Proud Day," 30. The "bobalition" broadsides appearing in the first decades of the nineteenth century also represent an intriguing instance of such stagey treatments of underclass and African American modes of celebratory enfranchisement. The Library Company of Philadelphia holds a significant number of these broadsides.
86. *New York Commercial Advertiser*, November 25, 1803, at the New Theatre; *New York American*, November 25, 1820, at the Anthony-Street Theatre.
87. *New York Morning Chronicle*, May 12, 1804.
88. *New York Public Advertiser*, June 23, 1812; performance on June 24, 1812; the *New York Columbian*, June 29, 1814 advertises a performance with the same subtitle.
89. *Baltimore Patriot*, February 17, 1821, at the Pavilion Gardens on February 22, 1821; the *Newport Mercury*, August 8, 1824, announces a performance that evening of the "Capture of André and the preservation of West Point, taken from Dunlap's patriotic play of *The Glory of Columbia*"; *Baltimore Patriot* October 19, 1829, at the Baltimore Theatre and Circus.
90. *New York Morning Chronicle*, November 24, 1803.
91. Ibid.
92. See Waldstreicher, *Perpetual Fetes*, 117–26; Max Cavitch, "The Man That Was Used Up: Poetry, Particularity, and the Politics of Remembering George Washington," *American Literature* 75.2 (2003): 247–74.
93. *New York Morning Chronicle*, November 24, 1803.
94. *New York can Watch-Tower*, July 4, 1804.

5   PANTOMIME AND BLACKFACE BANDITRY IN *THREE-FINGER'D JACK*

1. Evidence of *Three-Finger'd Jack*'s staging appears in a manuscript libretto in the Huntington Library's Larpent collection of play manuscripts (LA 1297), a published score by Samuel Arnold, a London-published wordbook and prospectus, and various playbills, reviews, and descriptions.
2. There is no known evidence of *Three-Finger'd Jack* in the slaveholding southern states. Errol Hill, *The Jamaican Stage, 1655–1900: Profile of a Colonial Theatre* (Amherst: University of Massachusetts Press, 1992), 100.

3. See William Earle, *Obi, or, the History of Threefingered Jack in a Series of Letters from a Resident in Jamaica to His Friend in England* (Worcester, MA: printed by Isaiah Thomas Jr., 1804).
4. See John Nathan Hutchins, *Hutchins Improved, Being an Almanack and Ephemeris... For the Year of Our Lord 1802* (New York: Printed and sold by Ming and Young, 1801).
5. Introduction, *Three-Finger'd Jack*, in Robert H. B. Hoskins and Eileen Southern, eds., *Music for London Entertainment 1660–1800*, vol. 4, Series D, Pantomime, Ballet and Social Dance (London: Stainer & Bell, 1996), xxiv.
6. The newspaper reviews appear in the *Morning Post*, July 2, 1800; *Morning Herald* July 1, 1800; *Whitehall Evening Post* July 1–3, 1800. The most detailed review appears in the *The Dramatic Censor, or, Monthly Epitome of Taste, Fashion, and Manners*, ed. Thomas Dutton, vol. 3 (London: J. Roach and C. Chapple, 1801), 13–29.
7. Heather S. Nathans, *Early American Theatre from the Revolution to Thomas Jefferson: Into the Hands of the People* (Cambridge: Cambridge University Press, 2003), 47.
8. Ibid., 144.
9. See John O'Brien, *Harlequin Britain: Pantomime and Entertainment, 1690–1760* (Baltimore: Johns Hopkins University Press, 2004).
10. See Douglas R. Egerton, *Gabriel's Rebellion: The Virginia Slave Conspiracies of 1800 and 1802* (Chapel Hill: University of North Carolina Press, 1993); and James Sidbury, *Ploughshares into Swords: Race, Rebellion, and Identity in Gabriel's Virginia, 1730–1810* (New York: Cambridge University Press, 1997).
11. *Weekly Museum* 13.32 (May 23, 1801): [2]; *New York American Citizen*, May 26, 1801.
12. *New York Daily Advertiser*, May 27, 1801.
13. *Poulson's American Daily Advertiser*, December 26, 1801; *Massachusetts Mercury*, April 2, 1802.
14. "The Drama. For the Portfolio," *The Port Folio* 2.12 (March 27, 1802): 89. Thanks to Matthew Pethers for bringing this review, archived at the Library Company of Philadelphia, to my attention.
15. Leonora Sansay, *Secret History; or, the Horrors of St. Domingo: In a Series of Letters, Written by a Lady at Cape Francois, to Colonel Burr... Principally During the Command of General Rochambeau.* (Philadelphia: Bradford & Inskeep, 1808).
16. Benjamin Moseley, *A Treatise on Sugar* (London: G. G. & J. Robinson, 1799), 197. Moseley also quotes Governor Dalling's Proclamation of January 13, 1781 (199–200).
17. Moseley, *Treatise*, 200–201. Quashee appears to be the same Maroon who cut off Jack's fingers; Moseley identifies Sam as "Captain Davy's son, he who shot a Mr. Thompson," which would make him, according to Mavis Campbell, Sam Grant, who had advanced by 1803 to the rank of "Major of Maroons, and Chief Commander at Charles Town." Mavis Christine Campbell, *The Maroons of Jamaica, 1655–1796: A History of Resistance, Collaboration and Betrayal* (South Hadley, MA: Bergin and Garvey, 1988), 205–206; Moseley, *Treatise*, 205.
18. Orlando Patterson, "Slavery and Slave Revolts: A Sociohistorical Analysis of the First Maroon War, 1655–1740," *Maroon Societies: Rebel Slave Communities in the Americas*, ed. Richard Price (Baltimore, MD: Johns Hopkins University Press, 1996), 246.
19. Campbell, *Maroons*, 127–28, quotes the Leeward Treaty.
20. Ibid., 131.
21. Robert Charles Dallas, *The History of the Maroons* (London: 1803), 1:97.
22. See Kevin Mulroy, *Freedom on the Border: The Seminole Maroons in Florida, the Indian Territory, Coahuila, and Texas* (Lubbock: Texas Tech University Press, 1993).
23. *Songs, Duets and Choruses in the Pantomimical Drama of Obi, or, Three-Finger'd Jack*, third edn (London: T. Woodfall, 1800), 13. Arnold's score, published after the play's

first season, indicates essentially the same chorus, although calling for elaborate repetition of the phrase "sing tinga ring sing terry."
24. Ibid., 4.
25. Ibid., 19.
26. John Fawcett, *Obi, or, Three-Finger'd Jack* (1800), 1.1 (5); Larpent manuscript LA 1297, quoted by permission of the Huntington Library, San Marino, California.
27. John Fawcett, *Obi, or, Three-Finger'd Jack* (London: T. Woodfall, 1800), 7.
28. Samuel Arnold, *The Overture, Songs, Chorusses & Appropriate Music in the Grand Pantomimical Drama Call'd Obi; or Three Finger'd Jack… Composed & Adapted to the Action by Saml. Arnold… With Selections from the Most Eminent Masters, Arranged for the Voice & Piano Forte. Op. 48* (London: J. Longman, Clementi & Co, 1800), 46.
29. Fawcett, *Obi, or, Three-Finger'd Jack*, 1.3 (7) (Larpent LA 1297).
30. Katherine Rowe, *Dead Hands: Fictions of Agency, Renaissance to Modern* (Stanford: Stanford University Press, 1999) makes more of such symbolism; thanks to Kacy Tillman for pointing out the broader implications of the severed hand.
31. Peter Linebaugh, *The London Hanged: Crime and Civil Society in the Eighteenth Century* (Cambridge: Cambridge University Press, 1992), 17.
32. John Gay, *The Beggar's Opera*, eds. Edgar V. Roberts and Edward Smith (Lincoln: University of Nebraska Press, 1969), 1.3.7–8.
33. John Gay, *Polly: An Opera: Being the Sequel of the Beggar's Opera*, ed. George Colman (London: Printed for T. Evans, 1777), 28. Since Colman's 1777 version does not number the scenes, I cite page numbers for ease of reference. See Peter P. Reed, "Conquer or Die: Staging Circum-Atlantic Revolt in *Polly* and *Three-Finger'd Jack*," *Theatre Journal* 59.2 (2007): 241–58, and Robert G. Dryden, "John Gay's *Polly*: Unmasking Pirates and Fortune Hunters in the West Indies," *Eighteenth-Century Studies* 34.4 (2001): 539–57.
34. Gay, *Polly*, 45–46.
35. Calhoun Winton, *John Gay and the London Theatre* (Lexington: University Press of Kentucky, 1993), 141. The etymology of the word reveals that the Maroon communities came first, and marooning as punishment followed. In the *Oxford English Dictionary*, the earliest noun form of Maroon refers to a "member of a community of fugitive black slaves or (subsequently) of their descendants," especially "those who settled in the mountains and forests of Surinam and the West Indies" (*OED*, s.v. "Maroon"). The verb's derivation from the noun implies that the act of abandonment somehow rendered the victim in the condition of a Maroon-as-fugitive-slave, rather than implying that the Maroons derive their status from the practice of nautical abandonment.
36. E. P. Thompson, *Whigs and Hunters: The Origin of the Black Act* (New York: Pantheon Books, 1975).
37. O'Brien, *Harlequin Britain*, 128.
38. Dale Cockrell, *Demons of Disorder: Early Blackface Minstrels and Their World* (New York: Cambridge University Press, 1997), 30–61.
39. Gay, *Polly*, 31.
40. Peter Linebaugh and Marcus Rediker, *The Many-Headed Hydra: Sailors, Slaves, Commoners, and the Hidden History of the Revolutionary Atlantic* (Boston: Beacon Press, 2000). Morano represents a radical turn from Linebaugh and Rediker's relatively productive, disciplined, and law-abiding proletariat. See Paul Gilroy, *The Black Atlantic: Modernity and Double Consciousness* (Cambridge, MA: Harvard University Press, 1993), 3; David Armitage, "The Red Atlantic," *Reviews in American History* 29.4 (2001): 479–86.
41. John Fawcett and Samuel Arnold, *Songs, Duets and Choruses in the Pantomimical Drama of Obi, or, Three-Finger'd Jack*, eleventh edn (London: Woodfall, 1809), and Fawcett,

Songs, Duets and Chorusses in the Pantomimical Drama of Obi, or, Three-Finger'd Jack. Invented by Mr. Fawcett, and Performed at the New Theatre, Philadelphia. ([Philadelphia]: 1810).

42. E. P. Thompson, "The Moral Economy of the English Crowd in the Eighteenth Century," *Past and Present* 50 (1971): 76–136.
43. W. T. Lhamon, "Optic Black: Naturalizing the Refusal to Fit," *Black Cultural Traffic: Crossroads in Global Performance and Popular Culture*, eds. Harry J. Elam and Kennell A. Jackson (Ann Arbor: University of Michigan Press, 2005), 111.
44. Daphne Brooks, *Bodies in Dissent: Spectacular Performances of Race and Freedom, 1850–1910* (Durham: Duke University Press, 2006), 8.
45. Arnold, *Overture*, 2.
46. See Richard Brinsley Sheridan, *A Short Account of the Situations and Incidents Exhibited in the Pantomime of Robinson Crusoe at the Theatre-Royal, Drury-Lane. Taken from the Original Story* (London: T. Becket, 1781); Fawcett and Arnold, *Songs*, 5.
47. *Songs, Duets and Choruses*, 2–4.
48. Ibid., 4.
49. Ibid. See Charles D. Martin, *The White African American Body: A Cultural and Literary Exploration* (New Brunswick, NJ: Rutgers University Press, 2002), for a chronologically broader study of similar fascinations with representations of the whitened African American body.
50. *Songs, Duets and Choruses*, 17–18; I follow Fawcett's spelling of "Jonkanoo" to refer to the play's character; "Jonkonnu," the spelling stabilized by Caribbean scholars and commentators in the twentieth century, indicates the variety of vernacular performance practices represented in print variously as "John Canoe," "Johnkannaus," "John Connú," and so on.
51. *Dramatic Censor*, 20.
52. Mikhail M. Bakhtin, *Rabelais and His World* (Cambridge, MA: MIT Press, 1968); see also Peter Stallybrass and Allon White, *The Politics and Poetics of Transgression* (Ithaca, NY: Cornell University Press, 1986).
53. *Songs, Duets and Choruses*, 16.
54. See Judith Bettelheim, "Jamaican Jonkonnu and Related Caribbean Festivals," *Africa and the Caribbean: The Legacies of a Link*, eds. Margaret E. Crahan and Franklin W. Knight (Baltimore: Johns Hopkins University Press, 1979), 80–100; Judith Bettelheim, "The Jonkonnu Festival: Its Relation to Caribbean and African Masquerades," *Jamaica Journal* 10.2–4 (1976): 20–27; Elizabeth A. Fenn, "A Perfect Equality Seemed to Reign: Slave Society and Jonkonnu," *North Carolina Historical Review* 65.2 (1988): 127–53. Peter Reed, "'There Was No Resisting John Canoe': Circum-Atlantic Transracial Performance," *Theatre History Studies* 27 (2007): 65–85; and Cockrell, *Demons of Disorder*, 38–41, points out Jonkonnu's relationships to other Atlantic vernacular acts.
55. Chapter twenty-two contains the intriguing descriptions of Jonkonnu and its social setting; Harriet A. Jacobs, *Incidents in the Life of a Slave Girl*, ed. Lydia Maria Francis Child (Boston: Published for the author, 1861).
56. Ibid., 180.
57. Robert Dirks, *The Black Saturnalia: Conflict and Its Ritual Expression on British West Indian Slave Plantations* (Gainesville: University of Florida Press, 1987), discusses Jonkonnu as a safety valve, echoing Frederick Douglass's scathing assessment of holiday festivities as "keeping down the spirit of insurrection." See Frederick Douglass, *Narrative of the Life of Frederick Douglass, an American Slave*, ed. William Lloyd Garrison (Boston: Anti-Slavery Office, 1845), 74.
58. Jacobs, *Incidents*, 180.

59. Roger D. Abrahams, *Singing the Master: The Emergence of African American Culture in the Plantation South* (New York: Pantheon Books, 1992), 111.
60. *Songs, Duets and Choruses*, 13.
61. James Clifford, *Routes: Travel and Translation in the Late Twentieth Century* (Cambridge, MA: Harvard University Press, 1997), 25.
62. Edward Long, *The History of Jamaica. Or, General Survey of the Antient and Modern State of That Island: With Reflections on Its Situation, Settlements, Inhabitants, Climate, Products, Commerce, Laws, and Government. In Three Volumes. Illustrated with Copper Plates* (London: Printed for T. Lowndes, in Fleet-Street, 1774), 2.424; K. Y. Daaku, *Trade and Politics on the Gold Coast, 1600–1720: A Study of the African Reaction to European Trade* (London: Clarendon, 1970), 17, treats Conny as not just legend, but as historical fact.
63. Isaac Mendes Belisario, *Sketches of Character, in Illustration of the Habits, Occupation, and Costume of the Negro Population in the Island of Jamaica* (Kingston, Jamaica: Published by the artist, at his residence, no. 21, King-Street, 1837) depicts later scenes of such characters.
64. Jonkonnu acts bear significant formal resemblances to traditional Anglo-European mummeries in Ireland and Newfoundland. See Dirks, *Black Saturnalia*, 176. Matthew Gregory Lewis, *Journal of a Residence among the Negroes in the West Indies* (London: John Murray, Albemarle Street, 1845), 25; entry for January 1, 1816.
65. Isaac Mendes Belisario, "Jaw-bone, or House John Canoe"; Belisario, *Sketches of Character*, vol. 1, pl. 3, facing p. 4; Lithograph, 14 ¾ x 10 ¼ in. (37.5 x 26 cm); Yale Center for British Art, Paul Mellon Collection.
66. Lewis, *Journal*, 24.
67. Erin Skye Mackie, "Welcome the Outlaw: Pirates, Maroons, and Caribbean Countercultures," *Cultural Critique* 59 (2005): 32.
68. Ibid., 32.
69. Lhamon, "Black Cultural Traffic," 114.
70. Lewis, *Journal*, 24.
71. Ibid.
72. Marvin Edward McAllister, *White People Do Not Know How to Behave at Entertainments Designed for Ladies and Gentlemen of Colour: William Brown's African and American Theater* (Chapel Hill: University of North Carolina Press, 2003), 127; McAllister follows the argument of John Daniel Collins, *American Drama in Antislavery Agitation, 1792–1861*, dissertation (The University of Iowa, 1963).
73. The historical scholarship on American slave revolts, evolving from Herbert Aptheker's foundational studies, has increasingly addressed the ratio of myth, paranoia, and credible information in the records of slave insurrections and conspiracies. See Herbert Aptheker, *American Negro Slave Revolts* (New York: 1943), and a two-part forum entitled "The Making of a Slave Conspiracy" in the *William and Mary Quarterly* (Third Series 58.4 and 59.1), featuring responses to Michael P. Johnson, "Denmark Vesey and His Co-Conspirators," *The William and Mary Quarterly* 58.4 (2001): 915–76.
74. McAllister, *White People Do Not Know*, 122–30. George Thompson, *A Documentary History of the African Theatre* (Evanston, IL: Northwestern University Press, 1998), 132, dates the playbill announcing *Three-Finger'd Jack* from 1823. Charles Rzepka, "Thomas De Quincey's 'Three-Fingered Jack': The West Indian Origins of the 'Dark Interpreter,'" *European Romantic Review* 8.2 (1997): 122.
75. James O'Rourke, "The Revision of *Obi; or, Three-Finger'd Jack* and the Jacobin Repudiation of Sentimentality," *Nineteenth-Century Contexts* 28.4 (2006): 285–303. Other contemporary renditions of Jack's story reveal the competing agendas at work in

abolitionist melodrama. Held in the British Library, *The Life and Adventures of Three-Fingered Jack, the Terror of Jamaica* (London: Printed for Orlando Hodgson, 1834), features on its frontispiece a spectacularly exotic and highly sexualized scene of Jack as a melodramatic hero. He is flanked by a topless (but demurely concealed) Rosa and an obi-man, who are all framed within a border constructed of African tribal and animal emblems, including a cartouche of a kneeling manacled slave that recalls the Wedgwood abolitionist medallions popular from the end of the eighteenth century. Special thanks to Vincent Brown for sharing his copy of this image.

76. Shane White, *Stories of Freedom in Black New York* (Cambridge, MA: Harvard University Press, 2002), 159–65; White revises the portrait of Aldridge's career found in Herbert Marshall and Mildred Stock, *Ira Aldridge, the Negro Tragedian* (Carbondale: Southern Illinois University Press, 1968).
77. Marshall and Stock, *Aldridge*, 250.
78. W. T. Lhamon, Jr., *Raising Cain: Blackface Performance from Jim Crow to Hip Hop* (Cambridge, MA: Harvard University Press, 1998); Eric Lott, *Love and Theft: Blackface Minstrelsy and the American Working Class* New York: Oxford University Press, 1993).
79. Joseph R. Roach, *Cities of the Dead: Circum-Atlantic Performance* (New York: Columbia University Press, 1996), 122.
80. *Alexandria Herald*, September 11, 1818.
81. Ibid.
82. Ibid.

6 CLASS, PATRONAGE, AND URBAN SCENES IN *TOM AND JERRY*

1. George Clinton Densmore Odell, *Annals of the New York Stage* (New York: Columbia University Press, 1927), 3:59; Odell cites the New York *Evening Post*. Francis Courtney Wemyss, *Twenty-Six Years of the Life of an Actor and Manager* (New York: Burgess, Stringer and Co., 1847), 86. Wemyss lists the English actors Price brought to America, calling him the "Star Giver General to the United States" for his labor-importation efforts.
2. Pierce Egan, *Life in London, or, the Day and Night Scenes of Jerry Hawthorn, Esq.: And His Elegant Friend Corinthian Tom: Accompanied by Bob Logic, the Oxonian, in Their Rambles and Sprees through the Metropolis* (London: Printed for Sherwood, Nealy, and Jones, 1820).
3. My discussions cross-reference various editions of the play. I rely primarily on the later and most detailed W. T. Moncrieff, *Tom and Jerry, or Life in London, an Operatic Extravaganza, in Three Acts* (London: J. Cumberland, n.d.) to fill in scenes only thinly described by abridged versions. Other London editions influenced later American performances. Copies of another edition, W. T. Moncrieff, *Tom and Jerry: Or, Life in London, an Operatic Extravaganza in Three Acts*, second edn. (London: T. Richardson, 1828), reside (pinned and pencil-marked) in the Harvard Theatre Collection's promptbook collection. Finally, W. T. Moncrieff, *Tom and Jerry, or, Life in London: A Burletta of Fun, Frolic, Fashion, and Flash, in Three Acts, as Performed at the Boston, New York and Philadelphia Theatres* (Philadelphia: 1824), held at the Library Company of Philadelphia, represents earlier American versions of the play.
4. See John Cowie Reid, *Bucks and Bruisers: Pierce Egan and Regency England* (London: Routledge and K. Paul, 1971).
5. Jane Moody, *Illegitimate Theatre in London, 1770–1840* (Cambridge: Cambridge University Press, 2000), 218.

6. A playbill at the Harvard Theatre Collection dated November 11, 1821, announces the play's first performance at the Adelphi Theatre for November 26, 1821.
7. Moncrieff, *Tom and Jerry* (Philadelphia: 1824), 3.7 (68).
8. Odell, *Annals*, 3:59.
9. Charles Durang, *The Philadelphia Stage... By Charles Durang. Partly Compiled from the Papers of His Father, the Late John Durang; with Notes by the Editors* (Philadelphia: 1854–57), vol. 6, chapter 10. Durang's account, originally published in the Philadelphia *Sunday Dispatch*, was later compiled in a scrapbook; copies exist in the Library Company of Philadelphia and the Harvard Theatre Collection.
10. See W. T. Lhamon, Jr., *Raising Cain: Blackface Performance from Jim Crow to Hip Hop* (Cambridge, MA: Harvard University Press, 1998), 30.
11. Boston *Commercial Gazette*, January 8, 1823.
12. Moncrieff, *Tom and Jerry* (Philadelphia: 1824), 2.6 (48).
13. Undated clipping accompanying an illustration of Billy Waters (Victoria and Albert Museum, E.1070–1921).
14. Moncrieff, *Tom and Jerry* (London: J. Cumberland, n.d), n.p.
15. Benjamin A. Baker, *A Glance at New York: A Local Drama in Two Acts* (New York: S. French, 1857), 15.
16. Boston *Commercial Gazette*, January 8, 1823; Moncrieff, *Tom and Jerry* (Philadelphia: 1824), 3.3 (59). The English editions of the play call for a "Comic Pas Deux" to a popular jig tune.
17. Moncrieff, *Tom and Jerry* (Philadelphia: 1824), 3.3 (58–59).
18. Ibid., 3.3 (62).
19. Joe Cowell, *Thirty Years Passed among the Players in England and America* (New York: Harper & Brothers, 1844), 63.
20. Ibid., 63.
21. Wemyss, *Twenty-Six Years*, 84–85.
22. The *American Federalist Columbian Centinel*, December 20, 1823, lists Boston's first performance of *Tom and Jerry*; three days later, the *Boston Daily Advertiser*, December 23, 1823, advertises the fourth performance of *Tom and Jerry* at the Washington Garden Circus.
23. Durang, *Philadelphia Stage*, vol. 6, chapter 10, describes *Tom and Jerry*'s appearance in Philadelphia. The Harvard Theatre Collection newspaper clipping indicates *Tom and Jerry*'s sequel on February 13, 1824, at the "Philadelphia Theatre," as an afterpiece— "For the 2d time, a new satirical, burlesque, operative Parody, called The Death of Life in London; or, the funeral of Tom & Jerry." A newspaper clipping at the Harvard Theatre Collection (hand-dated 1826) announces "the celebrated Burletta, called TOM AND JERRY" at the Boston Theatre, featuring Mr. Hamblin in his last night there.
24. William Burke Wood's account books, listed by date, carefully tally totals for the performances. The individual entries are traceable in the transcription of the account book published in Reese Davis James, *Old Drury of Philadelphia; a History of the Philadelphia Stage, 1800–1835* (Philadelphia: University of Pennsylvania Press, 1932).
25. Odell, *Annals*, 3:263, 685; Arthur Herman Wilson, *A History of the Philadelphia Theatre, 1835 to 1855* (Philadelphia: University of Pennsylvania Press, 1935), 185, 326–27; William Warland Clapp, *A Record of the Boston Stage* (Boston: Munroe, 1853), 477; John Brougham, *Life in New York; or Tom and Jerry on a Visit. A Comic Drama in Two Acts* (New York: S. French, 1856); Harvard Theatre Collection.
26. Richard M. Dorson, "Mose the Far-Famed and World-Renowned," *American Literature* 15.3 (1943): 288–300; see also Peter Buckley, *To the Opera House: Culture and Society in New York City, 1820–1860*, dissertation (SUNY Stony Brook, 1984).

27. Moncrieff, *Tom and Jerry* (London: J. Cumberland, n.d), 6.
28. Harriet Arbuthnot, *The Journal of Mrs. Arbuthnot, 1820–1832*, ed. Francis Bamford and the Duke of Wellington (London: Macmillan, 1950), 144.
29. Moncrieff, *Tom and Jerry* (London: T. Richardson, 1828), n.p.
30. Moncrieff, *Tom and Jerry* (Philadelphia: 1824), 2.6 (46).
31. Ibid., 1.2 (11–12).
32. Moncrieff, *Tom and Jerry* (London: T. Richardson, 1828), n.p.
33. Moncrieff, *Tom and Jerry* (Philadelphia: 1824), 2.6 (47).
34. Playbill, Adelphi Theatre, November 26, 1821, Harvard Theatre Collection.
35. Richard Daniel Altick, *The Shows of London* (Cambridge, MA: Belknap Press of Harvard University Press, 1978); Angela L. Miller, "The Panorama, the Cinema and the Emergence of the Spectacular," *Wide Angle* 18.2 (1996): 34–69.
36. Moncrieff, *Tom and Jerry* (Philadelphia: 1824), 1.4 (15–16).
37. Dana Arnold, *Re-Presenting the Metropolis: Architecture, Urban Experience and Social Life in London 1800–1840* (Burlington, VT: Ashgate, 2000), 31.
38. Peter Linebaugh, *The London Hanged: Crime and Civil Society in the Eighteenth Century* (Cambridge: Cambridge University Press, 1992), 363; V. A. C. Gatrell, *The Hanging Tree: Execution and the English People 1770–1868* (Oxford: Oxford University Press, 1994).
39. Tyler Anbinder, *Five Points: The 19th-Century New York City Neighborhood That Invented Tap Dance, Stole Elections, and Became the World's Most Notorious Slum* (New York: Free Press, 2001), 1, 42–47.
40. Murray Newton Rothbard, *The Panic of 1819: Reactions and Policies* (New York: Columbia University Press, 1962); Daniel S. Dupre, "The Panic of 1819 and the Political Economy of Sectionalism," *The Economy of Early America: Historical Perspectives and New Directions*, ed. Cathy D. Matson (University Park: The Pennsylvania State University Press, 2006), 263–93.
41. Paul Joseph Erickson, *Welcome to Sodom: The Cultural Work of City-Mysteries Fiction in Antebellum America*, dissertation (University of Texas, 2005), 6.
42. Elizabeth Blackmar, *Manhattan for Rent, 1785–1850* (Ithaca: Cornell University Press, 1989), 214.
43. George William Curtis, "Editor's Easy Chair," *Harper's New Monthly Magazine* 8.48 (May 1854): 845–46; a similar observation appears a year later, in the May 1855 *Harper's*, 836–37. See also William Shepard Walsh, *Curiosities of Popular Customs and of Rites, Ceremonies, Observances, and Miscellaneous Antiquities* (Philadelphia: J. P. Lippincott Company, 1898), 728.
44. A letter reprinted in Philadelphia and Boston discusses the planned performance on May 17, 1787, associating the play with the city's "season for general removal" (see, e.g., the *Massachusetts Centinel*, May 23, 1787, and *Philadelphia Independent Gazetteer*, June 4, 1787). Jeffrey H. Richards, *Drama, Theatre, and Identity in the American New Republic* (Cambridge: Cambridge University Press, 2005), 219–21.
45. *The Cries of London, as They Are Daily Exhibited in the Streets with an Epigram in Verse, Adapted to Each* (Philadelphia: Printed for Benjamin Johnson, 1805); Samuel Wood, *The Cries of New-York* (New York: Printed and sold by S. Wood, at the Juvenile bookstore, No. 362, Pearl-Street, 1808); *The Cries of Philadelphia* (Philadelphia: Johnson and Warner, 1810).
46. Samuel L. Mitchill, *The Picture of New-York, or, the Traveller's Guide through the Commercial Metropolis of the United States* (New York: I. Riley & Co., 1807), [iii].
47. Anne Newport Royall, *Sketches of History, Life, and Manners, in the United States* (New Haven: Printed for the author, 1826), 243.

48. See Stuart M. Blumin, *The Emergence of the Middle Class: Social Experience in the American City, 1760–1900* (New York: Cambridge University Press, 1989); David M. Henkin, *City Reading: Written Words and Public Spaces in Antebellum New York* (New York: Columbia University Press, 1998).
49. James W. Cook, "Dancing across the Color Line: A Story of Mixtures and Markets in New York's Five Points," *Common-place* 4.1 (2003): 3. http://www.common-place.org/vol-04/no-01/cook.
50. See Richard Briggs Stott, *Workers in the Metropolis: Class, Ethnicity, and Youth in Antebellum New York City* (Ithaca: Cornell University Press, 1990).
51. Patricia Cline Cohen, Timothy J. Gilfoyle and Helen Lefkowitz Horowitz, eds., *The Flash Press: Sporting Male Weeklies in 1840s New York* (Chicago: University of Chicago Press, 2008), 13.
52. "Dramatis Personae," Moncrieff, *Tom and Jerry* (Philadelphia: 1824), n.p.
53. Durang, *Philadelphia Stage*, vol. 2, chapter 5.
54. Ibid.
55. Wemyss, *Twenty-Six Years*, 72.
56. Ibid., 116.
57. Moncrieff, *Tom and Jerry* (Philadelphia: 1824), 2.6 (46).
58. *Baltimore Patriot*, May 17, 1823.
59. Durang, *Philadelphia Stage*, vol. 6, chapter 10.
60. *American Federalist Columbian Centinel*, January 7, 1824.
61. Ibid.
62. Ibid.
63. Moncrieff, *Tom and Jerry* (London: J. Cumberland, n.d.), [5]; emphasis in the original.
64. George Thompson, *A Documentary History of the African Theatre* (Evanston, IL: Northwestern University Press, 1998), 131. Thompson reproduces a playbill from the Harvard Theatre Collection.
65. See Shane White, *Somewhat More Independent: The End of Slavery in New York City, 1770–1810* (Athens: University of Georgia Press, 1991); Shane White, *Stories of Freedom in Black New York* (Cambridge, MA: Harvard University Press, 2002), chapters 2 and 3.
66. Marvin Edward McAllister, *White People Do Not Know How to Behave at Entertainments Designed for Ladies and Gentlemen of Colour: William Brown's African and American Theater* (Chapel Hill: University of North Carolina Press, 2003), 4.
67. Thompson, *Documentary History*, 132. The Charleston *City Gazette*, e.g., includes numerous announcements of commercial activity at the Vendue Range.
68. Ibid.
69. Thomas F. De Voe, *The Market Book: Containing a Historical Account of the Public Markets of the Cities of New York, Boston, Philadelphia and Brooklyn, with a Brief Description of Every Article of Human Food Sold Therein, the Introduction of Cattle in America, and Notices of Many Remarkable Specimens* (New York: Printed for the author, 1862), 344–45.
70. McAllister, *White People Do Not Know*, 117.
71. See Michael Warner, Natasha Hurley, Luis Iglesias, Sonia Di Loreto, Jeffrey Scraba and Sandra Young, "A Soliloquy 'Lately Spoken at the African Theatre': Race and the Public Sphere in New York City, 1821," *American Literature* 73.1 (2001): 1–46.
72. White, *Stories of Freedom*, 68–69.
73. McAllister, *White People Do Not Know*, 7.
74. *New York National Advocate*, August 3, 1821.
75. McAllister, *White People Do Not Know*, 131.

76. *New York Commercial Advertiser*, August 17, 1822.
77. Thompson, *Documentary History*, chapter 8, discusses the "hooliganism at the African theatre"; see McAllister, *White People Do Not Know*, 143–50; Paul A. Gilje, *The Road to Mobocracy: Popular Disorder in New York City, 1763–1834* (Chapel Hill, NC: Published for the Institute of Early American History and Culture by the University of North Carolina Press, 1987), 157.
78. The *New Orleans Bee*, May 13, 1837, advertises *Life in New Orleans* for the benefit of N. H. Bannister, who received credit for writing it.
79. Henry A. Kmen, "Old Corn Meal: A Forgotten Urban Negro Folksinger," *Journal of American Folklore* 75.295 (1962): 29, documents these quotes from the *New Orleans Picayune* (May 12, 1837), the *New Orleans Bee* (May 13, 1837), and the *Louisiana Courier* (May 16, 1837).
80. Kmen, "Old Corn Meal," 31; Kmen cites the *New Orleans Picayune*, August 21, 1839.
81. Francis Cynric Sheridan, *Galveston Island; or, a Few Months Off the Coast of Texas: The Journal of Francis C. Sheridan, 1839–1840*, ed. Willis Winslow Pratt (Austin: University of Texas Press, 1954), 93–94.
82. *Georgia Telegraph*, February 16, 1841.
83. Dale Cockrell, *Demons of Disorder: Early Blackface Minstrels and Their World* (New York: Cambridge University Press, 1997), 50.
84. Meredith L. McGill, *American Literature and the Culture of Reprinting, 1834–1853* (Philadelphia: University of Pennsylvania Press, 2003).
85. *New-Hampshire Gazette*, June 7, 1842, reporting Old Corn Meal's death on May 20.
86. Sheridan, *Galveston Island*, 93–94.
87. *New Orleans Picayune*, August 28, 1839, quoted in Kmen, "Old Corn Meal," 31.
88. Caldwell managed the Camp Street Theatre immediately before the 1835 establishment of the St. Charles Theatre, where Old Corn Meal acted. John Smith Kendall, *The Golden Age of the New Orleans Theater* (Baton Rouge: Louisiana State University Press, 1952), 30, 112–13.
89. *The New Orleans Bee*, March 27, 1835.
90. The Library Company of Philadelphia has collected a number of Clay's prints, which appeared in London and Philadelphia in 1828 and 1830, and which Thomas Hood constructed into a narrative published in *The New Comic Annual for 1831* (London: Hurst, Chance, and Co., St. Paul's Church-Yard, 1830), 223–37.
91. W. T. Lhamon, Jr., *Jump Jim Crow: Lost Plays, Lyrics, and Street Prose of the First Atlantic Popular Culture* (Cambridge, MA: Harvard University Press, 2003), 52–61, discusses the intertextual influences on *Bone Squash Diavolo* and includes an edition of the play.
92. Kmen, "Old Corn Meal," 33; Kmen identifies a "skit" entitled "Old Corn Meal," but the *New Orleans Bee* (February 23, 1836) announcing Rice's *Bone Squash Diavolo* does not seem to indicate that. It simply places Rice's name next to the phrase "Old Corn Meal," which could indicate either Old Corn Meal performing alongside Rice or Rice performing as Old Corn Meal.
93. *Bone Squash Diavolo: A Burletta*, in Lhamon, *Jump Jim Crow*, 179.
94. According to Eric Lott, *Love and Theft: Blackface Minstrelsy and the American Working Class* (New York: Oxford University Press, 1993), 4, minstrelsy commits "small but significant crimes against settled ideas of racial demarcation"; even so, the significant actors remain the white working-class audiences who "enter the haunted realms of racial fantasy."
95. Charles Dickens, *American Notes for General Circulation*, ed. Patricia Ingham (London: Penguin, 2000), 95–96.

96. Marian Hannah Winter, "Juba and American Minstrelsy," *Inside the Minstrel Mask: Readings in Nineteenth-Century Blackface Minstrelsy*, eds Annemarie Bean, James Vernon Hatch, and Brooks McNamara (Hanover, NH: Wesleyan University Press, 1996), 228–29. Winter's article originally appeared in *Dance Index* 6.2 (1947), 28–47.
97. Dickens, *American Notes*, 95–96.
98. Ibid., 96.
99. Ibid.

7 SLAVE REVOLT AND CLASSICAL BLACKNESS IN *THE GLADIATOR*

1. Bird's manuscript "Secret Record," unnumbered page 4; folder 182, Robert Montgomery Bird Papers, Rare Book and Manuscript Library, University of Pennsylvania.
2. *New York Evening Post*, September 27, 1831; newspaper clipping in Robert Montgomery Bird Papers, Rare Book and Manuscript Library, University of Pennsylvania.
3. Robert Montgomery Bird, *The Gladiator*, ed. Jeffrey H. Richards (New York: Penguin Books, 1997), 1.1 (198).
4. Curtis Dahl, *Robert Montgomery Bird* (New York: Twayne Publishers, 1963), 19, 56.
5. Bird, *Gladiator*, 2.3 (196–97), applies all of the epithets in one scene; numerous labels appear throughout the script in a veritable orgy of class disdain. Bird's informal manuscript notes on Roman customs taken from his reading of the first volume of Adam Ferguson's *History of the Progress and Termination of the Roman Republic* (first printed in 1783, and with numerous transatlantic reprints through the 1830s) seem fascinated with Roman hierarchies. Unpaginated bound volume of *The Gladiator* in Folder 181, Robert Montgomery Bird Papers, Rare Book and Manuscript Library, University of Pennsylvania.
6. Bruce A. McConachie, *Melodramatic Formations: American Theatre and Society, 1820–1870* (Iowa City: University of Iowa Press, 1992), 91–118, argues for the simultaneous victimization and scapegoating of the low.
7. Introduction to *The Gladiator*, in Jeffrey H. Richards, *Early American Drama* (New York: Penguin Books, 1997), 168.
8. *New York Evening Post*, September 27, 1831; newspaper clipping in Robert Montgomery Bird Papers, Rare Book and Manuscript Library, University of Pennsylvania.
9. McConachie, *Melodramatic Formations*, 117.
10. Richard Moody, *Edwin Forrest, First Star of the American Stage* (New York: Knopf, 1960), 62, 66. Moody describes the Park Theatre debut, June 23, 1826, and notes that Forrest chose Othello again for his Bowery debut, November 6, 1826.
11. Richards, *Early American Drama*, 166.
12. Moody, *Edwin Forrest*, 88–91.
13. See ibid., 10–65.
14. George Clinton Densmore Odell, *Annals of the New York Stage* (New York: Columbia University Press, 1927), 3:75 and 3:254–55.
15. Reese Davis James, *Old Drury of Philadelphia; a History of the Philadelphia Stage, 1800–1835* (Philadelphia: University of Pennsylvania Press, 1932), 60.
16. McConachie, *Melodramatic Formations*, 77–78.
17. Ibid., 84; Odell, *Annals*, 2:310.
18. John Hanners, *"It Was Play or Starve": Acting in the Nineteenth-Century American Popular Theatre* (Bowling Green, Ohio: Bowling Green State University Popular Press, 1993), 20.
19. Dahl, *Robert Montgomery Bird*, 58; McConachie, *Melodramatic Formations*, 88.

20. Bird, *Gladiator*, 2.1 (185).
21. Ibid.; emphasis in the original
22. McConachie, *Melodramatic Formations*, 116.
23. *Brooklyn Eagle*, December 26, 1846.
24. "Mr. Forrest's Second Reception in England," *The United States Magazine and Democratic Review*, April 1845, 385.
25. Peter Buckley, *To the Opera House: Culture and Society in New York City, 1820–1860*, dissertation (SUNY Stony Brook, 1984), 294–409; Richard Butsch, "Bowery B'hoys and Matinee Ladies: The Re-Gendering of Nineteenth-Century American Theater Audiences," *American Quarterly* 46.3 (1994): 374–405.
26. James, *Old Drury*, 63.
27. Buckley, *To the Opera House*, 151.
28. Sean Wilentz, *Chants Democratic: New York City and the Rise of the American Working Class, 1788–1850* (New York: Oxford University Press, 1984), 257; Tyler Anbinder, *Five Points: The 19th-Century New York City Neighborhood That Invented Tap Dance, Stole Elections, and Became the World's Most Notorious Slum* (New York: Free Press, 2001), 14–20. Before 1813, as Anbinder notes, the neighborhood did not even exist: it was a five-acre pond known as "the Collect."
29. David R. Roediger, *The Wages of Whiteness: Race and the Making of the American Working Class*, revised edn (New York: Verso, 1999), 24.
30. McConachie, *Melodramatic Formations*, 105–106.
31. The manuscripts are part of the Robert Montgomery Bird Papers, Rare Book and Manuscript Library, University of Pennsylvania. The longest (and apparently latest) extant draft of *The Gladiator* is an inconsistently paginated bound manuscript in folder 178 of the Robert Montgomery Bird Papers. The manuscript has numerous significant passages marked, some for deletion and others for emendation. Editions of the play in Clement Edgar Foust, *The Life and Dramatic Works of Robert Montgomery Bird* (New York: Knickerbocker Press, 1919), 299–440, and more recently in Jeffrey Richards's *Early American Drama* indicate most of these potential cuts. Bird's manuscripts reveal a wonderfully messy process of collaborative, self-conscious, and frequently hesitant revision.
32. Bird, *Gladiator*, 4.2 (216–17).
33. Ibid., 3.1 (199).
34. An undated, unidentified clipping of a review (signed by "J. C. N.") in the Robert Montgomery Bird Papers, Rare Book and Manuscript Library, University of Pennsylvania, finds this passage a particularly persuasive demonstration of Bird's "poetic talent."
35. Bird, *Gladiator*, 2.3 (193).
36. Bird's notes on Roman history sources in the bound volume's manuscript notes in folder 181, Robert Montgomery Bird Papers, Rare Book and Manuscript Library, University of Pennsylvania.
37. Bird, *Gladiator*, 3.2 (204).
38. (Portland, Maine) *Eastern Argus*, June 3, 1831.
39. Bird, *Gladiator*, 2.1 (187).
40. Ibid. (188).
41. Ibid.
42. Ibid., 1.1 (177).
43. Ibid., 3.1 (199).
44. Ibid., 5.1 (223); the evocative phrase occurs in a passage marked for deletion.
45. Ibid.
46. Ibid.

47. Ibid. (224).
48. Ibid.
49. Ibid., 5.7 (242).
50. Ibid.
51. Ginger Strand, "'My Noble Spartacus': Edwin Forrest and Masculinity on the Nineteenth-Century Stage," *Passing Performances: Queer Readings of Leading Players in American Theater History*, eds Robert A. Schanke and Kim Marra (Ann Arbor, MI: University of Michigan Press, 1998), 19–40.
52. Robert Montgomery Bird, *The City Looking Glass. A Philadelphia Comedy, in Five Acts*, ed. Arthur Hobson Quinn (New York: Printed for the Colophon, 1933), xiv–xv.
53. Caroline Winterer, *The Culture of Classicism: Ancient Greece and Rome in American Intellectual Life, 1780–1910* (Baltimore: Johns Hopkins University Press, 2002); John C. Shields, *The American Aeneas: Classical Origins of the American Self* (Knoxville: University of Tennessee Press, 2001).
54. The discussions on American "neo-romanism" remain indebted to Bernard Bailyn, *The Ideological Origins of the American Revolution* (Cambridge: Belknap Press of Harvard University Press, 1967); J. G. A. Pocock, *The Machiavellian Moment: Florentine Political Thought and the Atlantic Republican Tradition* (Princeton, NJ: Princeton University Press, 1975).
55. Edwin A. Miles, "The Young American Nation and the Classical World," *Journal of the History of Ideas* 35.2 (1974): 259–74; Eran Shalev, "Empire Transformed: Britain in the American Classical Imagination, 1758–1783," *Early American Studies* 4.1 (2006): 112–46.
56. Eric Thomas Slauter, "Neoclassical Culture in a Society with Slaves: Race and Rights in the Age of Wheatley," *Early American Studies* 2.1 (2004): 81–122.
57. Jason Shaffer, "Making 'an Excellent Die': Death, Mourning, and Patriotism in the Propaganda Plays of the American Revolution," *Early American Literature* 41.1 (2006): 4.
58. Stanley M. Burstein, "The Classics and the American Republic," *The History Teacher* 30.1 (1996): 39; Bailyn, *Ideological Origins*, 24.
59. Bird, *Gladiator*, 2.3 (193).
60. Ibid., 4.1 (213). The line appears in the act's new opening scene after Bird's marks indicated cutting the entire first scene.
61. *Brooklyn Eagle*, December 26, 1846.
62. Bird, *Gladiator*, 1.1 (179).
63. New York *Mercantile Advertiser*, October 10, 1831; a review in the Philadelphia *Pennsylvanian*, October 29, 1831, notes Forrest's "sullen desolation" in the same scene. Hand-labeled newspaper clippings in Robert Montgomery Bird Papers, Rare Book and Manuscript Library, University of Pennsylvania.
64. Bird, *Gladiator*, 1.1 (181–82). The negotiation over the cost of human flesh seems to have garnered no special (written) attention in the manuscript.
65. See Joy S. Kasson, "Mind in Matter in History: Viewing the Greek Slave," *The Yale Journal of Criticism* 11.1 (1998): 79–83; Joseph R. Roach, "Slave Spectacles and Tragic Octoroons: A Cultural Genealogy of Antebellum Performance," *Theatre Survey* 33 (1992): 167–87.
66. Quoted in Julie K. Ellison, "Cato's Tears," *ELH* 63.3 (1996): 593.
67. Ellison, Ralph. *Invisible Man*. [1952] New York: Random House, 2002, chapter 10.
68. Ibid., 596; see also Laura J. Rosenthal, "Juba's Roman Soul: Addison's Cato and Enlightenment Cosmopolitanism," *Studies in the Literary Imagination* 32.2 (1999): 63–76.

69. The account, which I discuss in chapter six, appears in Charles Dickens, *American Notes for General Circulation*, ed. Patricia Ingham (London: Penguin, 2000), 95–96.
70. Tyrone Power, *Impressions of America; During the Years 1833, 1834, and 1835* (Philadelphia: Carey, Lea & Blanchard, 1836), 1:45.
71. Michael Kaplan, "New York City Tavern Violence and the Creation of a Working-Class Male Identity," *Journal of the Early Republic* 15.4 (1995): 606–607.
72. Odai Johnson, *Absence and Memory in Colonial American Theatre: Fiorelli's Plaster* (New York: Palgrave Macmillan, 2006), 229.
73. Trevor G. Burnard, "Slave Naming Patterns: Onomastics and the Taxonomy of Race in Eighteenth-Century Jamaica," *Journal of Interdisciplinary History* 31.3 (2000): 335.
74. Advertisements in the *Pennsylvania Gazette*, April 15, 1756, and August 11, 1757; quoted in Billy G. Smith and Richard Wojtowicz, *Blacks Who Stole Themselves: Advertisements for Runaways in the Pennsylvania Gazette, 1728–1790* (Philadelphia: University of Pennsylvania Press, 1989), entries 48 and 55.
75. *New England Galaxy*, November 19, 1831; newspaper clipping in Robert Montgomery Bird Papers, Rare Book and Manuscript Library, University of Pennsylvania. Bird gripes about the Galaxy's review in his manuscript notes.
76. Ibid.
77. "The Drama," *New-York Mirror* (February 24, 1827): 245.
78. Ibid.
79. "Mr. Forrest's Second Reception," 386.
80. Bird, "Secret Record," 5.
81. "Mr. Forrest's Second Reception," 386.
82. "The Drama," *The New-York Mirror: a Weekly Gazette of Literature and the Fine Arts*, February 24, 1827, 245.
83. "The Drama," *The Critic: A Weekly Review of Literature, Fine Arts, and the Drama*, May 30, 1829, 57.
84. Montrose Moses, *The Fabulous Forrest; the Record of an American Actor* (Boston: Little, Brown and Company, 1929), 41; Moody, *Edwin Forrest*, 29–34. James Rees, *The Life of Edwin Forrest: With Reminiscences and Personal Recollections* (Philadelphia: Peterson and Brothers, 1874), 78, concurs.
85. Artemus Ward, *Artemus Ward, His Book with Many Comic Illustrations* (New York: Carleton, 1862), 112. Hans Nathans explains that the "Essence of old Virginny" was a standard minstrel show dance; "dances featuring such jumps and leaps were called 'essences' from about the (eighteen)fifties on, of which the 'Essence of Old Virginny,' frequently accompanied by Emmett's tune 'Root, Hog or Die,' is the best known example. Hans Nathan, *Dan Emmett and the Rise of Early Negro Minstrelsy* (Norman: University of Oklahoma Press, 1962), 93.
86. James, *Old Drury*, 65.
87. Playbill, Bowery Theatre, May 22 1834; Harvard Theatre Collection. Reproduced in W. T. Lhamon, Jr., *Jump Jim Crow: Lost Plays, Lyrics, and Street Prose of the First Atlantic Popular Culture* (Cambridge, MA: Harvard University Press, 2003), 68–69.
88. James and Eliphalet Brown, "Dancing for eels; A scene from the new play of New-York As It Is, as played at the Chatham Theatre, N.Y." (1848); Library of Congress Prints and Photographs Division [LC-USZC4-632].
89. Eric Lott, *Love and Theft: Blackface Minstrelsy and the American Working Class* (New York: Oxford University Press, 1993), 38–62; W. T. Lhamon, Jr., *Raising Cain: Blackface Performance from Jim Crow to Hip Hop* (Cambridge, MA: Harvard University Press, 1998), 1–55.
90. William Lloyd Garrison, "The Insurrection," *The Liberator* (September 3, 1831): 143.

91. Bird, "Secret Record," 2.
92. Ibid., 2–3. The phrase "vis, et amor sceleratus habendi" ("violence and the vicious love of possessions") is a quotation of Ovid's *Metamorphosis*, 1.131.
93. Bird, *Gladiator*, 2.2 (191); David Walker, *Walker's Appeal, in Four Articles; Together with a Preamble, to the Coloured Citizens of the World, but in Particular, and Very Expressly, to Those of the United States of America*, third edn (Boston: D. Walker, 1830). See also Kenneth S. Greenberg, *Nat Turner: A Slave Rebellion in History and Memory* (New York: Oxford University Press, 2002).

EPILOGUE: ESCAPE ARTISTS AND SPECTATORIAL MOBS

1. Jonas B. Phillips, *Jack Sheppard, or the Life of a Robber! Melodrama in Three Acts Founded on Ainsworth's Novel* (1839). The unpaginated Harvard Theatre Collection manuscript [TS 4336.123] may be the only one in existence. Phillips authored a number of other popular melodramas in the 1830s, such as *The Evil Eye* and *Camillus, or, The Self-Exiled Patriot*.
2. George Clinton Densmore Odell, *Annals of the New York Stage* (New York: Columbia University Press, 1927), 3.370 and 85, attributes *Life in New York* to Phillips. W. T. Lhamon, Jr., *Jump Jim Crow: Lost Plays, Lyrics, and Street Prose of the First Atlantic Popular Culture* (Cambridge, MA: Harvard University Press, 2003), 67–69, reproduces a playbill that evidently advertises the same play, but does not name an author.
3. Odell, *Annals*, 4.367.
4. Peter Linebaugh, *The London Hanged: Crime and Civil Society in the Eighteenth Century* (Cambridge: Cambridge University Press, 1992), 7–41. Linebaugh's first chapter discusses Jack Sheppard's 1720s career.
5. *A Narrative of all the Robberies, Escapes, &c. of John Sheppard* (London: Printed and sold by John Applebee, 1724), 3.
6. Ibid.
7. Ibid., 16.
8. *Authentic Memoirs of the Life and Surprising Adventures of John Sheppard: Who Was Executed at Tyburn, November the 16th, 1724*, second edn (London: Printed for Joseph Marshall, 1724), 62.
9. Ibid., 63.
10. *Narrative of the Robberies, Escapes, &c.*, 28–29.
11. Linebaugh, *The London Hanged*, 37.
12. John Thurmond, *Harlequin Sheppard, a Night Scene in Grotesque Characters: As It Is Perform'd at the Theatre-Royal in Drury-Lane.* (London: Printed and sold by J. Roberts, and A. Dodd, 1724). Christopher Hibbert, *The Road to Tyburn: The Story of Jack Sheppard and the Eighteenth Century Underworld* (London: Longmans, 1957), 17, says it appeared "within a fortnight of his death."
13. *The History of the Remarkable Life of John Sheppard, Containing a Particular Account of His Many Robberies and Escapes* (London: Applebee, 1724), 24, reprinted in Richard Holmes, ed., *Defoe on Sheppard and Wild* (London: Harper Perennial, 2002), 1–44.
14. *History of the Remarkable Life of John Sheppard*, 24.
15. Hibbert, *The Road to Tyburn*, 16–17.
16. *History of the Remarkable Life of John Sheppard*, 24, reprinted in Holmes, ed., *Defoe on Sheppard and Wild*, 1–44.
17. "Prologue at the Opening of the Theatre in Drury Lane" (1747), lines 53–54, in Samuel Johnson, *The Yale Edition of the Works of Samuel Johnson*, eds E. L. McAdam and George Milne, vol. 6, Poems (New Haven: Yale University Press, 1964).

18. Hibbert, *Road to Tyburn*, 213, claims that "an eyewitness estimated the numbers of the crowd, probably without exaggeration, at 200,000."
19. *London Weekly Journal or British Gazetteer*, November 21, 1724.
20. Ibid. Hibbert, *Road to Tyburn*, 226–31; such actions were fairly common in the eighteenth century; see Peter Linebaugh, "The Tyburn Riot against the Surgeons," *Albion's Fatal Tree: Crime and Society in Eighteenth-Century England*, eds Douglas Hay, Peter Linebaugh, John G. Rule, E. P. Thompson, and Cal Winslow (New York: Peregrine Books, 1975), 65–117.
21. William Harrison Ainsworth, *Jack Sheppard: A Romance* (London: Nottingham Society, 1839).
22. Matthew Buckley, "Sensations of Celebrity: Jack Sheppard and the Mass Audience," *Victorian Studies* 44.3 (2002): 426–29.
23. Reviews appear, e.g., in the *London Morning Herald* and *Examiner* on October 28, 1839, and the *London Times* on October 29, 1839; see also Buckley, "Sensations of Celebrity," 423–63.
24. Playbills in the Harvard Theatre Collection advertise performances at the Adelphi Theatre on December 28–30, 1840, and the week of January 4, 1841. Martin Meisel, *Realizations: Narrative, Pictorial, and Theatrical Arts in Nineteenth-Century England* (Princeton, NJ: Princeton University Press, 1983), 271.
25. Elliott Vanskike, "Consistent Inconsistencies: The Transvestite Actress Madame Vestris and Charlotte Bronte's Shirley," *Nineteenth-Century Literature* 50.4 (1996): 467, describes Vestris's on- and offstage acts as "ambiguous and threatening." See also William Worthen Appleton, *Madame Vestris and the London Stage* (New York: Columbia University Press, 1974).
26. Odell, *Annals*, 4.367.
27. Dramatis personae, Phillips, *Jack Sheppard*, n.p.
28. Ibid., 2.4.
29. Ibid., 3.1.
30. Linebaugh, *The London Hanged*, 244.
31. John Baldwin Buckstone, *Jack Sheppard; a Drama in Four Acts* (New York: Samuel French, 1853), 3.1 (47). This edition, although published later, has the most detailed descriptions of the scenes. Moreover, French's acting editions, edited by Philadelphia manager Francis Courtney Wemyss, purported to reflect American performance practices, printing in this case Buckstone's play as acted at the Bowery Theatre in November 1853. The variety of related scripts suggests that American audiences could have been familiar with performances based on its descriptions.
32. Keith Hollingsworth, *The Newgate Novel, 1830–1847: Bulwer, Ainsworth, Dickens and Thackeray* (Detroit, MI: Wayne State University Press, 1963), 139–40.
33. The Harvard Theatre Collection also holds copies of John Baldwin Buckstone, *Jack Sheppard; a Drama, in Four Acts* (London: Chapman and Hall, [1840]); and T. L. Greenwood, *Jack Sheppard, or, the House-Breaker of the Last Century: A Romantic Drama in Five Acts Dramatised from Harrison Ainsworth's Novel* (London: J. Cumberland, 1840) marked for American productions; HTC promptbooks.
34. Buckstone, *Jack Sheppard*, 4.8 (90).
35. Phillips, *Jack Sheppard*, 2.5.
36. Paul A. Gilje, *The Road to Mobocracy: Popular Disorder in New York City, 1763–1834* (Chapel Hill, NC: Published for the Institute of Early American History and Culture by the University of North Carolina Press, 1987), 247–48.
37. Peter Buckley, *To the Opera House: Culture and Society in New York City, 1820–1860*, Dissertation (SUNY Stony Brook, 1984), 92–93 and 181.

38. Ibid., 151.
39. Ibid., 14–16; estimates of the numbers of spectators at the speeches ranges from eight thousand to twenty-five thousand, while the rioters numbered around sixty-five hundred persons. See also Richard Moody, *The Astor Place Riot* (Bloomington, IN: Indiana University Press, 1958).
40. *New York Herald*, May 16, 1849, collects the "Opinions of the Press on the Late Occurrences in Astor Place." Lawrence W. Levine, *Highbrow/Lowbrow: The Emergence of Cultural Hierarchy in America* (Cambridge, MA: Harvard University Press, 1988), 268, note 62.
41. Buckley, *To the Opera House*, 9.
42. George Lippard, *New York: Its Upper Ten and Lower Million* (Cincinnati, OH: H. M. Rulison, 1853); see also David S. Reynolds, *George Lippard, Prophet of Protest: Writings of an American Radical, 1822–1854* (New York: P. Lang, 1986); Shelley Streeby, "Opening up the Story Paper: George Lippard and the Construction of Class," *boundary 2* 24.1 (1997): 177–203; Dennis Berthold, "Class Acts: The Astor Place Riots and Melville's 'the Two Temples,'" *American Literature* 71.3 (1999): 429–61.
43. Levine, *Highbrow/Lowbrow*, 68.
44. George Templeton Strong, *Diary of George Templeton Strong*, eds Allan Nevins and Milton Halsey Thomas, 4 vols (New York: Macmillan, 1952), 1:351–53; Dennis Berthold, "Melville, Garibaldi, and the Medusa of Revolution," *American Literary History* 9.3 (1997): 425–59, examines the associations of the riots with the European revolutions.
45. Buckley, *To the Opera House*, 26.
46. W. T. Lhamon, Jr., "The Blackface Lore Cycle," in *Raising Cain: Blackface Performance from Jim Crow to Hip Hop* (Cambridge, MA: Harvard University Press, 1998), 56–115.

Works Cited

Abrahams, Roger D. *Singing the Master: The Emergence of African American Culture in the Plantation South*. New York: Pantheon Books, 1992.
Adams, Robert. *The Narrative of Robert Adams, an American Sailor, Who Was Wrecked on the Western Coast of Africa, in the Year 1810, Was Detained Three Years in Slavery by the Arabs of the Great Desert, and Resided Several Months in the City of Tombuctoo. With a Map, Notes and an Appendix*. Boston: Wells and Lilly, 1817.
Agnew, Jean-Christophe. *Worlds Apart: The Market and the Theater in Anglo-American Thought, 1550–1750*. Cambridge: Cambridge University Press, 1986.
Ahlquist, Karen. *Democracy at the Opera: Music, Theater, and Culture in New York City, 1815–60*. Urbana: University of Illinois Press, 1997.
Ainsworth, William Harrison. *Jack Sheppard: A Romance*. London: Nottingham Society, 1839.
Altick, Richard Daniel. *The Shows of London*. Cambridge, MA: Belknap Press of Harvard University Press, 1978.
Americanischer Haus- Und Wirthschafts-Calender Auf Das 1781ste Jahr Christi. Philadelphia, 1780.
Anbinder, Tyler. *Five Points: The 19th-Century New York City Neighborhood That Invented Tap Dance, Stole Elections, and Became the World's Most Notorious Slum*. New York: Free Press, 2001.
Anderson, Benedict R. *Imagined Communities: Reflections on the Origin and Spread of Nationalism*. London: Verso, 1983.
Appleton, William Worthen. *Madame Vestris and the London Stage*. New York: Columbia University Press, 1974.
Aptheker, Herbert. *American Negro Slave Revolts*. New York, 1943.
Arbuthnot, Harriet. *The Journal of Mrs. Arbuthnot, 1820–1832*. Ed. Francis Bamford and the Duke of Wellington. London: Macmillan, 1950.
Armitage, David. "The Red Atlantic." *Reviews in American History* 29.4 (2001): 479–86.
Armstrong, Nancy. *Desire and Domestic Fiction: A Political History of the Novel*. New York: Oxford University Press, 1987.
Arnold, Dana. *Re-Presenting the Metropolis: Architecture, Urban Experience and Social Life in London 1800–1840*. Burlington, VT: Ashgate, 2000.
Arnold, Samuel. *The Overture, Songs, Chorusses & Appropriate Music in the Grand Pantomimical Drama Call'd Obi; or Three Finger'd Jack... Composed & Adapted to the Action by Saml. Arnold... With Selections from the Most Eminent Masters, Arranged for the Voice & Piano Forte. Op. 48*. London: J. Longman, Clementi & Co, 1800.
Authentic Memoirs of the Life and Surprising Adventures of John Sheppard: Who Was Executed at Tyburn, November the 16th, 1724. Second edn. London: Printed for Joseph Marshall, 1724.
Baepler, Paul Michel. *White Slaves, African Masters: An Anthology of American Barbary Captivity Narratives*. Chicago: University of Chicago Press, 1999.
———. "The Barbary Captivity Narrative in American Culture." *Early American Literature* 39.2 (2004): 217–46.

Baer, Marc. *Theatre and Disorder in Late Georgian London*. Oxford: Oxford University Press, 1992.
Bailyn, Bernard. *The Ideological Origins of the American Revolution*. Cambridge: Belknap Press of Harvard University Press, 1967.
Baker, Benjamin A. *A Glance at New York: A Local Drama in Two Acts*. New York: S. French, 1857.
Bakhtin, Mikhail M. *Rabelais and His World*. Cambridge, MA: MIT Press, 1968.
Bank, Rosemarie K. *Theatre Culture in America, 1825–1860*. Cambridge: Cambridge University Press, 1997.
Barish, Jonas A. *The Antitheatrical Prejudice*. Berkeley: University of California Press, 1981.
"Beggar's Opera—Midnight Hour." *The American Monthly Magazine and Critical Review* November 1817: 62.
Belisario, Isaac Mendes. *Sketches of Character, in Illustration of the Habits, Occupation, and Costume of the Negro Population in the Island of Jamaica*. Kingston, Jamaica: Published by the artist, at his residence, no. 21, King-Street, 1837.
Benes, Peter. "Night Processions: Celebrating the Gunpowder Plot in England and New England." *New England Celebrates: Spectacle, Commemoration, and Festivity*. Ed. Peter Benes. Concord, MA: Boston University, 2002. 9–28.
Benson, Egbert. *Vindication of the Captors of Major André*. New York: Published by Kirk & Mercein, at the office of the Edinburgh and Quarterly reviews, no. 22 Wall-Street. T. & W. Mercein, printers, 1817.
Berlin, Ira. *Many Thousands Gone: The First Two Centuries of Slavery in North America*. Cambridge, MA: Belknap Press of Harvard University Press, 1998.
Berthold, Dennis. "Melville, Garibaldi, and the Medusa of Revolution." *American Literary History* 9.3 (1997): 425–59.
———. "Class Acts: The Astor Place Riots and Melville's 'the Two Temples.'" *American Literature* 71.3 (1999): 429–61.
Bettelheim, Judith. "The Jonkonnu Festival: Its Relation to Caribbean and African Masquerades." *Jamaica Journal* 10.2–4 (1976): 20–27.
———. "Jamaican Jonkonnu and Related Caribbean Festivals." *Africa and the Caribbean: The Legacies of a Link*. Eds Margaret E. Crahan and Franklin W. Knight. Baltimore: Johns Hopkins University Press, 1979. 80–100.
Bickerstaff, Isaac, and Charles Dibdin. *The Captive, a Comic Opera; as It Is Perform'd at the Theatre-Royal in the Hay-Market*. London: Printed for W. Griffin, 1769.
Bird, Robert Montgomery. *The City Looking Glass. A Philadelphia Comedy, in Five Acts*. Ed. Arthur Hobson Quinn. New York: Printed for the Colophon, 1933.
———. *The Gladiator*. 1831. Ed. Jeffrey H. Richards. New York: Penguin Books, 1997.
Blackmar, Elizabeth. *Manhattan for Rent, 1785–1850*. Ithaca: Cornell University Press, 1989.
Blumin, Stuart M. *The Emergence of the Middle Class: Social Experience in the American City, 1760–1900*. New York: Cambridge University Press, 1989.
Boswell, James. *Boswell's London Journal, 1762–1763*. Ed. Frederick Albert Pottle. New Haven: Yale University Press, 1991.
Boudreau, Kristin. "Early American Criminal Narratives and the Problem of Public Sentiments." *Early American Literature* 32.3 (1997): 249–69.
Bourdieu, Pierre. *Distinction: A Social Critique of the Judgement of Taste*. Cambridge, MA: Harvard University Press, 1984.
Bouton, Terry. *Taming Democracy: "The People," The Founders, and the Troubled Ending of the American Revolution*. Oxford: Oxford University Press, 2007.

Brooks, Daphne. *Bodies in Dissent: Spectacular Performances of Race and Freedom, 1850–1910*. Durham: Duke University Press, 2006.
Brougham, John. *Life in New York; or Tom and Jerry on a Visit. A Comic Drama in Two Acts*. New York: S. French, 1856.
Brown, Jared. *The Theatre in America during the Revolution*. Cambridge: Cambridge University Press, 1995.
Brown, John. *Barbarossa; a Tragedy: With Alterations and Amendments. As Performed at the Theatre in Boston*. Printed at the Apollo Press, in Boston, for David West, no. 36, Marlboro Street, and John West, no. 75, Cornhill, 1794.
Buckley, Matthew. "Sensations of Celebrity: Jack Sheppard and the Mass Audience." *Victorian Studies* 44.3 (2002): 423–63.
Buckley, Peter. "To the Opera House: Culture and Society in New York City, 1820–1860." Dissertation. SUNY Stony Brook, 1984.
Buckstone, John Baldwin. *Jack Sheppard; a Drama, in Four Acts*. London: Chapman and Hall [1840].
———. *Jack Sheppard; a Drama in Four Acts*. New York: Samuel French, 1853.
Burge, James C. *Lines of Business: Casting Practice and Policy in the American Theatre, 1752–1899*. New York: P. Lang, 1986.
Burke, Peter. *Popular Culture in Early Modern Europe*. New York: New York University Press, 1978.
Burnard, Trevor G. "Slave Naming Patterns: Onomastics and the Taxonomy of Race in Eighteenth-Century Jamaica." *Journal of Interdisciplinary History* 31.3 (2000): 325–46.
Burstein, Stanley M. "The Classics and the American Republic." *The History Teacher* 30.1 (1996): 29–44.
Bussard, Robert L. "The 'Dangerous Class' of Marx and Engels: The Rise of the Idea of the Lumpenproletariat." *History of European Ideas* 8.6 (1987): 675–92.
Butler, Judith. *Gender Trouble: Feminism and the Subversion of Identity*. New York: Routledge, 1990.
Butsch, Richard. "Bowery B'hoys and Matinee Ladies: The Re-Gendering of Nineteenth-Century American Theater Audiences." *American Quarterly* 46.3 (1994): 374–405.
Campbell, Mavis Christine. *The Maroons of Jamaica, 1655–1796: A History of Resistance, Collaboration and Betrayal*. South Hadley, MA: Bergin and Garvey, 1988.
Carey, Mathew. *A Short Account of Algiers*. Second edn. Philadelphia: Printed for Mathew Carey, no. 118, Market-Street, 1794.
Castiglia, Christopher. *Bound and Determined: Captivity, Culture-Crossing, and White Womanhood from Mary Rowlandson to Patty Hearst*. Chicago: University of Chicago Press, 1996.
Castle, Terry. *Masquerade and Civilization: The Carnivalesque in Eighteenth-Century English Culture and Fiction*. Stanford: Stanford University Press, 1986.
Cavitch, Max. "The Man That Was Used Up: Poetry, Particularity, and the Politics of Remembering George Washington." *American Literature* 75.2 (2003): 247–74.
Certeau, Michel de. *The Practice of Everyday Life*. Berkeley: University of California Press, 1984.
Chartier, Roger. *The Order of Books: Readers, Authors, and Libraries in Europe between the Fourteenth and Eighteenth Centuries*. Stanford: Stanford University Press, 1994.
Chetwood, William Rufus. *A General History of the Stage, from Its Origin in Greece Down to the Present Time*. London: Printed for W. Owen, 1749.
Clapp, William Warland. *A Record of the Boston Stage*. Boston: Munroe, 1853.
Clifford, James. *The Predicament of Culture: Twentieth-Century Ethnography, Literature, and Art*. Cambridge, MA: Harvard University Press, 1988.

Clifford, James. *Routes: Travel and Translation in the Late Twentieth Century.* Cambridge, MA: Harvard University Press, 1997.

Cobbett, William. *A Kick for a Bite, or, Review Upon Review with a Critical Essay on the Works of Mrs. S. Rowson: In a Letter to the Editor, or Editors, of the American Monthly Review.* Philadelphia: Printed by T. Bradford, 1795.

———. *A Little Plain English, Addressed to the People of the United States, on the Treaty, Negociated with His Britannic Majesty, and on the Conduct of the President Relative Thereto.* Philadelphia: Published by Thomas Bradford, printer, bookseller, and stationer, no. 8, South Front Street, 1795.

Cockrell, Dale. *Demons of Disorder: Early Blackface Minstrels and Their World.* New York: Cambridge University Press, 1997.

Cohen, Daniel A. *Pillars of Salt, Monuments of Grace: New England Crime Literature and the Origins of American Popular Culture, 1674–1860.* New York: Oxford University Press, 1993.

Cohen, Patricia Cline, Timothy J. Gilfoyle, and Helen Lefkowitz Horowitz, eds. *The Flash Press: Sporting Male Weeklies in 1840s New York.* Chicago: University of Chicago Press, 2008.

Collins, John Daniel. "American Drama in Antislavery Agitation, 1792–1861." Dissertation. The University of Iowa, 1963.

The Continental Almanac, for the Year of Our Lord, 1781. Philadelphia: Printed and sold by Francis Bailey, in Market-Street, between Third and Fourth-Streets, 1780.

Cook, James W. "Dancing across the Color Line: A Story of Mixtures and Markets in New York's Five Points." *Common-place* 4.1 (2003).

Cooke, William. *Memoirs of Charles Macklin, Comedian, with the Dramatic Characters, Manners, Anecdotes, &c. of the Age in Which He Lived.* Second edn. London: J. Asperne, 1806.

Cowell, Joe. *Thirty Years Passed among the Players in England and America.* New York: Harper & Brothers, 1844.

Cray, Robert E., Jr. "Major John Andre and the Three Captors: Class Dynamics and Revolutionary Memory Wars in the Early Republic, 1780–1831." *Journal of the Early Republic* 17.3 (1997): 371–97.

The Cries of London, as They Are Daily Exhibited in the Streets with an Epigram in Verse, Adapted to Each. Philadelphia: Printed for Benjamin Johnson, 1805.

The Cries of London, as They Are Daily Exhibited in the Streets with an Epigram in Verse, Adapted to Each. Embellished with Sixty-Two Elegant Cuts. To Which Is Added, a Description of the Metropolis in Verse. London: Printed for E. Newbery, 1796.

The Cries of Philadelphia. Philadelphia: Johnson and Warner, 1810.

Curtis, George William. "Editor's Easy Chair." *Harper's New Monthly Magazine* 8.48 (May 1854): 845–46.

Daaku, K. Y. *Trade and Politics on the Gold Coast, 1600–1720: A Study of the African Reaction to European Trade.* London: Clarendon, 1970.

Dahl, Curtis. *Robert Montgomery Bird.* New York: Twayne Publishers, 1963.

Dallas, Robert Charles. *The History of the Maroons.* London, 1803.

Davidson, Cathy N. *Revolution and the Word: The Rise of the Novel in America.* New York: Oxford University Press, 1986.

———. *Revolution and the Word: The Rise of the Novel in America.* Expanded edn. New York: Oxford University Press, 2004.

Davis, Susan G. *Parades and Power: Street Theatre in Nineteenth-Century Philadelphia.* Philadelphia: Temple University Press, 1986.

De Voe, Thomas F. *The Market Book: Containing a Historical Account of the Public Markets of the Cities of New York, Boston, Philadelphia and Brooklyn, with a Brief Description of Every*

Article of Human Food Sold Therein, the Introduction of Cattle in America, and Notices of Many Remarkable Specimens. New York: Printed for the author, 1862.

Defoe, Daniel. *A Narrative of all the Robberies, Escapes, &c. of John Sheppard.* London: Printed and sold by John Applebee, 1724.

Denning, Michael. "Beggars and Thieves: *The Beggar's Opera* and the Ideology of the Gang." *Literature and History* 8.1 (1982): 41–55.

A Dialogue between Elizabeth Smith, and John Sennet. [Boston], 1773.

Dickens, Charles. *American Notes for General Circulation.* Ed. Patricia Ingham. London: Penguin, 2000.

Diebels, Mary Chrysostom. *Peter Markoe (1752?–1792): A Philadelphia Writer.* Washington, D.C.: The Catholic University of America Press, 1944.

Dillon, Elizabeth Maddock. "*Slaves in Algiers*: Race, Republican Genealogies, and the Global Stage." *American Literary History* 16.3 (2004): 407–36.

Dirks, Robert. *The Black Saturnalia: Conflict and Its Ritual Expression on British West Indian Slave Plantations.* Gainesville: University of Florida Press, 1987.

Dorson, Richard M. "Mose the Far-Famed and World-Renowned." *American Literature* 15.3 (1943): 288–300.

Douglass, Frederick. *Narrative of the Life of Frederick Douglass, an American Slave.* Ed. William Lloyd Garrison. Boston: Anti-Slavery Office, 1845.

"The Drama." *The New-York Mirror: A Weekly Gazette of Literature and the Fine Arts*, February 24, 1827: 245.

"The Drama." *The Critic: A Weekly Review of Literature, Fine Arts, and the Drama*, May 30, 1829: 57.

The Dramatic Censor, or, Monthly Epitome of Taste, Fashion, and Manners. Ed. Thomas Dutton. Vol. 3. London: J. Roach and C. Chapple, 1801.

Dryden, John. *Don Sebastian, King of Portugal: A Tragedy Acted at the Theatre Royal.* London: Printed for Jo. Hindmarsh, at Golden Ball in Cornhill, 1690.

Dryden, Robert G. "John Gay's *Polly*: Unmasking Pirates and Fortune Hunters in the West Indies." *Eighteenth-Century Studies* 34.4 (2001): 539–57.

Dudden, Faye E. *Women in the American Theatre: Actresses and Audiences, 1790–1870.* New Haven: Yale University Press, 1994.

Dugaw, Dianne. *Warrior Women and Popular Balladry, 1650–1850.* New York: Cambridge University Press, 1989.

———. *Deep Play: John Gay and the Invention of Modernity.* Newark, DE: University of Delaware Press, 2001.

Dunlap, William. *André: A Tragedy, in Five Acts: As Performed by the Old American Company, New-York, March 30, 1798. To Which Are Added Authentic Documents Respecting Major Andre; Consisting of Letters to Miss Seward, the Cow Chace, Proceedings of the Court Martial, &C.* New York: Printed by T. & J. Swords, no. 99 Pearl-street, 1798.

———. *Blue Beard: A Dramatic Romance by G. Colman, the Younger, as Altered for the New York Theatre, with Additional Songs.* New York: Printed and published by D. Longworth, at the Shakspeare gallery, 1802.

———. *The Glory of Columbia; Her Yeomanry. A Play in Five Acts. The Songs, Duets, and Chorusses, Intended for the Celebration of the Fourth of July, at the New-York Theatre.* New York: Printed and published by D. Longworth, at the Shakspeare-Gallery, 1803.

———. *The Glory of Columbia; Her Yeomanry! A Play in Five Acts.* New York: Published by David Longworth at the Dramatic Repository, Shakspeare-Gallery, 1817.

———. *A History of the American Theatre.* New York: J. & J. Harper, 1832.

———. *Diary of William Dunlap, 1766–1839; the Memoirs of a Dramatist, Theatrical Manager, Painter, Critic, Novelist, and Historian.* Ed. Dorothy C. Barck. New York: B. Blom, 1969.

Dupre, Daniel S. "The Panic of 1819 and the Political Economy of Sectionalism." *The Economy of Early America: Historical Perspectives and New Directions*. Ed. Cathy D. Matson. University Park: The Pennsylvania State University Press, 2006. 263–93.

Durang, Charles. *The Philadelphia Stage...By Charles Durang. Partly Compiled from the Papers of His Father, the Late John Durang; with Notes by the Editors*. Philadelphia, 1854–57.

Durang, John. *The Memoir of John Durang, American Actor, 1785–1816*. Ed. Alan Seymour Downer. Pittsburgh: University of Pittsburgh Press, 1966.

The Dying Groans of Levi Ames, Who Was Executed at Boston, the 21st of October, 1773, for Burglary. [Boston], 1773.

The Dying Penitent; or, the Affecting Speech of Levi Ames. [Boston]: Sold opposite the Court-House in Queen-Street, 1773.

Earle, William. *Obi, or, the History of Threefingered Jack in a Series of Letters from a Resident in Jamaica to His Friend in England*. Worcester, MA: Printed by Isaiah Thomas Jr., 1804.

Egan, Pierce. *Life in London, or, the Day and Night Scenes of Jerry Hawthorn, Esq.: And His Elegant Friend Corinthian Tom: Accompanied by Bob Logic, the Oxonian, in Their Rambles and Sprees through the Metropolis*. London: Printed for Sherwood, Nealy, and Jones, 1820.

Egerton, Douglas R. *Gabriel's Rebellion: The Virginia Slave Conspiracies of 1800 and 1802*. Chapel Hill: University of North Carolina Press, 1993.

Ekirch, Roger A. *Bound for America: The Transportation of British Convicts to the Colonies, 1718–1775*. Oxford: Oxford University Press, 1987.

Eliot, Andrew. *Christ's Promise to the Penitent Thief. A Sermon Preached the Lord's-Day before the Execution of Levi Ames, Who Suffered Death for Burglary, Oct. 21, 1773. Aet. 22*. Boston: Printed and sold by John Boyle, next door to the Three Doves in Marlborough-Street, 1773.

Ellison, Julie K. "Cato's Tears." *Elh* 63.3 (1996): 571–601.

Empson, William. *Some Versions of Pastoral*. London: Chatto and Windus, 1935.

Erickson, Paul Joseph. "Welcome to Sodom: The Cultural Work of City-Mysteries Fiction in Antebellum America." Dissertation. University of Texas, 2005.

Espy, M. Watt, and John Ortiz Smykla. *Executions in the United States, 1608–1987: The Espy File*. Ann Arbor, MI: Inter-University Consortium for Political and Social Research, 1987.

An Exhortation to Young and Old to Be Cautious of Small Crime, Lest They Become Habitual, and Lead Them before Thay Are Aware into Those of the Most Heinous Nature. Occasioned by the Unhappy Case of Levi Ames, Executed on Boston-Neck, October 21st, 1773, for the Crime of Burglary. [Boston], 1773.

Faller, Lincoln B. *Turned to Account: The Forms and Functions of Criminal Biography in Late Seventeenth- and Early Eighteenth-Century England*. New York: Cambridge University Press, 1987.

Fawcett. *Songs, Duets and Chorusses in the Pantomimical Drama of Obi, or, Three-Finger'd Jack. Invented by Mr. Fawcett, and Performed at the New Theatre, Philadelphia*. [Philadelphia], 1810.

Fawcett, John. *Obi, or, Three-Finger'd Jack*. 1800.

Fawcett, John, and Samuel Arnold. *Songs, Duets and Choruses in the Pantomimical Drama of Obi, or, Three-Finger'd Jack*. Eleventh edn. London: Woodfall, 1809.

Fenn, Elizabeth A. "A Perfect Equality Seemed to Reign: Slave Society and Jonkonnu." *North Carolina Historical Review* 65.2 (1988): 127–53.

Fennell, James. *An Apology for the Life of James Fennell, Written by Himself*. Philadelphia: Published by Moses Thomas, No. 52, Chesnut-Street, J. Maxwell, printer, 1814.

Ferris, Lesley. *Acting Women: Images of Women in Theatre.* New York: New York University Press, 1989.
A Few Lines Wrote Upon the Intended Execution of Levi Ames, for Burglary, and Being Sent to Him for His Improvement, Are Now Published at His Desire. [Boston], 1773.
Fisher, Samuel Rowland. *Journal of Samuel Rowland Fisher, of Philadelphia, 1779–1781.* Ed. Anna Wharton Morris, 1928.
Flaherty, David H. "Crime and Social Control in Provincial Massachusetts." *The Historical Journal* 24.2 (1981): 339–60.
Fliegelman, Jay. *Declaring Independence: Jefferson, Natural Language, and the Culture of Performance.* Stanford: Stanford University Press, 1993.
Foner, Eric. *Free Soil, Free Labor, Free Men: The Ideology of the Republican Party before the Civil War.* New York: Oxford University Press, 1970.
Foss, John. *A Journal, of the Captivity and Sufferings of John Foss, Several Years a Prisoner in Algiers.* Newburyport [Mass.]: Printed by A. March, Middle-Street, 1798.
Foucault, Michel. *Discipline and Punish: The Birth of the Prison.* New York: Pantheon Books, 1977.
Foust, Clement Edgar. *The Life and Dramatic Works of Robert Montgomery Bird.* New York: Knickerbocker Press, 1919.
Gannett, Deborah Sampson. *Address Delivered with Applause at the Federal-Street Theatre, Boston, Four Successive Nights of the Different Plays, Beginning March 22, 1802 and after, at Other Principal Towns, a Number of Nights Successively at Each Place.* Dedham [MA]: Printed and sold by H. Mann, for Mrs. Gannet, at the Minerva office, 1802.
Garber, Marjorie B. *Vested Interests: Cross-Dressing and Cultural Anxiety.* New York: Routledge, 1992.
Gatrell, V. A. C. *The Hanging Tree: Execution and the English People 1770–1868.* Oxford: Oxford University Press, 1994.
Gay, John. *Polly: An Opera: Being the Sequel of the Beggar's Opera.* Ed. George Colman. London: Printed for T. Evans, 1777.
———. *The Beggar's Opera.* Eds Edgar V. Roberts and Edward Smith. Lincoln: University of Nebraska Press, 1969.
Geertz, Clifford. *The Interpretation of Cultures: Selected Essays.* New York: Basic Books, 1973.
Genette, Gérard. *Paratexts: Thresholds of Interpretation.* Cambridge: Cambridge University Press, 1997.
Gilfoyle, Timothy J. *A Pickpocket's Tale: The Underworld of Nineteenth-Century New York.* New York: W.W. Norton, 2006.
Gilje, Paul A. *The Road to Mobocracy: Popular Disorder in New York City, 1763–1834.* Chapel Hill, NC: Published for the Institute of Early American History and Culture by the University of North Carolina Press, 1987.
Gilroy, Paul. *The Black Atlantic: Modernity and Double Consciousness.* Cambridge, MA: Harvard University Press, 1993.
Goffman, Erving. *The Presentation of Self in Everyday Life.* Garden City, NY: Doubleday, 1959.
Goring, Paul. "'John Bull, Pit, Box, and Gallery, Said No!': Charles Macklin and the Limits of Ethnic Resistance on the Eighteenth-Century London Stage." *Representations* 79 (2002): 61–81.
Goudie, Sean X. *Creole America: The West Indies and the Formation of Literature and Culture in the New Republic.* Philadelphia: University of Pennsylvania Press, 2006.
Gough, Hugh, and David Dickson, eds. *Ireland and the French Revolution.* Dublin: Irish Academic Press, 1989.

Gould, Philip. *Barbaric Traffic: Commerce and Antislavery in the Eighteenth-Century Atlantic World*. Cambridge: Harvard University Press, 2003.

Graydon, Alexander. *Memoirs of a Life, Chiefly Passed in Pennsylvania, within the Last Sixty Years, with Occasional Remarks Upon the General Occurrences, Character and Spirit of That Eventful Period*. Harrisburgh, PA: Printed by John Wyeth, 1811.

Greenberg, Kenneth S. *Nat Turner: A Slave Rebellion in History and Memory*. New York: Oxford University Press, 2002.

Greenblatt, Stephen, Walter Cohen, Jean E. Howard, Katharine Eisaman Maus, and Andrew Gurr, eds. *The Norton Shakespeare*. Second edn. New York: W.W. Norton, 2008.

Greenwood, T. L. *Jack Sheppard, or, the House-Breaker of the Last Century: A Romantic Drama in Five Acts Dramatised from Harrison Ainsworth's Novel*. London: J. Cumberland, 1840.

Gross, John J. *Shylock: A Legend and Its Legacy*. New York: Simon & Schuster, 1992.

Gurr, Ted Robert. *Violence in America*. 2 vols. Newbury Park, CA: Sage Publications, 1989.

Gustafson, Sandra M. *Eloquence is Power: Oratory and Performance in Early America*. Chapel Hill: Published for the Omohundro Institute of Early American History and Culture, Williamsburg, Virginia, by the University of North Carolina Press, 2000.

Habermas, Jürgen. *The Structural Transformation of the Public Sphere: An Inquiry into a Category of Bourgeois Society*. Cambridge, MA: MIT Press, 1989.

Hadley, Elaine. "The Old Price Wars: Melodramatizing the Public Sphere in Early-Nineteenth-Century England." *PMLA* 107.3 (1992): 524–37.

Hanners, John. *"It Was Play or Starve": Acting in the Nineteenth-Century American Popular Theatre*. Bowling Green, Ohio: Bowling Green State University Popular Press, 1993.

Hay, Douglas. "Property, Authority, and the Criminal Law." *Albion's Fatal Tree: Crime and Society in Eighteenth-Century England*. Eds Douglas Hay, Peter Linebaugh, John G. Rule, E. P. Thompson, and Cal Winslow. New York: Peregrine Books, 1975. 17–63.

Henkin, David M. *City Reading: Written Words and Public Spaces in Antebellum New York*. New York: Columbia University Press, 1998.

Henry, Brian. *Dublin Hanged: Crime, Law Enforcement and Punishment in Late Eighteenth-Century Dublin*. Blackrock: Irish Academic Press, 1994.

Hibbert, Christopher. *The Road to Tyburn: The Story of Jack Sheppard and the Eighteenth Century Underworld*. London: Longmans, 1957.

Highfill, Philip H., Kalman A. Burnim, and Edward A. Langhans, eds. *A Biographical Dictionary of Actors, Actresses, Musicians, Dancers, Managers & Other Stage Personnel in London, 1660–1800*. Carbondale: Southern Illinois University Press, 1973.

Hill, Errol. *The Jamaican Stage, 1655–1900: Profile of a Colonial Theatre*. Amherst: University of Massachusetts Press, 1992.

Hiltner, Judith R. "'Like a Bewildered Star': Deborah Sampson, Herman Mann, and 'Address, Delivered with Applause.'" *Rhetoric Society Quarterly* 29.2 (1999): 5–24.

Hirsch, Adam Jay. *The Rise of the Penitentiary: Prisons and Punishment in Early America*. New Haven: Yale University Press, 1992.

The History of the Remarkable Life of John Sheppard, Containing a Particular Account of His Many Robberies and Escapes. London: Applebee, 1724.

Hobsbawm, E. J. *Bandits*. New York: New Press, distributed by W.W. Norton, 2000.

Hogeland, William. *The Whiskey Rebellion: George Washington, Alexander Hamilton, and the Frontier Rebels Who Challenged America's Newfound Sovereignty*. New York: Scribner, 2006.

Hollingsworth, Keith. *The Newgate Novel, 1830–1847: Bulwer, Ainsworth, Dickens and Thackeray*. Detroit, MI: Wayne State University Press, 1963.

Holmes, Richard, ed. *Defoe on Sheppard and Wild*. London: Harper Perennial, 2002.

Hoole, William Stanley. *The Ante-Bellum Charleston Theatre*. Tuscaloosa, AL: University of Alabama Press, 1946.
Hoskins, Robert H. B., and Eileen Southern, eds. *Music for London Entertainment 1660–1800*. Vol. 4, Series D, Pantomime, Ballet and Social Dance. London: Stainer & Bell, 1996.
Hutchins, John Nathan. *Hutchins Improved, Being an Almanack and Ephemeris… For the Year of Our Lord 1802*. New York: Printed and sold by Ming and Young, 1801.
Ignatiev, Noel. *How the Irish Became White*. New York: Routledge, 1995.
Jacobs, Harriet A. *Incidents in the Life of a Slave Girl*. Ed. Lydia Maria Francis Child. Boston: Published for the author, 1861.
James, Reese Davis. *Old Drury of Philadelphia; a History of the Philadelphia Stage, 1800–1835*. Philadelphia: University of Pennsylvania Press, 1932.
Johnson, Claudia D. *Church and Stage: The Theatre as Target of Religious Condemnation in Nineteenth Century America*. Jefferson, NC: McFarland & Co., 2008.
Johnson, Michael P. "Denmark Vesey and His Co-Conspirators." *The William and Mary Quarterly* 58.4 (2001): 915–76.
Johnson, Odai. *Absence and Memory in Colonial American Theatre: Fiorelli's Plaster*. New York: Palgrave Macmillan, 2006.
Johnson, Odai, William J. Burling, and James A. Coombs, eds. *The Colonial American Stage, 1665–1774: A Documentary Calendar*. Madison, NJ: Fairleigh Dickinson University Press, 2001.
Johnson, Samuel. *The Yale Edition of the Works of Samuel Johnson*. Eds E. L. McAdam and George Milne. Vol. 6, Poems. New Haven: Yale University Press, 1964.
Jones, Gareth Stedman. *Languages of Class: Studies in English Working Class History, 1832–1982*. New York: Cambridge University Press, 1983.
Kaplan, Michael. "New York City Tavern Violence and the Creation of a Working-Class Male Identity." *Journal of the Early Republic* 15.4 (1995): 591–617.
Kasson, John F. *Rudeness and Civility: Manners in Nineteenth-Century Urban America*. New York: Hill and Wang, 1990.
Kasson, Joy S. "Mind in Matter in History: Viewing the Greek Slave." *The Yale Journal of Criticism* 11.1 (1998): 79–83.
Kaul, Mythili, ed. *Othello: New Essays by Black Writers*. Washington, D.C.: Howard University Press, 1997.
Kealey, Linda. "Patterns of Punishment: Massachusetts in the Eighteenth Century." *The American Journal of Legal History* 30.2 (1986): 163–86.
Kendall, John Smith. *The Golden Age of the New Orleans Theater*. Baton Rouge: Louisiana State University Press, 1952.
Kidson, Frank. *The Beggar's Opera, Its Predecessors and Successors*. Cambridge: The University Press, 1922.
Kinservik, Matthew J. *Disciplining Satire: The Censorship of Satiric Comedy on the Eighteenth-Century London Stage*. London: Associated University Presses, 2002.
Kmen, Henry A. "Old Corn Meal: A Forgotten Urban Negro Folksinger." *Journal of American Folklore* 75.295 (1962): 29–34.
Koschnik, Albrecht. *Let a Common Interest Bind Us Together: Associations, Partisanship, and Culture in Philadelphia, 1775–1840*. Charlottesville: University of Virginia Press, 2007.
Lambert, Frank. *The Barbary Wars: American Independence in the Atlantic World*. New York: Hill and Wang, 2005.
Larkin, Samuel. *The Columbian Songster and Freemason's Pocket Companion. A Collection of the Newest and Most Celebrated Sentimental, Convivial, Humourous, Satirical, Pastoral, Hunting, Sea and Masonic Songs, Being the Largest and Best Collection Ever Published*

in America. Portsmouth, New-Hampshire: Printed by J. Melcher, for S. Larkin, at the Portsmouth book-store, 1798.

The Last Words and Dying Speech of Levi Ames, Who Was Executed at Boston, on Thursday the 21st Day of October, 1773. For Burglary. Boston: Printed and sold at the shop opposite the Court-House in Queen-Street, 1773.

Lehuu, Isabelle. *Carnival on the Page: Popular Print Media in Antebellum America.* Chapel Hill: University of North Carolina Press, 2000.

Leiner, Frederick C. *The End of Barbary Terror: America's 1815 War against the Pirates of North Africa.* Oxford: Oxford University Press, 2006.

Levine, Lawrence W. *Highbrow/Lowbrow: The Emergence of Cultural Hierarchy in America.* Cambridge, MA: Harvard University Press, 1988.

Lewis, Matthew Gregory. *Journal of a Residence among the Negroes in the West Indies.* London: John Murray, Albemarle Street, 1845.

Lewis, Paul. "Attaining Masculinity: Charles Brockden Brown and Woman Warriors of the 1790s." *Early American Literature* 40.1 (2005): 37–55.

Lhamon, W. T., Jr. *Raising Cain: Blackface Performance from Jim Crow to Hip Hop.* Cambridge, MA: Harvard University Press, 1998.

———. *Jump Jim Crow: Lost Plays, Lyrics, and Street Prose of the First Atlantic Popular Culture.* Cambridge, MA: Harvard University Press, 2003.

———. "Optic Black: Naturalizing the Refusal to Fit." *Black Cultural Traffic: Crossroads in Global Performance and Popular Culture.* Eds Harry J. Elam and Kennell A. Jackson. Ann Arbor: University of Michigan Press, 2005. 111–40.

Linebaugh, Peter. "The Tyburn Riot against the Surgeons." *Albion's Fatal Tree: Crime and Society in Eighteenth-Century England.* Eds Douglas Hay, Peter Linebaugh, John G. Rule, E. P. Thompson, and Cal Winslow. New York: Peregrine Books, 1975. 65–117.

———. *The London Hanged: Crime and Civil Society in the Eighteenth Century.* Cambridge: Cambridge University Press, 1992.

Linebaugh, Peter, and Marcus Rediker. *The Many-Headed Hydra: Sailors, Slaves, Commoners, and the Hidden History of the Revolutionary Atlantic.* Boston: Beacon Press, 2000.

Lippard, George. *New York: Its Upper Ten and Lower Million.* Cincinnati, OH: H. M. Rulison, 1853.

Long, Edward. *The History of Jamaica. Or, General Survey of the Antient and Modern State of That Island: With Reflections on Its Situation, Settlements, Inhabitants, Climate, Products, Commerce, Laws, and Government. In Three Volumes. Illustrated with Copper Plates.* London: Printed for T. Lowndes, in Fleet-Street, 1774.

Looby, Christopher. *Voicing America: Language, Literary Form, and the Origins of the United States.* Chicago: University of Chicago Press, 1996.

Lott, Eric. *Love and Theft: Blackface Minstrelsy and the American Working Class.* New York: Oxford University Press, 1993.

Mackie, Erin Skye. "Welcome the Outlaw: Pirates, Maroons, and Caribbean Countercultures." *Cultural Critique* 59 (2005): 24–62.

MacLean, William Jerry. "Othello Scorned: The Racial Thought of John Quincy Adams." *Journal of the Early Republic* 4.2 (1984): 143–60.

Mann, Herman. *The Female Review, or, Memoirs of an American Young Lady, Whose Life and Character Are Peculiarly Distinguished, Being a Continental Soldier, for Nearly Three Years, in the Late American War.* Dedham [Mass.]: Printed by Nathaniel and Benjamin Heaton, for the author, 1797.

Markoe, Peter. *The Algerine Spy in Pennsylvania: Or, Letters Written by a Native of Algiers on the Affairs of the United States in America, from the Close of the Year 1783 to the Meeting of the Convention*. Philadelphia: Printed and sold by Prichard & Hall, in Market between Front and Second Streets, 1787.

Marshall, Herbert, and Mildred Stock. *Ira Aldridge, the Negro Tragedian*. Carbondale: Southern Illinois University Press, 1968.

Martin, Charles D. *The White African American Body: A Cultural and Literary Exploration*. New Brunswick, NJ: Rutgers University Press, 2002.

Marx, Karl. *The Eighteenth Brumaire of Louis Bonaparte: With Explanatory Notes*. New York: International Publishers, 1963.

Masur, Louis P. *Rites of Execution: Capital Punishment and the Transformation of American Culture, 1776–1865*. New York: Oxford University Press, 1989.

Mather, Samuel. *Christ Sent to Heal the Broken Hearted. A Sermon, Preached at the Thursday Lecture in Boston, on October, 21st. 1773. When Levi Ames, a Young Man, under a Sentence of Death for Burglary, to Be Executed on That Day, Was Present to Hear the Discourse*. Boston: Printed and sold at William M'Alpine's printing office in Marlborough-Street, 1773.

Mathur, Maya. "An Attack of the Clowns: Comedy, Vagrancy, and the Elizabethan History Play." *Journal for Early Modern Cultural Studies* 7.1 (2007): 33–54.

McAllister, Marvin Edward. *White People Do Not Know How to Behave at Entertainments Designed for Ladies and Gentlemen of Colour: William Brown's African and American Theater*. Chapel Hill: University of North Carolina Press, 2003.

McConachie, Bruce A. *Melodramatic Formations: American Theatre and Society, 1820–1870*. Iowa City: University of Iowa Press, 1992.

McGill, Meredith L. *American Literature and the Culture of Reprinting, 1834–1853*. Philadelphia: University of Pennsylvania Press, 2003.

McIntosh, William A. "Handel, Walpole, and Gay: The Aims of the Beggar's Opera." *Eighteenth Century Studies* 7.4 (1974): 415–33.

McKenzie, Andrea. "Martyrs in Low Life? Dying 'Game' In Augustan England." *Journal of British Studies* 42.2 (2003): 167–205.

———. "The Real Macheath: Social Satire, Appropriation, and Eighteenth-Century Criminal Biography." *Huntington Library Quarterly* 69.4 (2006): 581–605.

Meisel, Martin. *Realizations: Narrative, Pictorial, and Theatrical Arts in Nineteenth-Century England*. Princeton, NJ: Princeton University Press, 1983.

Miles, Edwin A. "The Young American Nation and the Classical World." *Journal of the History of Ideas* 35.2 (1974): 259–74.

Miller, Angela L. "The Panorama, the Cinema and the Emergence of the Spectacular." *Wide Angle* 18.2 (1996): 34–69.

Mitchill, Samuel L. *The Picture of New-York, or, the Traveller's Guide through the Commercial Metropolis of the United States*. New York: I. Riley & Co., 1807.

Moncrieff, W. T. *Tom and Jerry, or Life in London, an Operatic Extravaganza, in Three Acts*. London: J. Cumberland, n.d.

———. *Tom and Jerry, or, Life in London: A Burletta of Fun, Frolic, Fashion, and Flash, in Three Acts, as Performed at the Boston, New York and Philadelphia Theatres*. Philadelphia, 1824.

———. *Tom and Jerry: Or, Life in London, an Operatic Extravaganza in Three Acts*. Second edn. London: T. Richardson, 1828.

Montgomery, Benilde. "White Captives, African Slaves: A Drama of Abolition." *Eighteenth-Century Studies* 27.4 (1994): 615–30.

Moody, Jane. *Illegitimate Theatre in London, 1770–1840*. Cambridge: Cambridge University Press, 2000.

Moody, Richard. *The Astor Place Riot*. Bloomington, IN: Indiana University Press, 1958.

———. *Edwin Forrest, First Star of the American Stage*. New York: Knopf, 1960.

Morgan, Gwenda, and Peter Rushton. *Eighteenth-Century Criminal Transportation: The Formation of the Criminal Atlantic*. New York: Palgrave Macmillan, 2003.

———. "Visible Bodies: Power, Subordination and Identity in the Eighteenth-Century Atlantic World." *Journal of Social History* 39.1 (2005): 39–64.

Moseley, Benjamin. *A Treatise on Sugar*. London: G. G. & J. Robinson, 1799.

Moses, Montrose. *The Fabulous Forrest; the Record of an American Actor*. Boston: Little, Brown and Company, 1929.

Mount, Thomas. *The Confession, &c. of Thomas Mount, Who Was Executed at Little-Rest, in the State of Rhode-Island, on Friday the 27th of May, 1791, for Burglary*. Ed. William Smith. [Newport, Rhode Island]: Printed and sold by Peter Edes, in Newport, 1791.

"Mr. Forrest's Second Reception in England." *The United States Magazine and Democratic Review* April 1845: 385.

Mullenix, Elizabeth Reitz. *Wearing the Breeches: Gender on the Antebellum Stage*. New York: St. Martin's Press, 2000.

Mulroy, Kevin. *Freedom on the Border: The Seminole Maroons in Florida, the Indian Territory, Coahuila, and Texas*. Lubbock: Texas Tech University Press, 1993.

A Narrative of all the Robberies, Escapes, &c. of John Sheppard. London: Printed and sold by John Applebee, 1724.

Nash, Gary B. "Poverty and Politics in Early American History." *Down and Out in Early America*. Ed. Billy G. Smith. University Park: Pennsylvania State University Press, 2004. 1–37.

———. *First City: Philadelphia and the Forging of Historical Memory*. Philadelphia: University of Pennsylvania Press, 2006.

Nathan, Hans. *Dan Emmett and the Rise of Early Negro Minstrelsy*. Norman: University of Oklahoma Press, 1962.

Nathans, Heather S. *Early American Theatre from the Revolution to Thomas Jefferson: Into the Hands of the People*. Cambridge: Cambridge University Press, 2003.

———. "A Much Maligned People: Jews on and Off the Stage in the Early American Republic." *Early American Studies* 2.2 (2007): 310–42.

The New Comic Annual for 1831. London: Hurst, Chance, and Co., St. Paul's Church-Yard, 1830.

Newman, Simon P. *Parades and the Politics of the Street: Festive Culture in the Early American Republic*. Philadelphia: University of Pennsylvania Press, 1997.

———. *Embodied History: The Lives of the Poor in Early Philadelphia*. Philadelphia: University of Pennsylvania Press, 2003.

The Nightingale. A Collection of the Most Elegant Songs, Now in Vogue. Amherst: Samuel Preston, 1797.

Nowatzki, Robert. "Paddy Jumps Jim Crow: Irish-Americans and Blackface Minstrelsy." *Éire-Ireland* 41.3 (2007): 162–84.

O'Brien, John. *Harlequin Britain: Pantomime and Entertainment, 1690–1760*. Baltimore: Johns Hopkins University Press, 2004.

Odell, George Clinton Densmore. *Annals of the New York Stage*. New York: Columbia University Press, 1927.

O'Keefe, John. *The Agreeable Surprise. A Comic Opera, in Two Acts*. [Boston]: Printed at the Apollo Press, in Boston, by Belknap and Hall, for William P. Blake, no. 59, Cornhill, and William T. Clap. No. 90, Newbury Street, 1794.

———. *Songs in the Comic Opera, Called, the Son-in-Law. By John O'Keefe, Esq. As Sung at the New Theatre, Philadelphia. Corrected and Revised by Mr. Rowson, Prompter*. [Philadelphia]: Printed for M. Carey, 1794.

O'Rourke, James. "The Revision of *Obi; or, Three-Finger'd Jack* and the Jacobin Repudiation of Sentimentality." *Nineteenth-Century Contexts* 28.4 (2006): 285–303.

Patterson, Orlando. "Slavery and Slave Revolts: A Sociohistorical Analysis of the First Maroon War, 1655–1740." *Maroon Societies: Rebel Slave Communities in the Americas.* Ed. Richard Price. Baltimore, MD: Johns Hopkins University Press, 1996: 246–291.

Pencak, William. *Jews and Gentiles in Early America: 1654–1800.* Ann Arbor: University of Michigan Press, 2005.

Pencak, William, Matthew Dennis, and Simon P. Newman, eds. *Riot and Revelry in Early America.* University Park, PA: Pennsylvania State University Press, 2002.

Phelan, Peggy. *Unmarked: The Politics of Performance.* New York: Routledge, 1993.

Phillips, Jonas B. *Jack Sheppard, or the Life of a Robber! Melodrama in Three Acts Founded on Ainsworth's Novel.* 1839.

Pocock, J. G. A. *The Machiavellian Moment: Florentine Political Thought and the Atlantic Republican Tradition.* Princeton, NJ: Princeton University Press, 1975.

Pollock, Thomas Clark. *The Philadelphia Theatre in the Eighteenth Century, Together with the Day Book of the Same Period.* Philadelphia: University of Pennsylvania Press, 1933.

Porter, Susan L. *With an Air Debonair: Musical Theatre in America, 1785–1815.* Washington: Smithsonian Institution Press, 1991.

Power, Tyrone. *Impressions of America; During the Years 1833, 1834, and 1835.* Philadelphia: Carey, Lea & Blanchard, 1836.

Price, Curtis Alexander, Gabriella Dideriksen, Judith Milhous, and Robert D. Hume. *Italian Opera in Late Eighteenth-Century London.* 2 vols. New York: Oxford University Press, 1995.

Ragussis, Michael. "Jews and Other 'Outlandish Englishmen': Ethnic Performance and the Invention of British Identity under the Georges." *Critical Inquiry* 26.4 (2000): 773–97.

Rediker, Marcus. *Between the Devil and the Deep Blue Sea: Merchant Seamen, Pirates, and the Anglo-American Maritime World, 1700–1750.* Cambridge, New York: Cambridge University Press, 1987.

———. *Villains of All Nations: Atlantic Pirates in the Golden Age.* Boston: Beacon Press, 2004.

Reed, Peter P. "Conquer or Die: Staging Circum-Atlantic Revolt in *Polly* and *Three-Finger'd Jack.*" *Theatre Journal* 59.2 (2007): 241–58.

———. "'There Was No Resisting John Canoe': Circum-Atlantic Transracial Performance." *Theatre History Studies* 27 (2007): 65–85.

Reel, Guy. *The National Police Gazette and the Making of the Modern American Man, 1879–1906.* New York: Palgrave Macmillan, 2006.

Rees, James. *The Life of Edwin Forrest: With Reminiscences and Personal Recollections.* Philadelphia: Peterson and Brothers, 1874.

Reid, John Cowie. *Bucks and Bruisers: Pierce Egan and Regency England.* London: Routledge and K. Paul, 1971.

Reinelt, Janelle G. "The Politics of Discourse: Performativity Meets Theatricality." *SubStance* 31.2 (2002): 201–15.

A Representation of the Figures Exhibited and Paraded through the Streets of Philadelphia, on Saturday, the 30th of September 1780. Philadelphia: Printed by John Dunlap [?], 1780.

Reynolds, David S. *George Lippard, Prophet of Protest: Writings of an American Radical, 1822–1854.* New York: P. Lang, 1986.

———. *Beneath the American Renaissance: The Subversive Imagination in the Age of Emerson and Melville.* New York: Knopf, 1988.

Reynolds, Frederick. *The Renegade; a Grand Historical Drama, in Three Acts. Interspersed with Music.* New York: Published by D. Longworth, at the Dramatic Repository, Shakspeare-Gallery, 1813.

Richards, Jeffrey H. *Theater Enough: American Culture and the Metaphor of the World Stage, 1607–1789*. Durham: Duke University Press, 1991.
———. *Early American Drama*. New York: Penguin Books, 1997.
———. *Drama, Theatre, and Identity in the American New Republic*. Cambridge: Cambridge University Press, 2005.
Richards, Leonard L. *Gentlemen of Property and Standing: Anti-Abolition Mobs in Jacksonian America*. New York: Oxford University Press, 1970.
Richardson, John. "John Gay, the Beggar's Opera, and Forms of Resistance." *Eighteenth-Century Life* 24.3 (2000): 19–30.
Rinehart, Lucy. "'Manly Exercises': Post-Revolutionary Performances of Authority in the Theatrical Career of William Dunlap." *Early American Literature* 36.2 (2001): 263–93.
Ritchey, David. *A Guide to the Baltimore Stage in the Eighteenth Century: A History and Day Book Calendar*. Westport, CT: Greenwood Press, 1982.
Roach, Joseph R. "Slave Spectacles and Tragic Octoroons: A Cultural Genealogy of Antebellum Performance." *Theatre Survey* 33 (1992): 167–87.
———. *Cities of the Dead: Circum-Atlantic Performance*. New York: Columbia University Press, 1996.
———. "Barnumizing Diaspora: The 'Irish Skylark' Does New Orleans." *Theatre Journal* 50.1 (1998): 39–51.
Roediger, David R. *The Wages of Whiteness: Race and the Making of the American Working Class*. New York: Verso, 1991.
———. *The Wages of Whiteness: Race and the Making of the American Working Class*. Revised edn. New York: Verso, 1999.
Rosenthal, Laura J. "Juba's Roman Soul: Addison's Cato and Enlightenment Cosmopolitanism." *Studies in the Literary Imagination* 32.2 (1999): 63–76.
Rothbard, Murray Newton. *The Panic of 1819: Reactions and Policies*. New York: Columbia University Press, 1962.
Rothman, David J. *The Discovery of the Asylum: Social Order and Disorder in the New Republic*. Boston: Little, 1971.
Rowe, Katherine. *Dead Hands: Fictions of Agency, Renaissance to Modern*. Stanford: Stanford University Press, 1999.
Rowson, Susanna Haswell. *Slaves in Algiers; or, a Struggle for Freedom: A Play, Interspersed with Songs, in Three Acts. By Mrs. Rowson. As Performed at the New Theatres, in Philadelphia and Baltimore*. Philadelphia: Printed for the author, by Wrigley and Berriman, no. 149, Chesnut-Street, 1794.
Royall, Anne Newport. *Sketches of History, Life, and Manners, in the United States*. New-Haven: Printed for the author, 1826.
Rudé, George F. E. *The Crowd in History: A Study of Popular Disturbances in France and England, 1730–1848*. New York: Wiley, 1964.
Rust, Marion. "'Daughters of America,' *Slaves in Algiers*: Activism and Abnegation Off Rowson's Barbary Coast." *Feminist Interventions in Early American Studies*. Ed. Mary Clare Carruth. Tuscaloosa: University of Alabama Press, 2006. 227–39.
———. *Prodigal Daughters: Susanna Rowson's Early American Women*. Chapel Hill: Omohundro Institute of Early American History and Culture, University of North Carolina Press, 2008.
Rzepka, Charles. "Thomas De Quincey's 'Three-Fingered Jack': The West Indian Origins of the 'Dark Interpreter.'" *European Romantic Review* 8.2 (1997): 117–38.
Said, Edward W. *Orientalism*. New York: Pantheon Books, 1978.
———. *Culture and Imperialism*. New York: Knopf, 1993.

Sansay, Leonora. *Secret History; or, the Horrors of St. Domingo: In a Series of Letters, Written by a Lady at Cape Francois, to Colonel Burr... Principally During the Command of General Rochambeau.* Philadelphia: Bradford & Inskeep, 1808.
Saxton, Alexander. *The Rise and Fall of the White Republic: Class Politics and Mass Culture in Nineteenth-Century America.* London: Verso, 2003.
Schechner, Richard. *Between Theater and Anthropology.* Philadelphia: University of Pennsylvania Press, 1985.
Scherr, Arthur. "'Sambos' and 'Black Cut-Throats': Peter Porcupine on Slavery and Race in the 1790's." *American Periodicals: A Journal of History, Criticism, and Bibliography* 13 (2004): 3–30.
Schueller, Malini Johar. *U.S. Orientalisms : Race, Nation, and Gender in Literature, 1790–1890.* Ann Arbor: University of Michigan Press, 1998.
Schultz, William Eben. *Gay's Beggar's Opera: Its Content, History, and Influence.* New Haven: Yale University Press, 1923.
Seilhamer, George Overcash. *History of the American Theatre.* Philadelphia: Globe Printing House, 1888.
Senelick, Laurence. *The Changing Room: Sex, Drag and Theatre.* New York: Routledge, 2000.
Shaffer, Jason. "Making 'an Excellent Die': Death, Mourning, and Patriotism in the Propaganda Plays of the American Revolution." *Early American Literature* 41.1 (2006): 1–27.
———. *Performing Patriotism: National Identity in the Colonial and Revolutionary American Theater.* Philadelphia: University of Pennsylvania Press, 2007.
Shalev, Eran. "Empire Transformed: Britain in the American Classical Imagination, 1758–1783." *Early American Studies* 4.1 (2006): 112–46.
Sheridan, Francis Cynric. *Galveston Island; or, a Few Months Off the Coast of Texas: The Journal of Francis C. Sheridan, 1839–1840.* Ed. Willis Winslow Pratt. Austin: University of Texas Press, 1954.
Sheridan, Richard Brinsley. *A Short Account of the Situations and Incidents Exhibited in the Pantomime of Robinson Crusoe at the Theatre-Royal, Drury-Lane. Taken from the Original Story.* London: T. Becket, 1781.
Sherman, Suzanne. *Comedies Useful: History of Southern Theater, 1775–1812.* Williamsburg, VA: Celest Press, 1998.
Shields, David S. *Civil Tongues and Polite Letters in British America.* Chapel Hill: Published for the Institute of Early American History and Culture, Williamsburg, Virginia, by University of North Carolina Press, 1997.
Shields, John C. *The American Aeneas: Classical Origins of the American Self.* Knoxville: University of Tennessee Press, 2001.
Sidbury, James. *Ploughshares into Swords: Race, Rebellion, and Identity in Gabriel's Virginia, 1730–1810.* New York: Cambridge University Press, 1997.
Slauter, Eric Thomas. "Neoclassical Culture in a Society with Slaves: Race and Rights in the Age of Wheatley." *Early American Studies* 2.1 (2004): 81–122.
Smith, Billy G., and Richard Wojtowicz. *Blacks Who Stole Themselves: Advertisements for Runaways in the Pennsylvania Gazette, 1728–1790.* Philadelphia: University of Pennsylvania Press, 1989.
A Solemn Farewell to Levi Ames, Being a Poem Written a Few Days before His Execution, for Burglary, Oct. 21, 1773. Boston: Printed and sold at Draper's printing-office, in Newbury-Street, 1773.
Songs, Duets and Choruses in the Pantomimical Drama of Obi, or, Three-Finger'd Jack. Third edn. London: T. Woodfall, 1800.

Sorensen, Janet. "Vulgar Tongues: Canting Dictionaries and the Language of the People in Eighteenth-Century Britain." *Eighteenth-Century Studies* 37.3 (2004): 435–54.

The Speech of Death to Levi Ames. Who Was Executed on Boston-Neck, October 21, 1773, for the Crime of Burglary. [Boston]: Printed by John Boyle, 1773.

Spraggs, Gillian. *Outlaws and Highwaymen: The Cult of the Robber in England from the Middle Ages to the Nineteenth Century*. London: Pimlico, 2001.

St. John de Crèvecoeur, J. Hector. *Letters from an American Farmer; and, Sketches of Eighteenth-Century America*. Ed. Albert E. Stone. New York: Penguin Books, 1983.

Stallybrass, Peter. "Marx and Heterogeneity: Thinking the Lumpenproletariat." *Representations* 31 (1990): 69–95.

Stallybrass, Peter, and Allon White. *The Politics and Poetics of Transgression*. Ithaca, NY: Cornell University Press, 1986.

Stansell, Christine. *City of Women: Sex and Class in New York, 1789–1860*. New York: Knopf, distributed by Random House, 1986.

Stern, Julia A. *The Plight of Feeling: Sympathy and Dissent in the Early American Novel*. Chicago: University of Chicago Press, 1997.

Stillman, Samuel. *Two Sermons: The First from Psalm CII. 19, 20. Delivered the Lords-Day before the Execution of Levi Ames, Who Was Executed at Boston, Thursday October 21, 1773*. Second edn. Boston: Printed and sold by J. Kneeland, in Milk-Street; sold also by Philip Freeman, in Union-Street, 1773.

Stott, Richard Briggs. *Workers in the Metropolis: Class, Ethnicity, and Youth in Antebellum New York City*. Ithaca: Cornell University Press, 1990.

Strand, Ginger. "'My Noble Spartacus': Edwin Forrest and Masculinity on the Nineteenth-Century Stage." *Passing Performances: Queer Readings of Leading Players in American Theater History*. Eds Robert A. Schanke and Kim Marra. Ann Arbor, MI: University of Michigan Press, 1998. 19–40.

Streeby, Shelley. "Opening up the Story Paper: George Lippard and the Construction of Class." *boundary 2* 24.1 (1997): 177–203.

Strong, George Templeton. *Diary of George Templeton Strong*. Eds Allan Nevins and Milton Halsey Thomas. 4 vols. New York: Macmillan, 1952.

Taylor, Diana. *The Archive and the Repertoire: Performing Cultural Memory in the Americas*. Durham: Duke University Press, 2003.

Theft and Murder! A Poem on the Execution of Levi Ames, Which Is to Be on Thursday, the 21st of October Inst. [Boston]: Sold [by Isaiah Thomas] near the Mill-Bridge: and at the printing office near the market, 1773.

Thompson, E. P. "The Moral Economy of the English Crowd in the Eighteenth Century." *Past and Present* 50 (1971): 76–136.

———. *Whigs and Hunters: The Origin of the Black Act*. New York: Pantheon Books, 1975.

———. *Customs in Common: Studies in Traditional Popular Culture*. New York: New Press, 1991.

Thompson, George. *A Documentary History of the African Theatre*. Evanston, IL: Northwestern University Press, 1998.

Thompson, Peter. *Rum Punch and Revolution: Taverngoing and Public Life in Eighteenth Century Philadelphia*. Philadelphia: University of Pennsylvania Press, 1999.

Thurmond, John. *Harlequin Sheppard, a Night Scene in Grotesque Characters: As It Is Perform'd at the Theatre-Royal in Drury-Lane*. London: Printed and sold by J. Roberts, and A. Dodd, 1724.

Tompkins, Jane P. *Sensational Designs: The Cultural Work of American Fiction, 1790–1860*. New York: Oxford University Press, 1985.

Trees, Andy. "Benedict Arnold, John Andre, and His Three Yeoman Captors: A Sentimental Journey or American Virtue Defined." *Early American Literature* 35.3 (2000): 246–73.

Turner, Victor Witter. *From Ritual to Theatre: The Human Seriousness of Play.* New York: Performing Arts Journal Publications, 1982.

Tyler, Royall. *The Algerine Captive, or the Life and Adventures of Doctor Updike Underhill, Six Years a Prisoner among the Algerines.* 2 vols. Walpole, Newhampshire: David Carlisle, jun., and sold at his bookstore, 1797.

———. *The Algerine Captive, or, the Life and Adventures of Doctor Updike Underhill, Six Years a Prisoner among the Algerines.* Ed. Caleb Crain. New York: Modern Library, 2002.

Vanskike, Elliott. "Consistent Inconsistencies: The Transvestite Actress Madame Vestris and Charlotte Bronte's Shirley." *Nineteenth-Century Literature* 50.4 (1996): 464–88.

Vitkus, Daniel J. *Turning Turk: English Theater and the Multicultural Mediterranean, 1570–1630.* New York: Palgrave Macmillan, 2003.

Vitkus, Daniel J., and N. I. Matar, eds. *Piracy, Slavery, and Redemption: Barbary Captivity Narratives from Early Modern England.* New York: Columbia University Press, 2001.

Wahrman, Dror. *The Making of the Modern Self: Identity and Culture in Eighteenth-Century England.* New Haven: Yale University Press, 2004.

Waldstreicher, David. *In the Midst of Perpetual Fetes: The Making of American Nationalism, 1776–1820.* Chapel Hill: Omohundro Institute of Early American History and Culture, University of North Carolina Press, 1997.

Walker, David. *Walker's Appeal, in Four Articles; Together with a Preamble, to the Coloured Citizens of the World, but in Particular, and Very Expressly, to Those of the United States of America.* Third edn. Boston: D. Walker, 1830.

Walsh, William Shepard. *Curiosities of Popular Customs and of Rites, Ceremonies, Observances, and Miscellaneous Antiquities.* Philadelphia: J. P. Lippincott Company, 1898.

Wanko, Cheryl. "Three Stories of Celebrity: *The Beggar's Opera* 'Biographies.'" *Studies in English Literature, 1500–1900* 38.3 (1998): 481–98.

Ward, Artemus. *Artemus Ward, His Book with Many Comic Illustrations.* New York: Carleton, 1862.

Warner, Michael. *The Letters of the Republic: Publication and the Public Sphere in Eighteenth-Century America.* Cambridge, MA: Harvard University Press, 1990.

Warner, Michael, Natasha Hurley, Luis Iglesias, Sonia Di Loreto, Jeffrey Scraba, and Sandra Young. "A Soliloquy 'Lately Spoken at the African Theatre': Race and the Public Sphere in New York City, 1821." *American Literature* 73.1 (2001): 1–46.

Watson, Charles S. *Antebellum Charleston Dramatists.* Tuscaloosa, AL: University of Alabama Press, 1976.

Wemyss, Francis Courtney. *Twenty-Six Years of the Life of an Actor and Manager.* New York: Burgess, Stringer and Co., 1847.

White, Shane. *Somewhat More Independent: The End of Slavery in New York City, 1770–1810.* Athens: University of Georgia Press, 1991.

———. "'It Was a Proud Day': African Americans, Festivals, and Parades in the North, 1741–1834." *The Journal of American History* 81.1 (1994): 13–50.

———. *Stories of Freedom in Black New York.* Cambridge, MA: Harvard University Press, 2002.

Wilentz, Sean. *Chants Democratic: New York City and the Rise of the American Working Class, 1788–1850.* New York: Oxford University Press, 1984.

Williams, Dan E. *Pillars of Salt: An Anthology of Early American Criminal Narratives.* Madison, WI: Madison House, 1993.

Williams, Daniel E. *Liberty's Captives: Narratives of Confinement in the Print Culture of the Early Republic.* Athens: University of Georgia Press, 2006.

Williams, Raymond. *The Long Revolution*. New York: Columbia University Press, 1961.

Wilmer, S. E. *Theatre, Society and the Nation: Staging American Identities*. Cambridge: Cambridge University Press, 2002.

Wilson, Arthur Herman. *A History of the Philadelphia Theatre, 1835 to 1855*. Philadelphia: University of Pennsylvania Press, 1935.

Wilson, Garff B. *Three Hundred Years of American Drama and Theatre, from Ye Bare and Ye Cubb to Hair*. Englewood Cliffs, NJ: Prentice-Hall, 1973.

Wilson, Gary E. "American Hostages in Moslem Nations, 1784–1796: The Public Response." *Journal of the Early Republic* 2.2 (1982): 123–41.

Winchester, Elhanan. *The Execution Hymn, Composed on Levi Ames, Who Is to Be Executed for Burglary, This Day, the 21st of October, 1773, Which Was Sung to Him and a Considerable Audience, Assembled at the Prison, on Tuesday Evening, the 19th of October, and, at the Desire of the Prisoner, Will Be Sung at the Place of Execution, This Day*. [Boston]: Sold by E. Russell, next the cornfield, Union-Street, 1773.

Winter, Marian Hannah. "Juba and American Minstrelsy." *Inside the Minstrel Mask: Readings in Nineteenth-Century Blackface Minstrelsy*. Eds Annemarie Bean, James Vernon Hatch and Brooks McNamara. Hanover, NH: Wesleyan University Press, 1996. 223–41.

Winterer, Caroline. *The Culture of Classicism: Ancient Greece and Rome in American Intellectual Life, 1780–1910*. Baltimore: Johns Hopkins University Press, 2002.

Winton, Calhoun. *John Gay and the London Theatre*. Lexington: University Press of Kentucky, 1993.

Wittke, Carl. "The Immigrant Theme on the American Stage." *The Mississippi Valley Historical Review* 39.2 (1952): 211–32.

Wood, Gordon S. *The Radicalism of the American Revolution*. New York: A.A. Knopf, 1992.

Wood, Samuel. *The Cries of New-York*. New-York: Printed and sold by S. Wood, at the Juvenile book-store, No. 362, Pearl-Street, 1808.

Wright, Richardson. *Revels in Jamaica, 1682–1838*. New York: Dodd Mead, 1937.

Young, Alfred F. *The Shoemaker and the Tea Party: Memory and the American Revolution*. Boston: Beacon Press, 1999.

———. *Masquerade: The Life and Times of Deborah Sampson, Continental Soldier*. New York: Alfred A. Knopf, distributed by Random House, 2004.

Index

abolitionism, 101, 121–2, 163, 172, 186, 212 n. 75
acrobatics, 49–51, 81, 83, 103
Adams, John Quincy, 64
Adelphi Theatre (London), *see* theatres and performance spaces
African Theatre (New York City), *see* theatres and performance spaces
Aldridge, Ira, 102, 122, 187
Alexander, Cato, 164–5
Algerine Captive, The (Tyler, 1797), 57, 62, 63, 69–70
Algerine Spy in Pennsylvania (Markoe, 1787), 67
All-Max (London, scene set in), 130–2, 142, 148
All-Max (New York City), *see* theatres and performance spaces
American Monthly Magazine and Critical Review, 36
American Notes (Dickens, 1842), 138, 148–50, 164
American Revolution, 13–14, 39, 57, 60, 79, 88–9, 91, 96–7, 129, 162; *see also The Glory of Columbia*
Americanischer Haus- und Wirthschafts-Calender (Philadelphia, 1780), 85–6, 205 n. 24
Ames, Levi, 27–8, 31–2, 39
  in *Algerine Captive*, 40
  and execution broadsides, 43–9
André (Dunlap, 1798), 4, 17, 79–81
André, Major John, 79; *see also André*; Dunlap, William; *The Glory of Columbia*
Appeal to the Coloured Citizens of the World (Walker, 1830), 173
Arabs of the Desert; or, Harlequin's Flight from Egypt (1799), 69
Arbuthnot, Harriet, 134
Arch Street Theatre (Philadelphia), *see* theatres and performance spaces
Arnold, Benedict, 83–9; *see also André*; Dunlap, William; *The Glory of Columbia*
Arnold, Samuel, 103
Astor Place Opera House (New York City), *see* theatres and performance spaces
Astor Place Riots, *see* riots or disturbance
audiences, 2, 9, 11, 22, 27, 58, 184
  of *The Beggar's Opera* 31–2, 36, 39
  of *The Gladiator* 151–6, 173

  of *The Glory of Columbia*, 88, 95, 100
  of *Jack Sheppard*, 177–8, 180
  of Levi Ames's execution, 44–6, 48–9
  of popular blackface, 169–70
  as self-conscious, 11–12
  of *Slaves in Algiers*, 70–1, 73
  of *Three-Finger'd Jack*, 115, 117, 121
  of *Tom and Jerry*, 129, 131, 140–1, 143, 149
  of Tom's execution, 124
  women in, 50–1

ballad opera, 5, 17, 19, 20, 29–36, 52, 73, 109–12, 134–6; *see also The Beggar's Opera*
ballads, 134–5, 140; *see also* music
Barbarossa, the Usurper of Algiers (Brown, 1794), 69, 202 n. 77
Barker, Robert, 135
Barlow, Joel, 75–6
Barnum, P. T., 14
Beggar's Opera, The (Gay, 1728), 1–4, 10, 17, 18, 19, 21, 28–31, 36–9, 42–3, 48, 50–2, 67, 72, 82, 101, 102, 109, 112, 127, 134–6, 137, 139, 175, 178–81, 186
  in Americas, 31–6
  censored, 34–5
  in cross-dress, 91
  and public disturbances, 1–4, 36
  see also ballad opera; *Polly*.
beggars and begging, 6, 8, 117, 129–30, 133–4, 137, 139–40, 170, 179–80
Belinda (Edgeworth, 1801), 164
Belisario, Isaac Mendes, 118–9
Belzebub, 185
Benito Cereno (Melville, 1855), 187
Bickerstaff, Isaac, 72, 200 n. 28, 203 n. 85
Bird, Robert Montgomery, "Secret Records," 172–3; *see also* Forrest, Edwin; *The Gladiator*
"Black Atlantic" (Gilroy), 111–12, 115, 116–22
blackface, 5, 11, 16, 18, 19–21, 67, 71, 84, 101–2, 109–15, 118, 122–3, 126, 133, 171–2, 175, 180, 217 n. 94
  and African American actors, 142–3
  and character types, 165
  and claims of authenticity, 130
  and Forrest, 152, 168–9
  in New Orleans 144–7

244 Index

blackface—*Continued*
  as self-consciously constructed, 120–32
  see also Obi; or, Three-Finger'd Jack; Polly;
    Rice, Thomas Dartmouth; *Tom and Jerry*
Blackwell's Island, 186
Blake, or the Huts of America (Delany, 1859–62), 187
Blue Beard (Colman, 1802), 68
"bobalition" broadsides, 208 n. 85
bodies
  as performing, 5, 28, 40, 43, 83, 105, 163, 186
  Forrest 167–71
  Sheppard, 177–9
  Stoker, 49–52
  see also Ames, Levi; captivity
Bone Squash Diavolo (Rice, 1835), 146–7
Bowery (New York City), 155, 169
Bowery Theatre (New York City), *see* theatres and performance spaces
broadsides, 3, 6, 17, 27–8, 43–9, 55, 84–6, 89, 128, 137, 177, 197 n. 76, 202 n. 77, 205 n. 24, 208 n. 85
Brown, Ann, 1–4, 10
Brown, William A., 102, 121, 141–4, 170; *see also* theatres and performance spaces
Brutus, or the Fall of Tarquin (Payne, 1818), 162
Bunker-Hill, or the Death of General Warren (Burk, 1797), 81
burlettas, 36, 128, 146–7, 214 n. 23
Byrne, Mr. (actor, Philadelphia), 68

Caldwell, James H., 15, 146, 217 n. 88
captivity (Algerian)
  and *Othello*, 64
  and playing Algerian, 66–71
  and religious or cultural conversion, 10, 54, 57, 61–7, 75–7
  see also Journal, of the Captivity and Sufferings (Foss, 1798); *Narrative of Robert Adams* (Adams, 1817); slavery; *Slaves in Algiers*
casting
  and extras or supernumeraries, 19, 22, 61
  "lines of business," 60
  star system, 16, 153–6, 213 n. 1
Cato (Addison, 1713), 56, 162, 163–4
Cato (aka Toby), 166–7
charivari (and related folk forms), 94; *see also* parades and processions
Charleston, South Carolina, scene set in, 142
Charlotte Temple (Rowson, 1791), 53, 56
Chatham Theatre (New York City), *see* theatres and performance spaces
Chestnut Street Theatre ("New Theatre," Philadelphia), *see* theatres and performance spaces

Christ's Promise to the Penitent Thief (Eliot, 1773), 27, 45–6, 48
Christ Sent to Heal the Broken Hearted (Mather, 1773), 27
circus, 14, 15, 20, 49–52, 82–3, 98, 103, 132, 143, 158, 168, 170, 198 n. 95, 208 n. 89, 214 n. 22
clowns, 20, 50, 82–3, 97
Cobbett, William, 55–6
Colman, George, Jr., 70
  and staging of Gay's *Polly*, 102, 109–12
colonial theatre, 13–15
comic opera, 72, 99, 135, 203 n. 85,
confessions, 17, 27, 38, 39–43, 49, 52, 74, 88–9
Confessions (Mount, 1791), 39–43, 51–2, 74, 88–9, 125, 175, 180, 181–2
Conny, John (African chief), 118, 212 n. 62
Contrast, The (Tyler, 1787), 88
convivial culture, 72–74
Cooper, Thomas Abthorpe, 80, 154
costuming, 2, 6, 19, 20, 23, 29, 63, 76, 84–5, 91, 97, 139, 168
  in *Jack Sheppard*, 176, 178–81
  in Jonkonnu, 117–20
  in *Slaves in Algiers*, 66–70
Courvoisier, B. F., 178
Covent Garden Theatre (London), *see* theatres and performance spaces
Cowboys (Revolutionary War), 88
Cowell, Joseph, 49–50, 132–3, 139
Cries of London, 65, 137
  and American adaptations, 137
  see also urban culture
criminal culture, 10–11, 28
  in America, 36–7
  as Atlantic subculture, 1–3, 28–31, 40–43, 92, 109–10, 134
  in *The Beggar's Opera*, 31–9
  in England, 29–31
  see also executions
cross-dress, 22, 90
  in *The Beggar's Opera*, 1, 34
  in *The Glory of Columbia*, 89–91, 94, 98
  in *Jack Sheppard*, 178–9, 180
  in *Polly*, 1, 109, 178, 180
  in *Slaves in Algiers*, 19, 56–7, 65
  in *Three-Finger'd Jack*, 101
  in *Tom and Jerry*, 130–1, 140, 142
  see also Keeley, Mary Ann; Sampson, Deborah; Vestris, Eliza
crowds, 2, 9–10, 27, 46–9, 50, 76–7, 87, 93, 95–6, 140, 170, 177, 183–4
  extras as, 22
  as "mobility," 11
  as "mobocracy," 184–5

and "moral economy" (Thompson), 36, 113
  see also audiences; riots or disturbance
Cruikshank, George and Robert, 127
cultural transmission, 13, 16–17, 21, 22–5, 31, 41, 84, 121–3, 136, 147, 149, 155–6, 161, 165, 182–3
  as "lore cycles," 186–8
  and "surrogation" (Roach), 3, 16, 54, 72, 122–3, 130, 169
Curtis, George William, 136

Dalling (Governor of Jamaica), 106
dance, 5, 19, 24, 94, 128, 138, 148–50
  "allemande de trois," (100)
  blackface, 101, 103, 107, 113, 116, 130–2, 147
  "breakdown," 148
  "Dancing for Eels" 169–71
  "Essence of Old Virginny," 168–9, 221 n. 85
  "pas de deux," 130, 214 n. 16
  "pas suel," 100
"Dancing for Eels" (Brown, James and Eliphalet, 1848), 169–70
Daniel, George, 130, 141
Darby's Return (Dunlap, 1789), 92
Darley, John, Jr., 67
DeCamp, Maria Theresa, 103
Dickens, Charles, 148–50, 187
Disappointment, The (1767), 92
Dixon, George Washington, 165
Don Sebastian, King of Portugal (Dryden, 1690), 72
Douglass, David, 13; and Jamaican slaves, 166
Douglass, Frederick, 211 n. 57
Dramatic Censor (London), 103, 115
Drury Lane Theatre (London), *see* theatres and performance spaces
Dunlap, William, 4, 15, 17–18, 53, 69, 79–83, 92, 95–6, 99–100, 103; *see also André*; *Darby's Return*; *The Glory of Columbia*
Dupuis, Robert, 63–4
Durang, Charles, 81, 83, 128, 139, 140
Durang, John, 67, 83
Dutch Broom Sellers, 145
Dutton, Charles, 105, 115
Dying Groans of Levi Ames (broadside, 1773), 44–5

effigy parades, 83–6
Eighteenth Brumaire of Louis Bonaparte (Marx, 1852), 6–7
equestrian performances, 98, 103, 143
ethnic identity and masking, 17, 19, 24, 53, 56, 58, 60–74, 91–3; *see also* blackface

"excarceration" (Linebaugh) 8, 17, 19, 36–9, 48, 50, 107, 123, 134, 175–6, 180, 183, 86
"Execution Hymn" (Winchester, 1773), 28, 44
executions
  as entertainment, 27–8, 43–52, 82–7, 92–3, 158, 181–3
  and print culture, 43–9
  see also Ames, Levi; *The Beggar's Opera*; *Confessions* (Mount, 1791); Sheppard, Jack; Stoker;

Fairfax, Mrs. George William (Sally), 163
Federal Street Theatre (Boston), *see* theatres and performance spaces
Female Review, The (Mann, 1797), 91
Fennell, James, 9–10
Fisher, Samuel Rowland, 83–4, 85–6
Five Points (New York City), 136, 138, 146, 164
flâneurie, 18, 19, 134, 136, 138, 140, 141, 164;
  see also Tom and Jerry
flash (criminal slang)
  dictionary, 41–2
  "flash gangs," 41–2, 49, 74, 176, 180
  see also Confessions (Mount, 1791)
Fly, William, 45
forgery, 65
Forrest, Edwin, 16, 18, 19, 184–5
  and blackface, 152–3, 167–71
  and physical performance, 153–6, 158, 160, 162
  see also The Gladiator; riots or disturbance
Francis, Mr. (actor, Philadelphia), 65
Franklin, Benjamin, 67
French Revolution (and theatre), 92, 161
Fulton Market, scene set in, 142

Garrick, David, 21, 65, 177
Garrison, William Lloyd, 152, 172, 174
Gay, John, 1, 20, 28–31, 177
General History of the Stage (Chetwood, 1749), 33
George Barnwell; or, the London Merchant (Lillo, 1731), 34
Gladiator, The (Bird, 1831), 4, 16, 18, 19, 114, 151–74
  and American slave revolt, 171–4
  and transnational revolution, 156–60
  see also Forrest, Edwin; neoclassical culture
Glance at New York in 1848, A (Baker, 1848), 130, 133, 155
Glory of Columbia; Her Yeomanry! (Dunlap, 1803), 4, 17, 19, 53, 79–100, 152
  and execution as spectacle; 82–3, 89–90, 91–4
  and Irish characters, 91–3

Glory of Columbia; Her Yeomanry! (Dunlap, 1803)—*Continued*
  and theatrical cross-dress, 89–91
  and yeomen, 87–96
  see also André, Major John; Arnold, Benedict
Graydon, Alexander, 35–6, 74, 86–7, 96–7
Great Expectations (Dickens, 1861), 181
Greek Slave, The (Powers, 1844), 163
Green, Mr. (actor, Philadelphia), 67

"Hail Columbia," *see* songs, by title
Hallam, Lewis, Jr., 14, 34, 61 (troupe), 103
Hallam, Lewis, Sr., 13
Hamblin, Thomas, 184
Harlequin Sheppard (Thurmond, 1724), 113, 177
harlequins, 69, 113, 177; Merry-Andrew, 120
Haymarket Little Theatre (London), *see* theatres and performance spaces
Haymarket Theatre (Boston), *see* theatres and performance spaces
highwaymen, 41, 88; *see also The Beggar's Opera*; Sheppard, Jack
History of the Maroons (Dallas, 1803), 106
Hodgkinson, John, 68, 103
Hogg, Miss (actor, New York City), 89
holidays
  as carnivalesque, 94, 115–17
  Election Day, 97
  Evacuation Day, 98
  Independence Day, 95, 98–100
  Pope Night, 83
  Training Day, 95, 97
  see also Jonkonnu; pantomime; Pinkster
Hutchins's *Almanack* (1802), 101
hybridity (cultural), 81, 118–20, 173

incarceration, 11, 27–8, 37–8, 40, 42–5, 51, 71, 75, 135, 137, 175–7, 181, 186–7; *see also* captivity; "excarceration"
Incidents in the Life of a Slave Girl (Jacobs, 1861), 117, 118
Incledon, Charles, 36
Indian Queen Tavern (Philadelphia), 76–7
Invisible Man (Ellison, 1952), 113, 164
Irving, Washington, 9
itinerant theatre, *see* theatrical circuits

Jack Sheppard (Ainsworth, 1839), 31, 177, 178, 181
Jack Sheppard (Buckstone, 1839), 31, 178–81, 223 n. 31, 33
Jack Sheppard, (Greenwood, 1840), 181
Jack Sheppard, or the Life of a Robber! (Phillips, 1839), 31, 175, 179, 181–4

Jamaica, 13, 33, 101, 101–6, 115–21, 166, 187
  English troupes in, 13, 33
  Leeward and Windward Treaties (1738/39), 105–6
  see also Jonkonnu
"Jaw-bone, or House John Canoe" (Belisario, 1837), 118–20
Jefferson, Joseph, 132, 139
John Street Theatre (New York City), *see* theatres and performance spaces
Jonkonnu
  festival, 115–20
  and Jonkanoo figure in *Three-Finger'd Jack*, 115–16
Journal, of the Captivity and Sufferings (Foss, 1798), 63, 75–6
Juba, *see* Lane, William Henry

Kean, Edmund, 139, 154
Kean, Thomas, 13, 33, 195 n. 32
Kearsley (Philadelphia Loyalist), 86–7
Keeley, Mary Ann, 178–80
Kemble, Charles, 103

"La Carmagnole," *see* songs, by title
labor and laborers, 1, 7–8, 9, 10–11, 15, 18, 19, 37–8, 62, 71–2, 76–7, 88, 97, 118–20, 137–8, 145, 147, 152–60, 163, 167–71, 175–6
  actors as, 25, 154
  see also beggars and begging; Forrest, Edwin; slavery; urban culture
Lane, William Henry ("Master Juba"), 148–50, 164
Last Words and Dying Speech of Levi Ames (broadside, 1773), 27, 42, 44
Laurence and Martin (dancers, New York City), 103
Lewis, Matthew G., 118, 120, 121
Liberty in Louisiana (Workman, 1803), 99
Life in London (Egan, 1820), 127
Life in New Orleans (1837), 144–7
Life in New York (Phillips, 1834), 133
Life in New York; or Tom and Jerry on a Visit (Brougham, 1856), 133
Life in Philadelphia (Burns, 1835), 133, 146
"Life in Philadelphia" (Clay, 1830), 146, 165, 217 n. 90
Lincoln's Inn Fields (London), *see* theatres and performance spaces
Lippard, George, 185
London Labour and the London Poor (Mayhew, 1851–1861), 138, 187
Long, Edward, 118
Love in a Village (Bickerstaff, 1762), 32
Ludlow, Noah, 15

Macklin, Charles, 91
Macready, William, 91, 184–5; *see also* riots or disturbance
Malefactor's Register, 31
Manfredi, Signor (acrobat, New York City), 81, 82
Market Book (De Voe, 1862), 138, 142
markets, and performance, 5, 8, 10, 142, 144, 168
maroons and marronage, 10, 104–6, 118, 120, 149, 209 n. 17, 210 n. 35
  in *Polly*, 109–13
  in *Three-Finger'd Jack*, 106, 108, 180
Marx, Karl
  and history repeating as farce, 7
  and "lumpenproletariat," 6–7, 21, 35, 71, 147, 157, 184–7, 189 n. 15
Mather, Cotton, 45
Mathews, Charles, 133, 154
May Day in Town, or New York in an Uproar (Tyler, 1787), 136–7
melodrama, 5, 15, 16, 18–21, 51, 114, 118, 122, 151–3, 158, 160, 163, 184
  and blackface, 168–9
militia musters, 10, 24, 80, 94–8
Modern Chivalry (Brackenridge, 1798), 92
Mogadore (Essaouira, Morocco), 63
moral lectures (plays as), 14, 34
Murray, Walter, 13, 33, 195 n. 32
museums, 14, 198 n. 100
music, 27, 44, 65, 71–3, 100, 101, 107, 110, 113–14, 116–17, 128, 134–5, 146–7, 177, 180, 187
  flash songs, 41–2, 180
  political or patriotic, 36, 67, 95–6
  songsters, 73–4
  see also ballad opera; melodrama; pantomime; songs, by title
Mysteries and Miseries of New York, The (Judson, 1848), 138

Narrative of all the Robberies, Escapes, &c. of John Sheppard (Defoe, 1724), 29, 114; *see also* Sheppard, Jack.
Narrative of Robert Adams (Adams, 1817), 63–4
Nassau Street Theatre (New York City), *see* theatres and performance spaces
neoclassical culture, 18, 19, 114
  as "Classical Blackness," 151–5, 160–74
New American Company, 34
New York by Gas-Light (Foster, 1850), 139
Newgate Calendars, 29, 31, 177
Newgate Prison, 38, 176–7, 179, 181
"Nix My Dolly Pals, Fake Away," *see* songs, by title

Noah, Mordecai M., 15, 141, 142; *National Advocate* (New York City), 141;

Obi; or, The History of Threefingered Jack (Earle, 1800/1804), 101
Obi; or, Three-Finger'd Jack (Fawcett, 1801), 16, 18, 19, 101–26, 129, 147, 159, 160, 171, 179, 180, 185, 186, 187, 212 n. 75
  at Brown's African Theatre, New York City, 143
  as melodrama, 152–3
  and obeah, 101, 104, 107–9, 114–15, 179
  see also blackface; Jonkonnu; maroons and marronage; "optic black"; pantomime
Octoroon, The (Boucicault, 1859), 163
Old American Company, 80
Old Corn Meal (performer, New Orleans), 144–7
Oliver Twist (Dickens, 1838), 177
Olympic Theatre (London), *see* theatres and performance spaces
one-man performance (of full play), 34
"optic black" (Lhamon), 113–15
  and "optic white" (Ellison), 113, 164
"orientalism" (Said), 17, 60, 67–70, 165, 200 n. 28
Osborne, Mrs. (actor, Virginia), 34
Otello (Rice, 1844), 64

Padlock, The (Bickerstaff, 1768), 111, 122
panoramic displays, 99, 128, 135
pantomime; 5, 18, 19, 20, 68–9, 81, 85, 92, 100, 107–9, 110–11, 113–16, 121–3, 147, 177;
  see also Three-Finger'd Jack
parades and processions, 10, 17, 69, 80–7, 94–8, 105
Park Theatre (New York City), *see* theatres and performance spaces
Parker, Mrs. (actor, Virginia and Maryland), 34
partisanship, 55–6, 61, 65, 95–6, 199 n. 11
Peale, Charles Willson, 84–5
Pelissier, Mr. (musician), 99
performance; and theatricality, 10–12, 185–8
  and "print-performance culture," 22–5
  and stage aesthetics, 19–22
  see also "structures of feeling"; "theatrical formations"; vernacular culture
Picture of New-York, The (Mitchill, 1807), 137
Pinkster, 97–8, 208 n. 83
poetry, 27, 48, 157, 219 n. 34
Polly (Gay, 1729), 1, 10, 16, 18, 19, 32, 34, 101
  London performances of (1777), 102, 109
  and Native Americans, 109
  and piracy, 109–11
  and Waltham and Windsor Blacks, 110

Poor of New York, The (Boucicault, 1857), 187
Poor Soldier, The (O'Keefe, 1786), 73, 91, 92
Port Folio (Philadelphia), 104
Power, Tyrone, 164–5
Price, Stephen A., 127, 141, 143

Quincy, Josiah, 35

Renegade, The (Reynolds, 1812), 72
"Republican Revolution" of 1800, 8
Rice, Thomas Dartmouth, 16, 133, 145–7, 169, 175, 186; *see also* blackface; *Bone Squash Diavolo*; *Life in Philadelphia*; *Otello*
Rickett's Circus (Philadelphia), *see* theatres and performance spaces
riots or disturbance, by date
  at Covent Garden Theatre (London, 1776), 1–4
  partisan singing (New York City 1798), 95–6
  Old Price Riots (London 1809), 9
  Fennell on (1814), 9
  at Park Theatre (New York City, 1817), 36
  at Brown's Theatre (New York City, 1823), 143–4
  "Anni Mirabili" (1830s), 183
  Astor Place Riots, 15, 184–6
ritual (performance as), 10, 24, 39, 41–2, 48, 51, 70, 72, 74, 77, 83–7, 95–7, 100–2, 105–7, 116–17, 122, 127, 133, 136–7, 145, 149, 154, 166, 184
Robinson Crusoe; or, Harlequin Friday (Sheridan, 1781), 113–14
"rogue performances"
  defined, 4–7, 185, 187
  in theatre, 12–16
"Rogue's March," *see* Songs, by title
Rowson, Susanna Haswell, 53–6, 57, 67; *see also Charlotte Temple*; *Slaves in Algiers*
Rowson, William, 73, 202 n. 64

Saint Domingue (Haitian Revolution), 102, 114
Sampson, Deborah, 90–1
Secret History; or, the Horrors of St. Domingo (Sansay, 1808), 104
Sennet, John (criminal), 46
sermons, 27–8, 45–6, 193 n. 1
servants, 7–9, 18, 21, 35, 37, 97, 118
Shaw, Mrs. (actor, New York City), 179
She Would Be A Soldier (Noah, 1819), 91
Sheppard, Jack, 4, 19, 29–31, 38, 175–7; *see also The Beggar's Opera*; "excarceration"; *Harlequin Sheppard*; *Jack Sheppard*
Sheridan, Francis Cynric, 145–6
Short Account of Algiers (Carey, 1794), 66
Shotaway; or, the Insurrection of the Caribs, of St. Domingo (1823), 143

Simpson, Edmund, 132
"Sixteen-String Jack," 31
Sketches of History, Life, and Manners, in the United States (Royall, 1826), 138
slavery, 102, 118–20
  and naming practices, 165–67
  and revolts, 13, 103–4, 105, 121, 125, 152, 172–3, 187
  and runaways, 6, 104–6, 123–6, 149, 166–7, 187
  and theatre, 166
  see also captivity; *Three-Finger'd Jack*
Slaves in Algiers (Rowson, 1794), 53–78
  and Algerians on display, 66–71
  and drinking songs, 72–4
  and transnational characters, 53–4, 65–6, 71–7
Smith, Elizabeth (criminal), 46
Smith, I. (or Joshua, co-conspirator with Arnold), 85
Smith, Mr. (actor, New York City), 142
Smith, Sol, 15
"social banditry" (Hobsbawn), 29–30
soldiers, *see The Glory of Columbia*; parades and processions
songs, by title
  "America, Commerce, and Freedom," 67
  "Hail Columbia," 95
  "La Carmagnole," 95–6
  "Nix My Dolly Pals, Fake Away," 180
  "Rogue's March," 85, 86–7, 95
  "Thus I Stand Like the Turk," 134
  "Youth's the Season," 30
  see also music
"spectacular opacity" (Brooks), 113
"Spectre" (pseudonym), 49–51
St. Giles (London), scenes set in, 39, 129, 135, 140, 177
Stoker (acrobat, Philadelphia), 49–50, 198 n. 95
Strand, the (London), 129
Strong, George Templeton, 185
"structures of feeling" (Williams), 25, 66, 136, 165, 187

tableaux, 70, 99, 105
tarring and feathering, 86–7
taverns and pubs, 10–11, 30, 76–7, 140, 164–6
theatres and performance spaces, by city; *see also* circus; taverns and pubs; theatrical circuits
  Annapolis, Maryland, 34
  Baltimore, Maryland, 69
  Boston, Massachusetts, Adelphia Theatre, 133; Federal Street Theatre, 14, 69, 104; Haymarket Theatre, 14
  Charleston, South Carolina, 13, 34, 69

London: Adelphi Theatre, 127–30, 135, 178–9; Covent Garden Theatre, 1, 9; Drury Lane Theatre, 114, 177; Haymarket Little Theatre, 1, 72, 101–2, 109, 128, 178; Lincoln's Inn Fields, 129; Olympic Theatre, 178
Macon, Georgia, 145
New Orleans: Camp Street Theatre, 217 n. 88; St. Charles Theatre, 144, 146–7
New York City: African ("Grove") Theatre, 102, 121, 141–4; All-Max (New York City), 148–9; Astor Place Opera House, 181, 184–5, 186, 188; Bowery Theatre, 15, 20, 133, 153–5, 169, 175, 179, 183–5, 218, n. 10, 223 n. 31; Chatham Theatre, 15, 153, 170; John Street Theatre, 14, 80, 103; Nassau Street Theatre, 33; Park Theatre, 11–12, 14, 36, 79–83, 89, 91, 101–3, 127, 132, 140–1, 148, 151, 152, 153, 155, 168, 184, 218, n. 10; West's Circus, 143–4
Newport, Rhode Island, 14, 69, 99
Philadelphia: Arch Street Theatre, 155; Chestnut Street Theatre ("New Theatre"), 14, 53, 53–61, 132, 139, 140; Circus (Walnut Street), 14, 49–50, 132; Rickett's Circus, 98, 103; Walnut Street Theatre, 14, 169
Portsmouth, New Hampshire, 34
Providence, Rhode Island, 69
Salem, Massachusetts, 69
Valley Forge, Pennsylvania, 162
Williamsburg, Virginia, 34
theatrical circuits, 13–16, 132, 153–4, 168, 169
"theatrical formation" (McConachie), 24
Theft and Murder! (broadside, 1773), 43, 48
Three-Finger'd Jack (Murray, 1830), 121–2, 152
"Thus I Stand Like the Turk," *see* songs, by title
Tom (Virginia bandit, 1818), 123–6
Tom and Jerry, or Life in London (Moncrieff, 1821); 18, 127–50
and American adaptations, 132–3, 136–9
and blackface; 130–2, 141–7
and sporting culture, 133–41
see also flâneurie; urban culture
transparencies, 84, 99
travesty, *see* cross-dress
Treatise on Sugar (Moseley, 1799), 105–6, 108, 114–15, 209 n. 16, n. 17
Turner, Nat, 152, 172–3
Two Sermons (Stillman, 1773), 27

Uncle Tom's Cabin (Stowe, 1852), 163, 186–7
underclasses
defined, 4–12
and "lateral sufficiency," 20
as "undersiders"; 18, 22, 36, 53, 127, 134, 138, 159, 186
see also criminal culture; ethnic identity and masking; flash gangs; Marx, Karl, "lumpenproletariat"
urban culture, 3, 5, 15–16, 18, 20, 55, 57, 82, 88, 109, 175–88; *see also* Tom and Jerry

Venice Preserved (Otway, 1682), 80
Verling, William, 34
vernacular culture, 3, 4, 15, 16, 19
and archives, 24–5
and criminals, 28, 42
and *The Glory of Columbia*, 79, 83–4, 86, 98–100
and *Three-Finger'd Jack*, 101–2, 115–26
and *Tom and Jerry*, 134–8
Vesey, Denmark, 125, 173
Vestris, Eliza, 178, 223 n. 25
Vindication of the Captors of Major André (Benson, 1817), 88
Virginia Company of Comedians, 34
Virginius (Knowles, 1820), 162
visual display, 83–6, 99, 108–9, 135
voyeurism 50, 127, 129–32, 135–9, 147–50; *see also* flâneurie

Walker, Mr. (actor, Maryland), 34
Walker, Thomas, 30
Wallack, Henry, 132, 139
Walnut Street Circus (Philadelphia), *see* theatres and performance spaces
Walnut Street Theatre (Philadelphia), *see* theatres and performance spaces
Waltham and Windsor Blacks, 110–11
Ward, Artemus, 168–9
Washington, George, 92, 162, 164–5
as *André*'s "General," 79
in stage displays, 99
Waters, Billy, 129–30, 144
waxworks, 51, 199, n. 100
Wemyss, Francis Courtney, 132, 139
West's Circus (New York City), *see* theatres and performance spaces
whiteface, 114, 118–20, 142–3; *see also* Jonkonnu
Whitman, Walt, 96, 155, 163
Williams, James (thief), 41

youth culture, 1–2, 30–31, 95–6, 129, 140–1, 175–6; *see also* urban culture
"Youth's the Season," *see* songs, by title

GPSR Compliance

The European Union's (EU) General Product Safety Regulation (GPSR) is a set of rules that requires consumer products to be safe and our obligations to ensure this.

If you have any concerns about our products, you can contact us on

ProductSafety@springernature.com

In case Publisher is established outside the EU, the EU authorized representative is:

Springer Nature Customer Service Center GmbH
Europaplatz 3
69115 Heidelberg, Germany

www.ingramcontent.com/pod-product-compliance
Lightning Source LLC
LaVergne TN
LVHW051915060526
838200LV00004B/160